International Workshop on Discovery Physics at the LHC

(Kruger2014)

AA008606

Journal of Physics: Conference Series Volume 623

Kruger, South Africa
1-6 December 2014

ISBN: 978-1-5108-0797-6
ISSN: 1742-6588

Printed from e-media with permission by:

Curran Associates, Inc.
57 Morehouse Lane
Red Hook, NY 12571

Some format issues inherent in the e-media version may also appear in this print version.

Printed by Curran Associates, Inc. (2015)

For permission requests, please contact the Institute of Physics
at the address below.

Institute of Physics
Dirac House, Temple Back
Bristol BS1 6BE UK

Phone: 44 1 17 929 7481
Fax: 44 1 17 920 0979

techtracking@iop.org

Additional copies of this publication are available from:

Curran Associates, Inc.
57 Morehouse Lane
Red Hook, NY 12571 USA
Phone: 845-758-0400
Fax: 845-758-2634
Email: curran@proceedings.com
Web: www.proceedings.com

Table of contents

Volume 623

International Workshop on Discovery Physics at the LHC (Kruger2014)
1–6 December 2014, Kruger, South Africa

Accepted papers received: 18 May 2015
Published online: 11 June 2015

Preface

011001

International Workshop on Discovery Physics at the LHC (Kruger2014) J Cleymans

011002

Peer review statement

Papers

012001

Two-loop renormalisation in UED models Ammar Abdalgabar and A S Cornell **1**

012002

Analysis of the Quark-Gluon Plasma by Heavy Quarks J Aichelin, V Ozvenchuck, T Gousset and P B Gossiaux **7**

012003

The LHCb Trigger System: Present and Future Johannes Albrecht (On behalf of the LHCb Collaboration) **14**

012004

Search for Supersymmetry at CMS Christian Autermann **19**

012005

The QGP dynamics in relativistic collisions E L Bratkovskaya, W Cassing, V P Konchakovski and V D Toneev **29**

012006

The dynamics of Composite Higgses Giacomo Cacciapaglia **40**

012007

Search for heavy resonances at CMS Otman Charaf **48**

012008

Vector-like Multiplets, Mixings and the LHC Aldo Deandrea **57**

012009

Longitudinal Properties of High Energy Collisions A Deppman, L Marques and J Cleymans **65**

012010

Recent Standard Model results from CMS Simon de Visscher (on behalf of the CMS collaboration) **73**

012011

Measurements of the top-quark mass and properties at CMS Marc Dünser (on behalf of the CMS Collaboration) **81**

012012

Searching for P- and CP-odd effects in HIC D Espriu **89**

012013

Results of Beyond-the-Standard Model 2HDM Higgs Searches using the ATLAS detector in Run-I Guillermo Nicolas Hamity **97**

012014

Performance of the ATLAS Tile Calorimeter Louise Heelan (on behalf of the ATLAS Collaboration) **105**

012015

NLO Heavy Quark Energy Loss in Strongly-Coupled Quark-Gluon Plasmas W A Horowitz **113**

012016

Radiation hardness of plastic scintillators for the Tile Calorimeter of the ATLAS detector H Jivan, B Mellado, E Sideras-Haddad, R Erasmus, S Liao, M Madhuku, G Peters and O Solvyanov **120_**

012017

Double Higgs production at FCC-he and prospects for measurements of self-coupling Mukesh Kumar, Xifeng Ruan, Alan S Cornell, Rashidul Islam and Bruce Mellado **127**

012018

SUSY Searches in the ATLAS Experiment Lawrence Lee **133**

012019

The Energy Momentum Tensor associated with Hard Parton production in Finite Time Ben Meiring and W A Horowitz **143**

012020

Search for the Higgs boson in fermionic channels using ATLAS detector Kazuya Mochizuki (on behalf of the ATLAS collaboration) **150**

012021

Jet Nuclear Modification Factor from the AdS/CFT Correspondence R Morad and W A Horowitz **158**

012022

Central exclusive quarkonia production in the forward region at LHCb S Katharina Müller **164**

012023

LHCb results in proton-nucleus collisions at the LHC Katharina Müller **172**

012024

The Upgrade of the ATLAS Tile Calorimeter Readout Electronics for Phase II Robert Graham Reed (On behalf of the ATLAS Tile Calorimeter System) **179**

012025

$B^0_{(s)} \to \mu^+\mu^-$ combination at the LHC Maximilian Schlupp **184**

012026

FAIR - Cosmic Matter in the Laboratory Horst Stöcker, Thomas Stöhlker and Christian Sturm **190**

012027

Search for the critical point of strongly interacting matter at the CERN SPS NA61/SHINE experiment Ludwik Turko (for the NA61/SHINE Collaboration) **200**

012028

CERN: A European laboratory for a global project Rüdiger Voss **206**

012029

Dark matter searches in CMS Nadir Daci (on behalf of the CMS Collaboration) **211**

Journal of Physics: Conference Series **623** (2015) 011001 doi:10.1088/1742-6596/623/1/011001

The third biannual conference on 'Discovery Physics at the LHC' was held on December 1-6 2014 at the Kruger Gate Hotel in South Africa. Over 100 participants attended from Austria, Australia, Belgium, Brazil, Canada, China, the Czech Republic, France, Germany, Italy, the Netherlands, Norway, Poland, South Africa, Switzerland, the UK and the USA. The latest results from the Large Hadron Collider as well the latest theoretical insights were presented. With the majestic Kruger National Park in the background this led to a very stimulating conference with many exchanges taking place. The proceedings reflect the high level of the conference. The financial contributions from the SA-CERN programme, the UCT-CERN Research Centre, the University of Johannesburg, the University of the Witwatersrand and iThemba L.A.B.S. are gratefully acknowledged.

Jean Cleymans

Chair of the Local Organizing Committee

Local Organizing Committee:

Z. Buthelezi

J. Cleymans (chair)

S. H. Connell

A. S. Cornell

T. Dietel

S. Förtsch

N. Haasbroek

A. Hamilton

W. A. Horowitz

B. Mellado

Z. Z. Vilakazi

S. Yacoob

Journal of Physics: Conference Series **623** (2015) 011001 doi:10.1088/1742-6596/623/1/011001

Group Photo of the conference Kruger2014: Discovery Physics at the LHC, 1- 6 December 2014

Leopard in the Kruger National Park, courtesy of Dr. Vincenzo Chiochia

Peer review statement

All papers published in this volume of *Journal of Physics: Conference Series* have been peer reviewed through processes administered by the proceedings Editors. Reviews were conducted by expert referees to the professional and scientific standards expected of a proceedings journal published by IOP Publishing.

International Workshop on Discovery Physics at the LHC (Kruger2014) IOP Publishing
Journal of Physics: Conference Series **623** (2015) 012001 doi:10.1088/1742-6596/623/1/012001

Two-loop renormalisation in UED models

Ammar Abdalgabar[1] **and A. S. Cornell**[2]

National Institute for Theoretical Physics; School of Physics and Mandelstam Institute for Theoretical Physics, University of the Witwatersrand, Wits 2050, South Africa

E-mail: [1]ammar.abdalgabar@students.wits.ac.za, [2]alan.cornell@wits.ac.za

Abstract. The evolution equations of the gauge and Yukawa couplings are derived for the two-loop renormalisation group equations in a five-dimensional SM compactified on a S^1/Z_2 to yield standard four space-time dimensions. Different possibilities can be discussed, however, we shall consider the limiting case in which all matter fields are localised on the brane. We will compare our two-loop results to the results found at one-loop level, and investigate the evolution of $\sin^2\theta_W$ in this scenario also.

1. Introduction

The renormalisation group evolution of gauge and Yukawa couplings is important in many contexts, and in some cases one needs to know this evolution quite precisely. Furthermore, in order to take full advantage of future experimental observations, it will be necessary to provide theoretical predictions for measured observables which are as accurate as possible, where one can hope that the values of these couplings will be deduced directly from experimental data.

Note that in order to obtain very precise predictions for the running of Yukawa couplings, one usually uses known masses of quarks and leptons [1], since it is the Yukawa interactions that give the fundamental fermions their masses after spontaneous symmetry breaking. Due to the observed hierarchy of the fermion masses, the corresponding values are usually defined at different scales. Therefore, one inevitably makes use of Renormalisation Group Equations (RGEs) to connect these scales. It should also be mentioned that contrary to leptons, quarks are not observed as free particles, so the pole mass which is usually associated with the physical mass of a particle, although being a gauge invariant and finite quantity, suffers from ambiguity [2]. As a consequence, theoretical uncertainty in the determination of the top Yukawa coupling, for example, is reduced, thus requiring more precise determinations of the corresponding RGEs.

So by computing couplings to two-loop accuracy we are able to set limits, such as on the prediction of the masses of heavy fermions [3]. But this is exasperated further when considering models with extra-dimensions, where we can have a range of heavy fermions and bosons, arising as Kaluza-Klein states, leading to runnings for our couplings now exhibiting a power law running, see Refs. [4] for example. Furthermore, this observed power law running can mean contributions from higher-order loops could be very significant when compared to our earlier one-loop calculations [4, 5]. Note that in this proceedings we shall restrict ourselves to a minimal Universal Extra-Dimensional (UED) model, where all matter fields are constrained to the brane, as a simpler testing ground for two-loop calculations. This shall be extended in an upcoming work to bulk cases also [6]. However, two-loop precision remains desirable here because fermions can be heavy, and this means that their Yukawa couplings, and other couplings, can be large.

As such, the precise value of the corresponding coupling is required to test whether the Standard Model (SM), or UED models, correctly describe nature. The paper is organised as follows: In section 2 we shall recall the one- and two-loop SM beta functions. In section 3, the one- and two-loop UED beta functions shall be presented. Section 4 concludes with some preliminary results.

2. The beta functions in the SM Model

The one- and two-loop beta functions in the SM have been known for decades now in various mass limits, see Refs. [7] for example, and are presented here in the form we shall use for completeness.

2.1. *Gauge sector*

The one-loop beta functions for the gauge couplings in SM model are given by:

$$(4\pi)^2 g_1^{-3}\beta_{g_1}^{(1)} = \frac{41}{10}, \quad (1)$$

$$(4\pi)^2 g_2^{-3}\beta_{g_2}^{(1)} = -\frac{19}{6}, \quad (2)$$

$$(4\pi)^2 g_3^{-3}\beta_{g_3}^{(1)} = -7. \quad (3)$$

The two-loop contribution to the SM gauge couplings are given by:

$$(4\pi)^4 g_1^{-3}\beta_{g_1}^{(2)} = \frac{199}{50}g_1^2 + \frac{27}{10}g_2^2 + \frac{44}{5}g_3^2 - \mathrm{Tr}\left(\frac{17}{5}Y_u^\dagger Y_u + Y_d^\dagger Y_d + 3Y_e^\dagger Y_e\right), \quad (4)$$

$$(4\pi)^4 g_2^{-3}\beta_{g_2}^{(2)} = \frac{9}{10}g_1^2 + \frac{35}{6}g_2^2 + 12g_3^2 - \mathrm{Tr}\left(3Y_u^\dagger Y_u + 3Y_d^\dagger Y_d + Y_e^\dagger Y_e\right), \quad (5)$$

$$(4\pi)^4 g_3^{-3}\beta_{g_3}^{(2)} = \frac{11}{10}g_1^2 + \frac{9}{2}g_2^2 - 26g_3^2 - 4\mathrm{Tr}\left(Y_u^\dagger Y_u + Y_d^\dagger Y_d\right). \quad (6)$$

2.2. *Yukawa sector*

The one-loop beta function for the Yukawa couplings in SM model are given by:

$$(4\pi)^2 Y_u^{-1}\beta_{Y_u}^{(1)} = \mathrm{Tr}\left(3Y_u^\dagger Y_u + 3Y_d^\dagger Y_d + Y_e^\dagger Y_e\right) - \left(\frac{17}{20}g_1^2 + \frac{9}{4}g_2^2 + 8g_3^2\right) + \frac{3}{2}\left(Y_u^\dagger Y_u - Y_d^\dagger Y_d\right) \quad (7)$$

$$(4\pi)^2 Y_d^{-1}\beta_{Y_d}^{(1)} = \mathrm{Tr}\left(3Y_u^\dagger Y_u + 3Y_d^\dagger Y_d + Y_e^\dagger Y_e\right) - \left(\frac{1}{4}g_1^2 + \frac{9}{4}g_2^2 + 8g_3^2\right) + \frac{3}{2}\left(Y_d^\dagger Y_d - Y_u^\dagger Y_u\right) \quad (8)$$

$$(4\pi)^2 Y_e^{-1}\beta_{Y_e}^{(1)} = \mathrm{Tr}\left(3Y_u^\dagger Y_u + 3Y_d^\dagger Y_d + Y_e^\dagger Y_e\right) - \left(\frac{9}{4}g_1^2 + \frac{9}{4}g_2^2\right) + \frac{3}{2}Y_e^\dagger Y_e. \quad (9)$$

The two-loop contribution to the SM Yukawa couplings are given by:

$$\begin{aligned}(4\pi)^4 Y_u^{-1}\beta_{Y_u}^{(2)} = {} & 11(Y_d^\dagger Y_d)^2 - 5(Y_d^\dagger Y_d)(Y_u^\dagger Y_u) + \left(5Y_d^\dagger Y_d - 9Y_u^\dagger Y_u\right)\mathrm{Tr}\left(3Y_u^\dagger Y_u + 3Y_d^\dagger Y_d + Y_e^\dagger Y_e\right) \\ & 6(Y_u^\dagger Y_u)^2 - 9\mathrm{Tr}\left(3(Y_u^\dagger Y_u)^2 + 3(Y_d^\dagger Y_d)^2 - \frac{2}{3}(Y_d^\dagger Y_d)(Y_u^\dagger Y_u) + (Y_e^\dagger Y_e)^2\right) \\ & + Y_u^\dagger Y_u\left(\frac{223}{40}g_1^2 + \frac{135}{8}g_2^2 + 32g_3^2\right) - Y_d^\dagger Y_d\left(\frac{43}{40}g_1^2 - \frac{9}{8}g_2^2 + 32g_3^2\right) \\ & + \frac{1187}{600}g_1^4 - \frac{23}{4}g_2^4 - 108g_3^4 - \frac{9}{20}g_1^2g_2^2 - \frac{19}{15}g_1^2g_3^2 + 9g_2^2g_3^2 + \frac{15}{4}\mathrm{Tr}(Y_e^\dagger Y_e)\left(g_1^2 + g_2^2\right) \\ & + 5\mathrm{Tr}(Y_u^\dagger Y_u)\left(\frac{17}{20}g_1^2 + \frac{9}{4}g_2^2 + 8g_3^2\right) + 5\mathrm{Tr}(Y_d^\dagger Y_d)\left(\frac{1}{4}g_1^2 + \frac{9}{4}g_2^2 + 8g_3^2\right), \end{aligned} \quad (10)$$

Journal of Physics: Conference Series **623** (2015) 012001 doi:10.1088/1742-6596/623/1/012001

$$
\begin{aligned}
(4\pi)^4 Y_d^{-1}\beta_{Y_d}^{(2)} &= 11(Y_u^\dagger Y_u)^2 - 5(Y_d^\dagger Y_d)(Y_u^\dagger Y_u) + \left(5Y_u^\dagger Y_u - 9Y_d^\dagger Y_d\right)\mathrm{Tr}\left(3Y_u^\dagger Y_u + 3Y_d^\dagger Y_d + Y_e^\dagger Y_e\right) \\
& 6(Y_d^\dagger Y_d)^2 - 9\mathrm{Tr}\left(3(Y_u^\dagger Y_u)^2 + 3(Y_d^\dagger Y_d)^2 - \frac{2}{3}(Y_d^\dagger Y_d)(Y_u^\dagger Y_u) + (Y_e^\dagger Y_e)^2\right) \\
& + Y_d^\dagger Y_d\left(\frac{187}{40}g_1^2 + \frac{135}{8}g_2^2 + 32g_3^2\right) - Y_u^\dagger Y_u\left(\frac{79}{40}g_1^2 - \frac{9}{8}g_2^2 + 32g_3^2\right) \\
& -\frac{127}{600}g_1^4 - \frac{23}{4}g_2^4 - 108g_3^4 - \frac{27}{20}g_1^2g_2^2 + \frac{13}{15}g_1^2g_3^2 + 9g_2^2g_3^2 + \frac{15}{4}\mathrm{Tr}(Y_e^\dagger Y_e)\left(g_1^2+g_2^2\right) \\
& + 5\mathrm{Tr}(Y_u^\dagger Y_u)\left(\frac{17}{20}g_1^2 + \frac{9}{4}g_2^2 + 8g_3^2\right) + 5\mathrm{Tr}(Y_d^\dagger Y_d)\left(\frac{1}{4}g_1^2 + \frac{9}{4}g_2^2 + 8g_3^2\right), \qquad (11)
\end{aligned}
$$

$$
\begin{aligned}
(4\pi)^4 Y_e^{-1}\beta_{Y_e}^{(2)} &= 6(Y_e^\dagger Y_e)^2 - 9\mathrm{Tr}\left(3(Y_u^\dagger Y_u)^2 + 3(Y_d^\dagger Y_d)^2 - \frac{2}{3}(Y_d^\dagger Y_d)(Y_u^\dagger Y_u) + (Y_e^\dagger Y_e)^2\right) \\
& -9Y_e^\dagger Y_e\mathrm{Tr}\left(3Y_u^\dagger Y_u + 3Y_d^\dagger Y_d + Y_e^\dagger Y_e\right) + Y_e^\dagger Y_e\left(\frac{387}{40}g_1^2 + \frac{135}{8}g_2^2\right) \\
& + 5\mathrm{Tr}(Y_u^\dagger Y_u)\left(\frac{17}{20}g_1^2 + \frac{9}{4}g_2^2 + 8g_3^2\right) + 5\mathrm{Tr}(Y_d^\dagger Y_d)\left(\frac{1}{4}g_1^2 + \frac{9}{4}g_2^2 + 8g_3^2\right) \\
& + \frac{15}{4}\mathrm{Tr}(Y_e^\dagger Y_e)\left(g_1^2+g_2^2\right) + \frac{1371}{200}g_1^4 - \frac{23}{4}g_2^4 + \frac{27}{20}g_1^2g_2^2. \qquad (12)
\end{aligned}
$$

3. The beta functions in a brane-localised UED model

In this scenario the SM $SU(3)_c \times SU(2)_L \times U(1)_Y$ gauge fields and the Higgs (H) propagate into the fifth dimension, y. As a consequence these fields will have Kaluza-Klein modes which contribute to the RGEs at energies $E > 1/R$. Different possibilities for the matter fields can be studied, however we shall study the limiting case with SM matter fields restricted to the $y = 0$ brane. As such there will be no additional Kaluza-Klein contributions of these matter fields to the RGEs.

3.1. Gauge sector

The one-loop beta functions for the gauge couplings in this five dimensional model are given by:

$$
\begin{aligned}
(4\pi)^2 g_1^{-3}\beta_{g_1}^{(1)} &= (S(t)-1)\left(\frac{1}{10} + \frac{8}{3}\eta\right), \qquad (13)\\
(4\pi)^2 g_2^{-3}\beta_{g_2}^{(1)} &= (S(t)-1)\left(-\frac{41}{6} + \frac{8}{3}\eta\right), \qquad (14)\\
(4\pi)^2 g_3^{-3}\beta_{g_3}^{(1)} &= (S(t)-1)\left(-\frac{21}{2} + \frac{8}{3}\eta\right). \qquad (15)
\end{aligned}
$$

where η is the number of fermion generations.

The two-loop contribution to the gauge couplings in this five dimensional model are given by:

$$
\begin{aligned}
(4\pi)^4 g_1^{-3}\beta_{g_1}^{(2)} &= (S(t)^2-1)\left(\frac{199}{50}g_1^2 + \frac{27}{10}g_2^2 + \frac{44}{5}g_3^2\right), \qquad (16)\\
(4\pi)^4 g_2^{-3}\beta_{g_2}^{(2)} &= (S(t)^2-1)\left(\frac{9}{10}g_1^2 + \frac{35}{6}g_2^2 + 12g_3^2\right), \qquad (17)\\
(4\pi)^4 g_3^{-3}\beta_{g_3}^{(2)} &= (S(t)^2-1)\left(\frac{11}{10}g_1^2 + \frac{9}{2}g_2^2 - 26g_3^2\right). \qquad (18)
\end{aligned}
$$

3.2. Yukawa sector

The one-loop beta functions for the Yukawa couplings in this five dimensional model are given by:

$$(4\pi)^2 Y_u^{-1}\beta_{Y_u}^{(1)} = 2(S(t)-1)\left(-\left(\frac{17}{20}g_1^2+\frac{9}{4}g_2^2+8g_3^2\right)+\frac{3}{2}\left(Y_u^\dagger Y_u - Y_d^\dagger Y_d\right)\right), \quad (19)$$

$$(4\pi)^2 Y_d^{-1}\beta_{Y_d}^{(1)} = 2(S(t)-1)\left(-\left(\frac{1}{4}g_1^2+\frac{9}{4}g_2^2+8g_3^2\right)+\frac{3}{2}\left(Y_d^\dagger Y_d - Y_u^\dagger Y_u\right)\right), \quad (20)$$

$$(4\pi)^2 Y_e^{-1}\beta_{Y_e}^{(1)} = 2(S(t)-1)\left(-\left(\frac{9}{4}g_1^2+\frac{9}{4}g_2^2\right)+\frac{3}{2}Y_e^\dagger Y_e\right). \quad (21)$$

The two-loop contribution to the Yukawa couplings in this five dimensional are given by:

$$\begin{aligned}(4\pi)^4 Y_u^{-1}\beta_{Y_u}^{(2)} &= 2(S(t)^2-1)\Big(6(Y_u^\dagger Y_u)^2 - 5(Y_d^\dagger Y_d)(Y_u^\dagger Y_u) + 11(Y_d^\dagger Y_d)^2 \\ &+ Y_u^\dagger Y_u\left(\frac{223}{40}g_1^2+\frac{135}{8}g_2^2+32g_3^2\right) - Y_d^\dagger Y_d\left(\frac{43}{40}g_1^2-\frac{9}{8}g_2^2+32g_3^2\right) \\ &+\frac{1187}{600}g_1^4-\frac{23}{4}g_2^4-108g_3^4-\frac{9}{20}g_1^2g_2^2-\frac{19}{15}g_1^2g_3^2+9g_2^2g_3^2\Big), \quad (22)\\ (4\pi)^4 Y_d^{-1}\beta_{Y_d}^{(2)} &= 2(S(t)^2-1)\Big(6(Y_d^\dagger Y_d)^2 - 5(Y_d^\dagger Y_d)(Y_u^\dagger Y_u) + 11(Y_u^\dagger Y_u)^2 \\ &+ Y_d^\dagger Y_d\left(\frac{187}{40}g_1^2+\frac{135}{8}g_2^2+32g_3^2\right) - Y_u^\dagger Y_u\left(\frac{79}{40}g_1^2-\frac{9}{8}g_2^2+32g_3^2\right) \\ &-\frac{127}{600}g_1^4-\frac{23}{4}g_2^4-108g_3^4-\frac{27}{20}g_1^2g_2^2+\frac{13}{15}g_1^2g_3^2+9g_2^2g_3^2\Big), \quad (23)\\ (4\pi)^4 Y_e^{-1}\beta_{Y_e}^{(2)} &= 2(S(t)^2-1)\Big(6(Y_e^\dagger Y_e)^2 + Y_e^\dagger Y_e\left(\frac{387}{40}g_1^2+\frac{135}{8}g_2^2\right) \\ &+\frac{1371}{200}g_1^4-\frac{23}{4}g_2^4+\frac{27}{20}g_1^2g_2^2\Big). \quad (24)\end{aligned}$$

That is, when the energy scale parameter $t = \ln(E/M_Z) > \ln(1/RM_Z)$ or $E > 1/R$, these equations shall be used. When the energy $0 < t < \ln(1/RM_Z)$ (that is $M_Z < E < 1/R$) the evolution of the gauge and Yukawa couplings at low energy are given by the usual four dimensional SM expressions, as was presented in section 2.

4. Result and Discussion

As expected the extra-dimension brings down the unification scale to a lower scale, see Fig. 1. At one-loop level the gauge and Yukawa coupling RGEs are disentangled, however, at the two-loop level the RGEs for the gauge and Yukawa couplings are entangled. As such, only a few percent change in the evolution of gauge couplings, due to the appearance of Yukawa couplings over the one-loop running, is observed. This also may change the running of g_3, due to the large size of Y_t.

In Fig. 2 we present the evolution of $\sin^2\theta_W$ in this brane constrained UED model. Once the new contributions from the extra-dimensions arise, the behaviour is changed until we reach the cut-off scale. One can see for $R^{-1} = 1$ TeV $\sin^2\theta_W$ can rise to ~ 0.4. This result may be useful, at least from a model building perspective, where there is no discernible difference from the one- and two-loop trajectories in this plot.

As such, for models with an extra-dimension the one-loop running of Yukawa couplings is clearly insufficient, since higher order corrections can be just as important at scales a few times above $1/R$. Although this type of large correction, which we require to claim unification, see Fig. 3, needs to be tested further to ensure that Y_t remains perturbative up to the unification

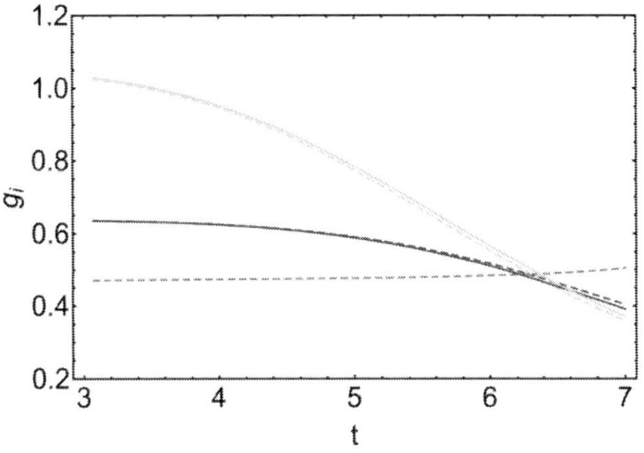

Figure 1. *(Colour online) Gauge couplings g_1 (red), g_2 (blue), g_3 (green) with all matter fields in the brane; for the compactification scales $R^{-1} = 2$ TeV as a function of the scale parameter t. Solid lines are the one-loop level runnings, dashed lines are two-loop level runnings.*

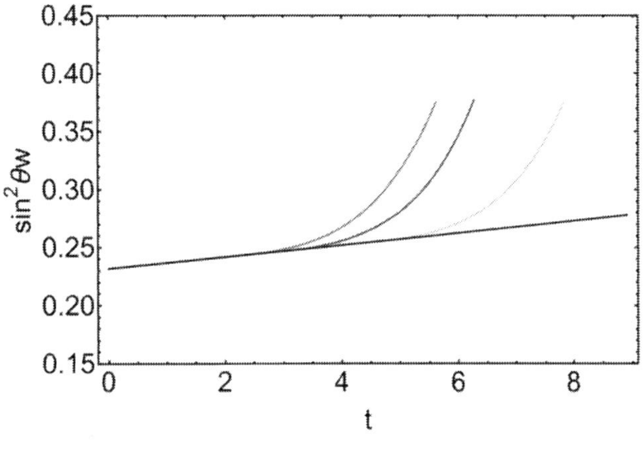

Figure 2. *$sin^2\theta_W$ with all matter fields in the brane; for three different values of the compactification scales R^{-1} = 1 TeV (red), 2 TeV (blue), 10 TeV (green) as a function of the scale parameter t.*

scale. This shall be more fully explored in an upcoming work, where the case of bulk propagating matter fields shall also be investigated [6].

Acknowledgments

We would like to thank our collaborator Aldo Deandrea for the helpful discussions. This work is supported by the National Research Foundation (South Africa).

References

[1] K. A. Olive *et al.* [Particle Data Group Collaboration], Chin. Phys. C **38**, 090001 (2014).

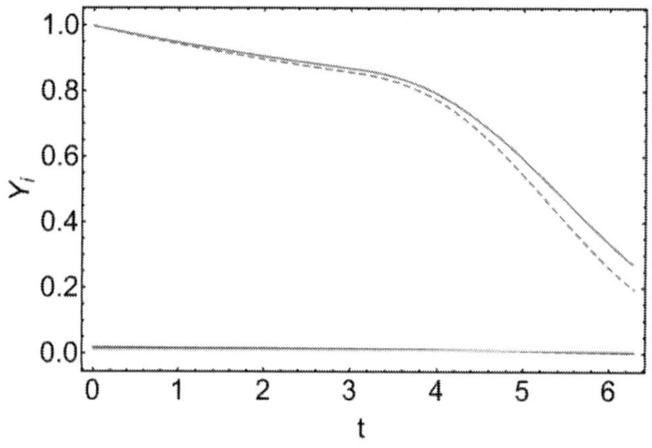

Figure 3. *(Colour online) Yukawa couplings Y_e (green), Y_d (blue), Y_u (red) with: all matter fields in the brane; for the compactification scales $R^{-1} = 2$ TeV as a function of the scale parameter t. As in Fig. 1, solid lines are the one-loop level runnings, dashed lines are two-loop level runnings.*

[2] M. Beneke and V. M. Braun, Nucl. Phys. B **426**, 301 (1994) [hep-ph/9402364]; I. I. Y. Bigi, M. A. Shifman, N. G. Uraltsev and A. I. Vainshtein, Phys. Rev. D **50**, 2234 (1994) [hep-ph/9402360].
[3] M. E. Machacek and M. T. Vaughn, Phys. Lett. B **103**, 427 (1981).
[4] A. S. Cornell and L. -X. Liu, Phys. Rev. D **83**, 033005 (2011) [arXiv:1010.5522 [hep-ph]]; A. S. Cornell, A. Deandrea, L. X. Liu and A. Tarhini, Phys. Rev. D **85**, 056001 (2012) [arXiv:1110.1942 [hep-ph]]; A. S. Cornell, A. Deandrea, L. -X. Liu and A. Tarhini, Mod. Phys. Lett. A **28**, no. 11, 1330007 (2013) [arXiv:1209.6239 [hep-ph]].
[5] L. -X. Liu and A. S. Cornell, PoS KRUGER **2010**, 045 (2010) [arXiv:1103.1527 [hep-ph]]; A. Abdalgabar, A. S. Cornell, A. Deandrea and A. Tarhini, Phys. Rev. D **88**, 056006 (2013) [arXiv:1306.4852 [hep-ph]]; A. Abdalgabar, A. S. Cornell, A. Deandrea and A. Tarhini, Eur. Phys. J. C **74**, no. 5, 2893 (2014) [arXiv:1307.6401 [hep-ph]]. A. Abdalgabar and A. S. Cornell, J. Phys. Conf. Ser. **455**, 012050 (2013) [arXiv:1305.3729 [hep-ph]].
[6] A. Abdalgabar, A. S. Cornell and M. Kumar, *in preparation.*
[7] M. Fischler and J. Oliensis, Phys. Rev. D **28**, 2027 (1983); J. van der Bij and M. J. G. Veltman, Nucl. Phys. B **231**, 205 (1984); J. J. van der Bij, Nucl. Phys. B **248**, 141 (1984).

International Workshop on Discovery Physics at the LHC (Kruger2014) IOP Publishing
Journal of Physics: Conference Series **623** (2015) 012002 doi:10.1088/1742-6596/623/1/012002

Analysis of the Quark-Gluon Plasma by Heavy Quarks

J. Aichelin [1] , V. Ozvenchuck, T. Gousset and PB. Gossiaux
SUBATECH, Laboratoire de Physique Subatomique et des Technologies Associées,
Université de Nantes - IN2P3/CNRS - Ecole des Mines de Nantes
4 rue Alfred Kastler, F-44072 Nantes, Cedex 03, France

Abstract. We discuss the present status of the tomography of a quark gluon plasma (QGP) with help of heavy quarks. We address the question whether collisional or radiative energy loss dominates, how different expansion scenarios of the medium influence the observables and which state of the expansion is impressed in the R_{AA} and v_2 observables.

1. Introduction

Hadrons have a finite radius. Therefore, if the density becomes too high one expects naively that the hadrons start to overlap and the constituents of the hadrons, the quarks and gluons, can move freely from one hadron to another. Thus hadrons loose their identity and form a new stage of matter - a plasma of unbound gluons and quarks which is in local thermal equilibrium. Because the density increases with temperature we expect such a transition at some given temperature in analogy with the boiling temperature of water when the phase transition from the liquid state to vapor occurs.

The phase transition from the hadronic to the partonic phase is also predicted by the fundamental theory for strongly interaction systems, the Quantum Chromo Dynamics (QCD). This follows from so called lattice gauge calculations [1], which are presently the only known way to predict expectation values of observables in strongly interacting systems. These calculations are done assuming an infinite system in global thermal equilibrium.

If one wants to do systematic studies of the QGP and its properties there is only one possibility, to create a QPG in heavy ion reactions. There the situation differs substantially from the conditions used for the lattice QCD calculations, i.e. the infinite partonic matter in equilibrium: In heavy ion reactions such a state can be only created for a very short time (of the order of $10^{-23}s$) and has an extension of a couple of 10^{-15} m. Then the system, which expands with almost the speed of light, forms after about 10 fm/c hadrons which are finally observed in the detector. Thus, it is quite difficult to prove - from the theoretical as well as from the experimental side - that in a such small system in a very short time a partonic plasma has be formed.

The problem is therefore to reconstruct from the observed hadrons the existence and the properties of such a QGP. The vast majority of the observed hadrons are formed from light (u,d,s) quarks which, unfortunately, tell us little about the formation of a plasma. It turned

[1] invited speaker

out that the multiplicity of light hadrons is very well described by statistical models [2]. This means that the system of light quarks or hadrons (made of light quarks) has reached a state of equilibrium (and the temperature is close to that predicted by lattice gauge calculations for the transition temperature). Therefore light hadrons do not contain information about the plasma properties at higher temperatures. One may argue that the spectra of light hadrons may give additional information. Unfortunately hadronic final state interactions are too strong and contain too many experimentally unknown cross sections in order to infer from the measured spectra the spectra at the moment when hadrons move out of equilibrium.

The light particle spectra are well described by so called event generators [3] which model the whole reaction on a computer. They assume that after a violent initial phase hadrons are formed which, if the energy density exceeds a given value, form a QGP. The expansion of the QGP is modelled either in viscous or in non viscous hydrodynamics. At a given energy density which is around $1GeV/fm^3$ [4] a sudden transition to the hadronic world takes place followed by a final state interactions among the hadrons. The problem is that the physics of the initial state is little known and leaves a lot of room for different assumptions. As a consequence, some model assume that the measured centrality dependence of the elliptic flow presents evidence for the need of viscous hydrodynamics [5] whereas other models describe the data equally well in an ideal hydrodynamical approach [6] assuming that not all particles take part in the hydrodynamical expansion.

Also the fast equilibration is not yet understood. Therefore other models [7] do not assume that a thermal equilibrium is established and describe the strongly interacting quark-gluon plasma by relativistic transport equations, derived from many-body Kadanoff-Baym equations, with a dynamical hadronization of partons to hadrons. This concept leads as well to a very satisfying description of a variety of data.

In a situation like this it is evident that one looks for possible observables which do not suffer from this memory erasing equilibrium phase. There are essentially two: High p_t hadrons which originate from jets as well as the p_t and v_2 distribution of heavy mesons which contain either a c or a b quark because neither jets nor heavy quarks come to an equilibrium with the plasma. Jets have the problem that the leading particle, i.e. that with the highest momentum, may change by interactions with the plasma. This makes the understanding of jets difficult.

Heavy quarks are produced in hard binary initial collisions between the incoming protons. Their production cross sections are known from pp collisions and can as well be calculated in pQCD calculations. Therefore the initial transverse momentum distribution of the heavy quarks is known. Comparing this distribution with that measured in heavy ion collisions allows defining $R_{AA} = (d\sigma_{AA}/dp_t^2)\ /\ (N_c d\sigma_{pp}/dp_t^2)$, where N_c is the number of the initial binary collisions between projectile and target. The deviation of R_{AA} from one measures the interaction of the heavy quark with the plasma because the hadron cross sections of heavy mesons are small. The heavy quark does not come to thermal equilibrium with the QGP therefore R_{AA} contains the information on the interaction of the heavy quark while it traverses the plasma. In addition, the distribution of heavy quarks at the moment of their creation is isotropic in azimuthal direction, therefore the elliptic flow $v_2 =< cos2(\phi - \phi_R) >$, where ϕ (ϕ_R) is the azimuthal angle of the emitted particle (reaction plane) is 0. The observed finite v_2 value of the observed heavy meson can only originate from interactions between light QGP constituents and the heavy quarks. The simultaneous description of R_{AA} and v_2 and their centrality dependence, presently the only observables for which data exist, give then the possibilities to understand the interactions inside the QGP.

Unfortunately the experimental results depend not only on the elementary interaction but also on the description of the expansion of the QGP [8]. Therefore the ultimate aim is to control the expansion by results on the light meson sector. This has not been achieved yet for the LHC and therefore it is difficult to asses the influence of the expansion on the observables. We use

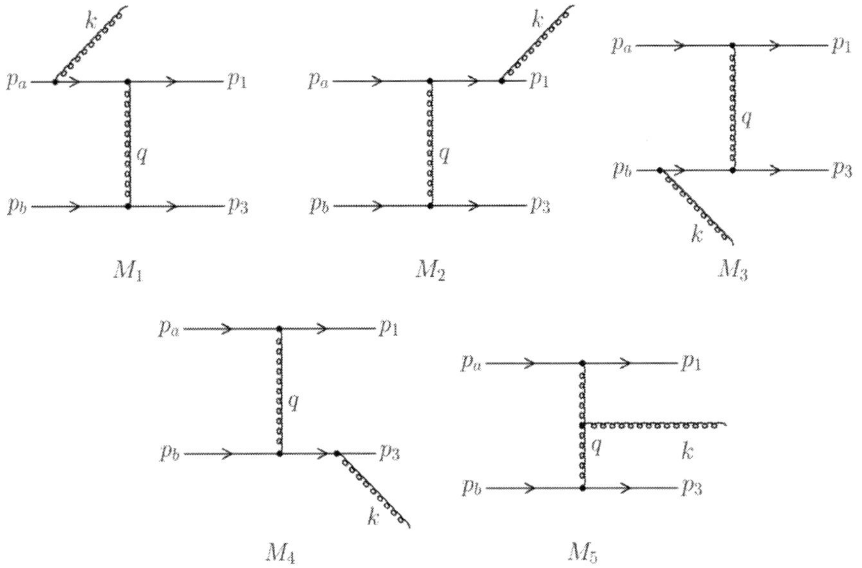

Figure 1. (Color online) The five matrix elements which contribute to the gluon bremsstrahlung.

here the approach from Kolb and Heinz which has reasonably well described the midrapidity light mesons at RHIC [9]. We adjust only the charged particle multiplicity to the value measured at LHC.

The R_{AA} of 0.2 values observed for large p_t heavy mesons are much smaller than originally expected. Early theoretical approaches based on perturbative QCD (pQCD) calculation gave much larger values and it has been doubted, whether pQCD is the right tool to describe this interaction. This early calculation, however, used ad hoc assumptions on the coupling constant α_s and the infrared regulator μ. With a standard choice μ and α_s an artificial K factor, an overall multiplication factor of the elastic cross section of around 10 [10, 11] had to be introduced to match the experimental data.

A while ago we advanced an approach for the collisional energy loss of heavy quarks in the QGP [12, 13, 14] in which a) μ has been fixed by the demand that more realistic calculations using the hard thermal loop approach give the same energy loss as our Born type pQCD calculation and b) the coupling constant is running and fixed by the sum rule advanced by Dokshitzer and later used by Peshier. Both these improvements increased the cross section especially for small momentum transfers and.

Later we included in addition the radiative energy loss [15, 16]. For this we have to consider the 5 Feynman diagrams displayed in fig.1. The commutation relation

$$T^b T^a = T^a T^b - i f_{abc} T^c \tag{1}$$

allows us regrouping the 5 matrix elements into 3 combinations, each of them being independently gauge invariant:

$$\begin{aligned} iM^{QED}_{h.q.} &= C_a i(M_1 + M_2) \\ iM^{QED}_{l.q.} &= C'_a i(M_3 + M_4) \end{aligned}$$

$$iM^{QCD} = C_c i(M_1 + M_3 + M_5). \tag{2}$$

h.q. (l.q.) mark the emission of the gluon from the heavy (light quark) line. C_a, C'_a and C_c are the color algebra matrix elements. The matrix elements labeled as QED are the bremsstrahlung diagrams already observed in Quantum Electrodynamics (QED), whereas that labeled QCD is the genuine diagram of Quantum Chromodynamics (QCD). The QCD diagram is the main objet of interest here because it dominates the energy loss of heavy quarks.

Collisional and radiative energy loss act quite differently as has been discussed in [17]. The cross section for collisional energy loss in our approach is rather large whereas that for radiational energy loss is small. However, in elastic collisions the energy transfer is small whereas in radiative collisions it is large. Fig. 2 shows the distribution of the number of elastic (left) and radiational collisions (right) of a heavy quark in a central heavy ion reaction of AuAu at $\sqrt{s} = 200MeV$. We see that there are on the average 40 times more elastic collisions than radiative collisions. This

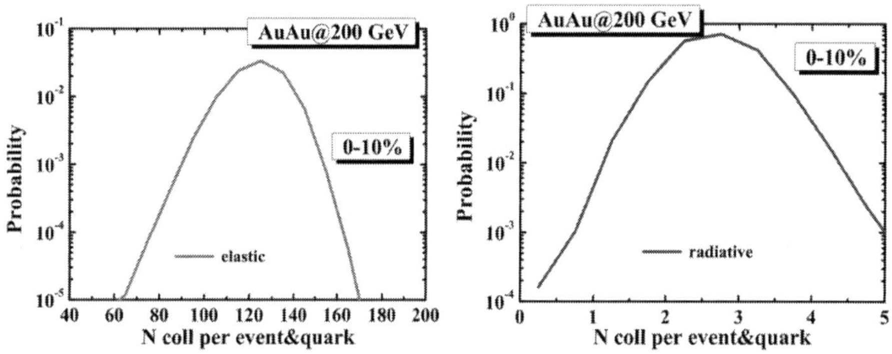

Figure 2. Distribution of the number of elastic collisions (left) and radiative collisions (right) a heavy quarks suffers in a central AuAu reaction at $\sqrt{s} = 200AGeV$.

quite different number of collisions triggered the hope that the form of the heavy meson spectra is different depending on the whether collisional and radiative energy loss dominates while the quark is traveling through the plasma. This hope has been in vain as is shown in Fig.3 which displays on the left hand side R_{AA}. If a nuclear reaction is nothing else than a superposition of pp collisions R_{AA} would be one. On the right side we display for the same reaction v_2, the second order Fourier coefficient of the azimuthal distribution, of heavy quarks with respect to the reaction plane. We display these quantities for two different scenarios. In scenario one we assume elastic collisions only and multiply the calculated cross section with an artificial factor K =1.5-1.8 to obtain the right energy loss. In scenario two we employ elastic and radiative collisions multiplied with K=0.8-1 to show how R_{AA} changes if one modifies slightly the cross sections. The expanding quark gluon plasma is modeled by the event generator EPOS II. We see first of all that R_{AA} differs strongly from one and therefore the heavy quarks interact strongly with the plasma (hadronic interactions of the D-mesons are neglected here). We observe furthermore that the form of R_{AA} is for both scenarios almost identical. Both calculations reproduce reasonably well the experimental results of the STAR collaboration [18]. We will come back to this later. Also the p_T dependence of v_2 is almost identical for the both scenarios and both scenarios give v_2 values close to the preliminary experimental results. On the basis of these two observables it is therefore impossible to discriminate experimentally between collisional and radiative energy loss of the heavy quarks.

The energy loss of heavy quarks in the expansing plasma depends on the one side on the elementary interaction of the heavy quarks with the plasma partons but on the other side as

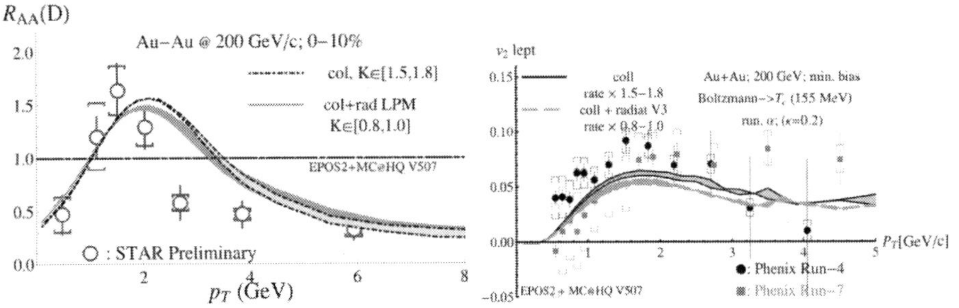

Figure 3. R_{AA} of D- mesons (left) and v_2 of non photonic electrons from the decay of heavy mesons (right) for central AuAu collisions at $\sqrt{s} = 200AGeV$. For the different scenarios see text.

well on how the plasma expansion is described. Presently in most of the models the plasma expansion is described by (viscous) hydrodynamics. These hydrodynamical models describe a multitude of experimental observables of hadrons consisting of light quarks. They are, however, mostly sensitive to the end of the plasma expansions where the chemical composition of these hadrons follows the predictions for a statistically equilibrated system. As we will see later, the energy loss of heavy quarks is created very early in the expansion of the plasma and therefore heavy quarks are sensitive to the early phase of the hydrodynamical expansion.

In Fig. 4 we display R_{AA} and v_2 for two different models for the expansion of the QGP, keeping the elementary interaction between heavy quarks and plasma partons identical. In this case we use only collisional energy loss with a K factor of 1.5-2 and compare the result to central STAR data for R_{AA} and central Phenix data for v_2. The violet curve shows the results for EPOS2 calculations [3] whereas the red curve is the result for the Kolb-Heinz hydrodynamical approach [9]. Whereas the v_2 value is rather independent of the expansion scenario, the R_{AA} of

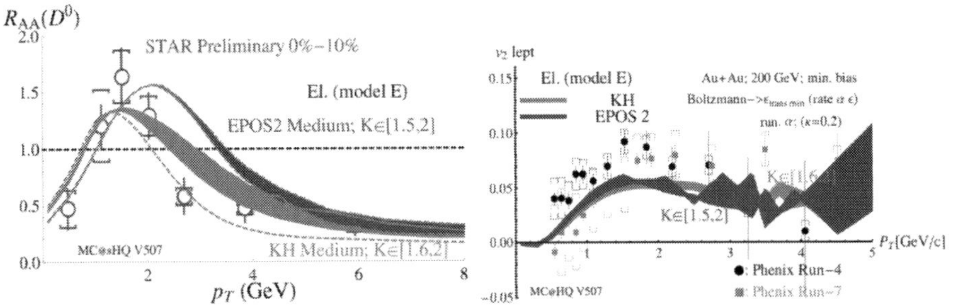

Figure 4. R_{AA} of D- mesons (left) and v_2 of non photonic electrons from the decay of heavy mesons (right) for central AuAu collisions at $\sqrt{s} = 200AGeV$ for different plasma expansion scenarios. For the different scenarios see text.

the both approaches differ substantially for p_T around 2 GeV. This is a consequence of the early creation of a radial flow in the EPOS2 approach [3]. In the Kolb-Heinz approach this creation is much softer. This comparison makes evident that heavy meson spectra do not only test the elementary interaction between heavy quarks and the QGP partons but also the expansion of the QGP itself. Therefore for a comprehensive understanding of a heavy ion reaction light and heavy quark observables cannot be entangled and have to be studied simultaneously.

One may ask the question how R_{AA} and v_2 are formed during the expansion of the QGP. To study this in more detail we display in Fig. 5 the time evolution of these variables for central AuAu reactions at $\sqrt{s}$ = 200 AGeV. We see that in our approach the R_{AA} is quite sensitive to the beginning of the expansion where the energy density is highest and therefore a lot of collisions take place. At 3 fm/c R_{AA} has almost obtained its asymptotic value. The elliptic flow, on the contrary, is more sensitive to the late phase of the expansion. Initially neither heavy nor light quarks have a v_2. The light quarks acquire it when hydrodynamics converts the initial special eccentricity of the interaction to an elliptic flow. This takes time. Only then the v_2 can be impressed on the heavy quarks by the collisional interaction with the light quarks. Therefore at 3 fm/c v_2 has not even half of the asymptotic value. If, as in our model, the partons of

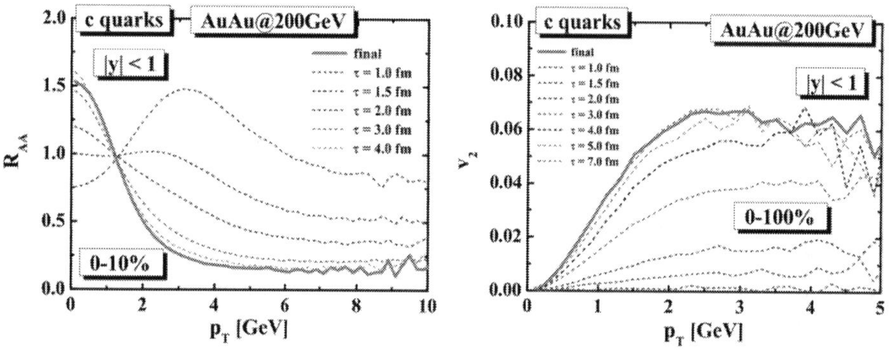

Figure 5. R_{AA} (left) and v_2 (right) of c quarks as a function from time. Final marks the time at which the c-quarks are converted into hadrons either by coalescence of by fragmentation.

the QGP are considered as massless when they interact with the heavy quarks, we can identify different time scales for the development of R_{AA} and v_2 of the heavy quarks. Whether this is also the case in models in which the plasma, with which the charm quarks interact, is described by nonequilibrium models like RSP [19, 20] or PHSD [21] remains to be seen. This observations opens in any case the opportunity to study the plasma expansion in detail, when the high precision data with new vertex detectors at the RHIC experiments will be published.

2. References

[1] S. Borsanyi *et al.*, J. Phys. G **38**, 124101 (2011).
[2] A. Andronic, P. Braun-Munzinger, K. Redlich and J. Stachel, J. Phys. G **38** (2011) 124081 [arXiv:1106.6321 [nucl-th]].
[3] K. Werner, I. Karpenko, T. Pierog, M. Bleicher and K. Mikhailov, Phys. Rev. C **82** (2010) 044904 [arXiv:1004.0805 [nucl-th]].
[4] J. Steinheimer, J. Aichelin and M. Bleicher, Phys. Rev. Lett. **110** (2013) 4, 042501 [arXiv:1203.5302 [nucl-th]].
[5] M. Luzum and P. Romatschke, Phys. Rev. C **78** (2008) 034915 [Erratum-ibid. C **79** (2009) 039903] [arXiv:0804.4015 [nucl-th]].
[6] J. Aichelin and K. Werner, J. Phys. G **37** (2010) 094006 [arXiv:1008.5351 [nucl-th]].
[7] W. Cassing and E.L. Bratkovskaya, Nucl. Phys. A831 (2009) 215, arXiv:0907.5331 [nucl-th].
[8] P. B. Gossiaux, S. Vogel, H. van Hees, J. Aichelin, R. Rapp, M. He and M. Bluhm, arXiv:1102.1114 [hep-ph].
[9] P. F. Kolb and U. W. Heinz, arXiv:nucl-th/0305084. published in Qurak Gluon Plasma, World Scientific
[10] G. D. Moore and D. Teaney, Phys. Rev. C **71** 064904 (2005).
[11] B. Svetitsky, Phys. Rev. D **37** (1988) 2484.
[12] P. B. Gossiaux, J. Aichelin Phys. Rev. **C78**, 014904 (2008), arXiv:0802.2525 [hep-ph]
[13] P. B. Gossiaux, R. Bierkandt and J. Aichelin, Phys. Rev. C **79** (2009) 044906 [arXiv:0901.0946 [hep-ph]].
[14] P. B. Gossiaux and J. Aichelin, J. Phys. G **36** (2009) 064028 [arXiv:0901.2462 [nucl-th]].

Journal of Physics: Conference Series **623** (2015) 012002 doi:10.1088/1742-6596/623/1/012002

[15] P. B. Gossiaux, J. Aichelin, T. Gousset and V. Guiho, J. Phys. G **37**, 094019 (2010) [arXiv:1001.4166 [hep-ph]].
[16] J. Aichelin, P. B. Gossiaux and T. Gousset, Phys. Rev. D **89** (2014) 7, 074018 [arXiv:1307.5270 [hep-ph]].
[17] M. Nahrgang, J. Aichelin, P. B. Gossiaux and K. Werner, Phys. Rev. C **90** (2014) 2, 024907 [arXiv:1305.3823 [hep-ph]].
[18] L. Adamczyk *et al.* [STAR Collaboration], Phys. Rev. Lett. **113** (2014) 14, 142301 [arXiv:1404.6185 [nucl-ex]].
[19] R. Marty and J. Aichelin, Phys. Rev. C **87** (2013) 3, 034912 [arXiv:1210.3476 [hep-ph]].
[20] R. Marty, E. Bratkovskaya, W. Cassing and J. Aichelin, arXiv:1412.5375 [hep-ph].
[21] T. Song, H. Berrehrah, D. Cabrera, J. M. Torres-Rincon, L. Tolos, W. Cassing and E. Bratkovskaya, arXiv:1503.03039 [nucl-th].

The LHCb Trigger System: Present and Future

Johannes Albrecht
TU Dortmund University
On behalf of the LHCb Collaboration

E-mail: johannes.albrecht@tu-dortmund.de

Abstract. The current LHCb trigger system consists of a hardware level, which reduces the LHC inelastic collision rate of 30 MHz to 1 MHz, at which the entire detector is read out. In a second level, implemented in a CPU farm, the event rate is reduced to about 5 kHz. The major bottleneck in LHCb's trigger efficiencies for hadronic heavy flavour decays is the hardware trigger. The LHCb experiment plans a major upgrade of the detector and DAQ system in the LHC shutdown of 2018. In this upgrade, a purely software based trigger system is being developed, which will have to process the full 30 MHz of inelastic collisions delivered by the LHC. Both the current trigger system and its planned upgrade are discussed in these proceedings.

1. Introduction

The LHCb detector at the LHC is a precision experiment dedicated to beauty and charm physics, covering a rapidity range of $2 < \eta < 5$ [1]. Ref. [2] summarizes the performance of the LHCb detector in Run 1 (2010 – 2012). The trigger system during Run 1 [3, 4] consisted of a fixed latency hardware level which reduced the visible bunch crossing rate to 1 MHz. At this rate, the whole detector was read out and a flexible software High-Level Trigger (HLT) was run to further reduce the rate to about 5 kHz, which was written to offline storage. This configuration allowed LHCb to record the largest beauty and charm hadron samples, at a very high signal purity.

For the second data taking period (Run 2, 2015 – 2018), the detector will be mostly unchanged with respect to Run 1, while the trigger software will undergo an incremental update. The software trigger will be split in two separate instances, the first one running synchronous with the LHC collisions while the second stage will be run asynchronously. This split allows to to perform a real-time calibration and alignment of the detector. The software trigger is thus able to select events based on a reconstruction with a quality almost identical to the offline processing.

The LHCb experiment plans major upgrades in preparation for the third LHC data taking period (Run 3, 2019++). As part of the detector upgrade, all of the front-end devices will be replaced in order to read the detector out at the full bunch crossing rate of 40 MHz [5]. Removing the readout bottleneck allows the complete event reconstruction at the full collision rate and hence an unprecedented trigger performance, especially for hadronic modes.

This note discusses the LHCb trigger system and its performance in Run 1 as well as the prospects and preparations for Run 2 and 3.

2. Run 1: Trigger Performance in 2009 – 2012

The trigger system of the LHCb experiment in Run 1 consists of two levels: the first level, implemented in hardware (L0), and the High Level Trigger (HLT), implemented in a CPU farm of about 29 000 logical cores. The LHCb trigger system and its performance during 2011 data taking has been described in detail in Ref. [4]. Here, the adjustments of the system for 2012 and its performance are discused. The LHCb detector was originally designed for an instantaneous luminosity of $\mathcal{L} = 2 \cdot 10^{32}\,\mathrm{cm}^{-2}\mathrm{s}^{-1}$. During the running period of 2012, this design luminosity was doubled to $\mathcal{L} = 4 \cdot 10^{32}\,\mathrm{cm}^{-2}\mathrm{s}^{-1}$, which was possible thanks to the excellent performance of both the detector hardware and the trigger system. In these conditions, the average number of visible interactions per bunch crossing is $\mu = 1.6$. One novel feature of the Run 1 trigger was that of event deferral: 20% of all L0 accepted events were stored on disk to be processed by the HLT during the LHC inter-fill time, which made an effective 25% of extra CPU available.

The trigger efficiency is evaluated using events that are reconstructed using the full offline software and selected with the final analysis selection for the respective channel. Thus, the trigger efficiency contains only the additional inefficiency due to simplifications used in the trigger, possible alignment inaccuracies, worse resolution than the offline reconstruction or harder cuts imposed by rate and processing time limitations. The method to evaluate the trigger performance and its implementation are discussed in Refs. [4, 6].

The rate of visible collisions in LHCb is about 13 MHz, which then is reduced by the L0 trigger to 1 MHz, at which the full detector is read out. The L0 trigger is implemented in custom hardware and has a latency of $4\,\mu$s. It triggers on high transverse momentum (p_T) muons and on large transverse energy (E_T) deposition in the calorimeter.

The High Level Trigger consists of two stages, HLT1 and HLT2. The first stage, HLT1, performs a partial event reconstruction and an inclusive selection of signal candidates. At the reduced rate of 80 kHz, HLT2 performs a full event reconstruction with only minor adjustments as compared to the offline reconstruction sequence. After this reconstruction, a set of inclusive and exclusive selections reduces the trigger rate to 5 kHz, which are saved for later offline analysis. The rates discussed above are average rates from the 2012 run of the LHC, in 2011 the HLT1 output rate was approximately 40 kHz and the HLT2 output rate was 3 kHz.

2.1. Hadron trigger performance
The L0 hardware trigger performance for hadronic modes is shown in Fig. 1. It is most efficient on two prong beauty decays (~40%), and least for four prong charm decays (~20%). The other modes lie in between.

Hadronic signals are generally selected in the HLT most efficiently by the inclusive beauty and charm trigger lines for both levels of the HLT. The HLT1 line selects good quality track candidates based on their p_T and displacement from the primary vertex. This trigger line gets the dominant part of the HLT1 bandwidth allocated. It is the dominant trigger line for most physics channels that do not contain leptons in the final state. The performance of HLT1 for hadronic signatures is shown in Fig. 2 as a function the p_T of the decaying charm or beauty hadron.

A multivariate selection is used to trigger B decays into charged hadrons in an inclusive selection based on two- three- and four-prong vertices. These trigger lines are based on a BDT classifier that uses discretized input variables [7] which ensures a fast and robust implementation. A crucial input to the BDT is the corrected mass, which takes the missing momentum transverse to the direction of flight into account. This allows the trigger to select heavy flavour decays even when some final state particles are not reconstructed. Fig. 3 shows the efficiency for the inclusive HLT2 trigger lines on a set of hadronic beauty decays.

2.2. Lepton trigger performance
The L0 hardware trigger selects muons based on their track segment reconstructed in the Muon chambers. The corresponding efficiency is shown in Fig. 4 for the decay $B^- \to J/\psi K^-$ with J/ψ ($\mu^+\mu^-$). The single muon trigger contributes the dominant part to the efficiency. The largest

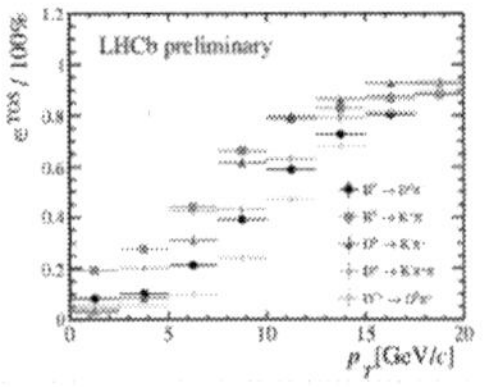

Figure 1. L0 hadron trigger performance: Trigger efficiency for beauty and charm decay modes, listed in the legend.

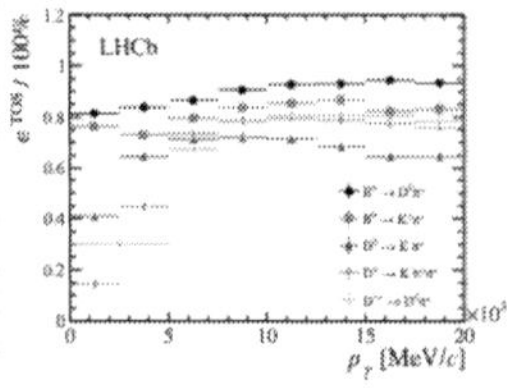

Figure 2. HLT1 inclusive trigger performance: efficiency for various channels as a function of p_{T} of the decaying hadron.

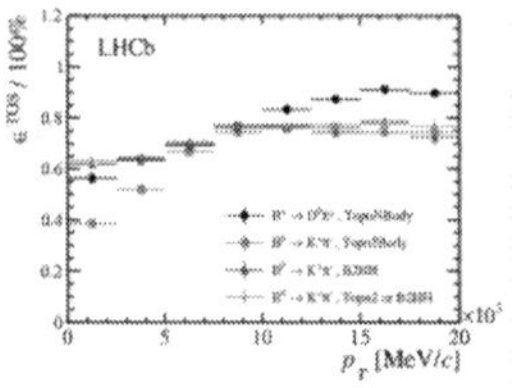

Figure 3. HLT2 inclusive beauty trigger performance as a function of p_{T} of the decaying hadron.

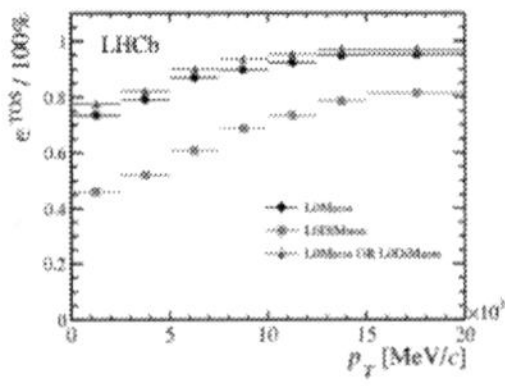

Figure 4. L0 muon trigger performance: Trigger efficiency for selected $B^- \to J/\psi K^-$ candidates.

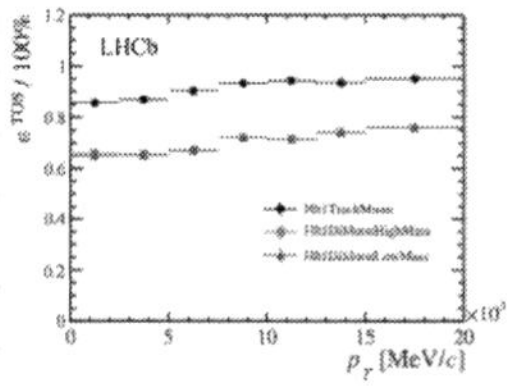

Figure 5. HLT1 muon trigger performance: Efficiency for $B^- \to J/\psi K^-$ candidates as function of B^+ p_{T}.

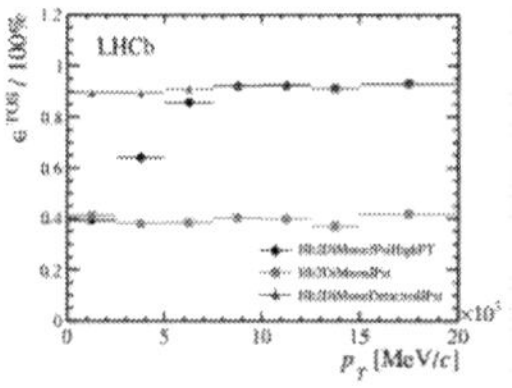

Figure 6. HLT2 muon trigger performance.

inefficiency originates in the tight muon identification requirements inside the L0 reconstruction algorithm. The L0 dimuon trigger selects a small fraction of additional candidates at lower transverse momenta. The combined efficiency for both L0 muon triggers, integrated over p_{T}, is 89%.

In the first part of the HLT, muons are selected both by an inclusive single track line (analogous to the hadron line discussed above) and by dimuon triggers. The dominant inefficiency for these lines originates in the online muon identification algorithms. The performance of HLT1 at selecting muonic signatures is shown in Fig. 5 as a function of p_{T} of the B^+ candidate.

In HLt2, several trigger lines select events with one or two identified muons. The muon identification procedure is identical to the one used in offline analysis. Single muon candidate events are selected with very tight p_{T} and vertex separation requirements. Dimuon candidate events are selected with either mass or vertex separation requirements, or with a combination of both. The efficiency of the HLT2 muon selections is shown in Fig. 6.

3. Run 2: Trigger preparation for 2015 – 2018

In 2015, the collision energy will be increased compared to 2012 (from 8 TeV to 13 TeV), and the bunch spacing will reach the design value of 25 ns. In these conditions the 2012 instantaneous luminosity of $4 \times 10^{32}\,\mathrm{cm}^{-2}\mathrm{s}^{-1}$ can be achieved at a lower pile-up (1 interaction/crossing instead of 1.6). The yield will further benefit from increased cross-sections for b- and c-hadron production. The requirements on the High Level Trigger will be partly relaxed by increasing the output rate for storing data to tape (12.5 kHz). A part of this data will be saved in a reduced data format for analysis without offline processing. On the other hand, the maximum read-out rate will be still limited to 1 MHz, and the new conditions will be challenging for the L0 trigger.

The computing resources will be adjusted to the 2015 conditions by doubling the Event Filter Farm CPU power and increasing the disk space for deferral. Changes will also be introduced in the architecture: both HLT levels will be split in two separate physical processes, allowing the second stage to run asynchronously. This allows a more efficient use of disk space and of the deferral mechanism, ultimately leading to an increase in the resources available for the High Level Trigger processing. A real time alignement and calibration can be performed on the output of HLT1 and used to improve the quality of the reconstruction in HLT2.

The track reconstruction sequence will be improved such that it allows to adopt multi-track trigger selections already in the first stage of the High Level Trigger.

4. Run 3: Trigger strategy from 2018 onwards

The LHC will provide LHCb with an instantaneous luminosity of $2 \times 10^{33}\mathrm{cm}^{-2}\mathrm{s}^{-1}$. This implies signal rates of about 300 kHz of beauty hadrons and 800 kHz for charm hadrons that are reconstructible inside the detector [8]. This represents a fundamental change in trigger requirements of LHCb. It is no longer the case that the trigger must reject background from signal, it must now categorise signal according to physics requirements. Such a strategy requires significantly more information in the trigger than can be provided by low-latency, hardware based solutions.

With the removal of the L0 hardware trigger, LHCb will be the first hadron collider experiment to deploy a trigger exclusively in software, using off-the-shelf hardware. The upgraded LHCb trigger will select and categorise events at the full collision rate. The advantages of such a system are clear. Software is easily modified, allowing an unprecedented flexibility as and when the LHCb physics programme changes. It also has the advantage that computing power is readily upgradeable and benefits from the ability to purchase more CPU power for the same price at a later stage. The deferral technique used in Run 1 will be leveraged in Run 2 and Run 3. Subdetector alignment and calibration will be performed while events are buffered to

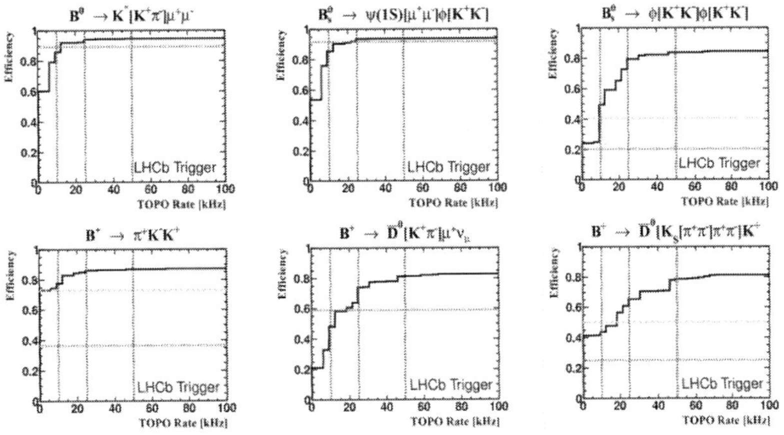

Figure 7. Efficiencies as a function of the output rate for several benchmark decay modes. Red lines indicate the efficiency during Run 1. Green lines indicate twice the Run 1 efficiency. Blue lines indicate three prospective output rates at 10, 25 and 50 kHz respectively.

disk, permitting the use of offline-quality reconstruction, and reducing the need for reprocessing of data at a later stage.

In the Run 3 trigger, all tracks will be made available at the earliest stages of the trigger and the full tracking sequence will be performed upfront [9]. In the trigger tracking sequence the same algorithms as in the offline sequence are used, but the sequencing and configuration are modified to reconstruct the most valuable tracks first and perform some basic candidate selection using these. Slower more specialised reconstruction algorithms only need to be executed later in the HLT algorithm flow.

After the tracking sequence, the remaining timing budget is available to apply offline-quality trigger selections. An inclusive, multivariate beauty signal selection is then performed directly on the 30 MHz of inelastic collision data. The performance of the inclusive trigger selections is summarised in Fig. 7. This inclusive trigger selection is complemented by a set of exclusive trigger selections, that are designed to maximize the efficiencies for specific core modes or to minimize systematic uncertainties. One example are trigger selections that do not bias the decay time of the selected mesons.

5. Summary

The LHCb trigger system has performed extraordinarily well in the first running phase of the LHC. It is designed to select charm and beauty hadrons in a large range of decay modes. The flexible design of the HLT, fully deployed in software, allows to quickly adjust to changes in running conditions and physics goals. Inclusive selections in the full trigger chain allow an efficient trigger for basically any beauty decay to charged tracks. Several innovative concepts have enabled this performance: the deferred triggering allows to optimise the trigger usage for mean instead of peak usage of the available computing resources. Multivariate selections allow the inclusive selection of beauty decays into charged tracks with high efficiency.

For the upcoming Run 2, several improvements of the trigger system are planned: the two software trigger levels will be decoupled which allows calibrations between them and thus a performance much closer to the one achieved offline. The use of event buffering for alignment and calibration permits selections that would otherwise only be available in the offline analysis environment.

The upgraded LHCb trigger represents a turning point in the design of hadron collider trigger systems. The approach to trigger purely in software running on commercially available hardware allows an inexpensive, flexible and scalable design. The upgraded trigger will perform offline-quality tracking at the full collision rate, which allows a number of very efficient inclusive and exclusive trigger selections. Subject to output bandwidth requirements, gains of up to a factor of four in efficiency are possible for several LHCb benchmark modes.

6. References

[1] A. A. Alves Jr. et al., The LHCb Detector at the LHC, JINST 3, S08005 (2008).
[2] R. Aaij *et al.* [LHCb Collaboration], LHCb Detector Performance, Int.J.Mod.Phys. A30 (2015) 1530022
[3] R. Aaij et al., The LHCb trigger and its performance in 2011, JINST 8, P04022 (2013).
[4] J. Albrecht et al., Performance of the LHCb High Level Trigger in 2012, J. Phys.: Conf. Ser. 513, 012001 (2014).
[5] I. Bediaga et al., LHCb Trigger and Online Upgrade Technical Design Report, CERN-LHCC-2014-016 (2014).
[6] S. Tolk, J. Albrecht, F. Dettori and A. Pellegrino, Data driven trigger efficiency determination at LHCb, LHCb-PUB-2014-039
[7] V. V. Gligorov and M. Williams, Efficient, reliable and fast high-level triggering using a bonsai boosted decision tree, JINST 8, P02013 (2013).
[8] C. Fitzpatrick and V. Gligorov, Anatomy of an upgrade event in the upgrade era, and implications for the LHCb trigger, LHCb-PUB-2014-027
[9] T. Head, J. Albrecht, and K. Dungs, Track finding scenarios for the software trigger of the upgraded LHCb experiment, LHCb-PUB-2014-028

Search for Supersymmetry at CMS

Christian Autermann
RWTH Aachen University, I. Physikalisches Institut B, 52056 Aachen, Germany
E-mail: christian.autermann@cern.ch

Abstract. CMS has searched for supersymmetry using the full dataset recorded until the end of 2012, at a center-of-mass energy of 8 TeV corresponding to an integrated luminosity of 20 fb^{-1}. The observations are consistent with the expectations from the standard model. CMS maintains a broad spectrum of search strategies that target a wide range of supersymmetric scenarios. A selection of results from very inclusive and almost model-independent searches to very specific and specialized analyses will be discussed in this article.

1. Introduction

Supersymmetry extends the standard model by adding a new SUSY-particle for each standard model particle. The superpartners differ in spin by $\frac{1}{2}$ and in mass, because the supersymmetry is broken. All other properties of the superpartners, such as charges and couplings, are identical, so that the corrections to the Higgs mass-squared parameter cancel at first order. The remaining correction terms depend on the superpartner masses but only logarithmically on the energy scale. As long as the superpartner mass-differences, in particular the 3rd generation squark masses, are not too large, no unnatural amount of fine-tuning is necessary. Supersymmetry provides a dark matter candidate particle, if the R-parity and thus the lepton- and baryon-number violating couplings are sufficiently small. The lightest stable supersymmetric particle (LSP) is typically the neutralino.

No direct evidence for supersymmetry has been observed at the CMS experiment [1] yet. Electroweak precision measurements of standard model parameters, like the $B_s \to \mu^+\mu^-$ branching fraction [2] or the Higgs mass measurements [3] constrain the supersymmetry parameter space. In combination with direct searches, models like the constrained minimal supersymmetric standard model (cMSSM) can be tested [4]. The Fittino cMSSM-fit shows that parameter points in best agreement with all constraints have particles masses of typically more than 1 TeV and p-values of less than 2.5%. With this justification the simplest cMSSM can therefore be assumed to be excluded at more than 95% confidence-level (CL).

As the details of the supersymmetry breaking mechanism are unknown and simplest models like the cMSSM have become uninteresting, search analyses are organized, optimized, and interpreted using simplified model spectra (SMS). The SMS model typical production processes of supersymmetric particles and their decays, where the involved sparticles masses are the only free parameters. The resulting cross section limits are typically presented in these mass planes. In the following, the various searches for supersymmetry are discussed in the order of specialization: First, inclusive analyses with very few model assumptions and therefore typically good sensitivity over wide parameter regions are presented, followed by increasingly specialized searches.

2. Inclusive searches

Inclusive searches for supersymmetry obtain generally good sensitivity over wide regions of the accessible SUSY phase-space, also because only a small amount of model dependent optimization of the analyses is necessary. Searches in the fully hadronic final state require the selected events only to have large missing transverse energy ($E_{\mathrm{T}}^{\mathrm{miss}}$) caused by the two LSPs. Alternatively, analyses attempt to identify the two cascade decay-chains through hemisphere algorithms, exploiting the mass hierarchy of the SUSY particles. Two exemplary analyses approaches are discussed in this section.

In the $E_{\mathrm{T}}^{\mathrm{miss}}$ +jets search [5] the data sample is divided into 36 statistically independent signal search regions that are combined into one likelihood-ratio test-statistic, rather than optimizing the selection for particluar signal scenarios. The signal regions are defined with respect to three jet multiplicity categories ($3-5$, $6-7$, and 8 or more jets), the scalar sum of jet transverse momenta (H_{T}) and the missing transverse jet-momentum ($H_{\mathrm{T}}^{\mathrm{miss}}$). $H_{\mathrm{T}}^{\mathrm{miss}}$ and $E_{\mathrm{T}}^{\mathrm{miss}}$ are similar, except that the $H_{\mathrm{T}}^{\mathrm{miss}}$ is calculated only from jets, while the $E_{\mathrm{T}}^{\mathrm{miss}}$ uses all calorimeter energy depositions, thus reducing the influence of noise and multiple simultaneous interactions.

An alternative approach [6] is based on the M_{T2} variable, which is a measure of the transverse momentum imbalance in an event and replaces the $E_{\mathrm{T}}^{\mathrm{miss}}$-like variables. Similarly as before, multiple statistically independent search regions are defined with respect to the value of the M_{T2} variable, the hadronic energy in the event H_{T}, the jet multiplicity, and the number of jets identified as originating from bottom quarks.

The precise estimation of the standard model backgrounds is the key to all search analyses and especially challenging in the fully hadronic final state, which therefore serves well as an example to discuss the different data-driven background estimation strategies. Typically the individual background sources are estimate from, and validated in, data control regions. The QCD-multijet background source is difficult to simulate with Monte Carlo (MC) generators, because of the large QCD cross-section and the rare probability for effects that lead to $E_{\mathrm{T}}^{\mathrm{miss}}$

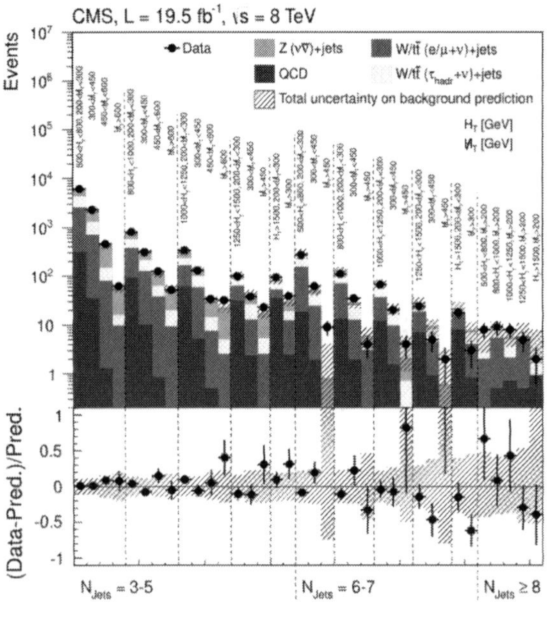

Figure 1. Summary of the observed number of events in each of the 36 search regions in comparison to the corresponding background prediction for the $H_{\mathrm{T}}^{\mathrm{miss}}$ +jets analysis [5]. The hatched region shows the total uncertainty of the background prediction. Good agreement, and therefore no sign for new physics, is observed. The different statistically independent bins are combined into one single likelihood-ratio test-statistic, maximizing the analysis sensitivity to the signal over a wide range of the examined phase-space.

in the final state. This background can be modeled by data events, where the jet momenta are adjusted in a first step, such that the $E_{\rm T}^{\rm miss}$ vanishes. At this stage, the sample of these events are comparable to generator level QCD-multijet simulation. In a second step, the jet momenta are smeared according to jet energy resolution templates, that also can be determined on data events and where especially non-Gaussian tails are of crucial importance. The resulting data-driven QCD-multijet estimation models the expected background with an uncertainty of typically better than 50%, depending on the signal region. Backgrounds with genuine $E_{\rm T}^{\rm miss}$ like $W \to \nu\tau_{\rm had}$ can be modeled using a muon data-sample, where the muon per event is replaced with a jet and the $E_{\rm T}^{\rm miss}$ is recalculated according to a tau-energy response template, that models the hadronic decay of tau leptons. The same data control sample can be used to describe the $t\bar{t} \to q\bar{q}b\bar{b}l\nu$ and $W \to l\nu$ backgrounds, where the lepton fails the acceptance or identification criteria. Finally, the $Z \to \nu\bar{\nu}$ background is of relevance and can be modeled through $Z \to \mu^+\mu^-$, $W \to \mu\nu$, or γ+jets events. The relative importance of all four standard model backgrounds depends on the signal phase-space region under study.

The observed numbers of events are found to be consistent with the expected standard model background evaluated from the data, as shown in Figure 1 for the $H_{\rm T}^{\rm miss}$ +jets analysis. The results are presented in the context of simplified models (Fig. 2, left) for the $H_{\rm T}^{\rm miss}$ +jets analysis and (Fig. 2, right) for the $M_{\rm T2}$ analysis), where final states are described by the pair production of new particles decaying to one, two, or more jets and a weakly interacting stable neutral particle, e.g. the lightest supersymmetric particle (LSP). Squark masses below 780 GeV and gluino masses of up to 1.1−1.3 TeV are excluded at 95% CL within the studied models depending on the neutralino masses and on the squark flavour.

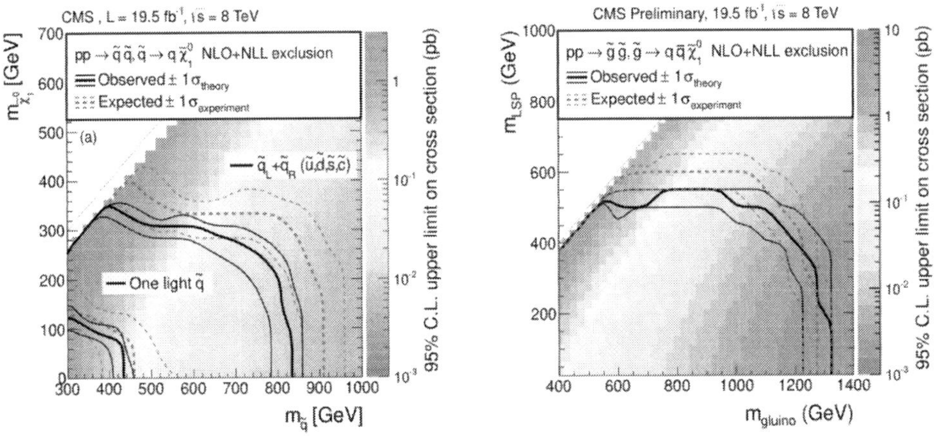

Figure 2. The observed 95% CL upper cross section limits together with the observed and expected exclusion contours on (left) the squark-squark production in the m(squark)-m(LSP) mass plane and (right) for gluino pair production, with $\tilde{g} \to q\bar{q}\tilde{\chi}_1^0$ in the m(gluino)-m(LSP) simplified model plane. The upper set of curves for the limits on squark pair-production corresponds to the scenario where the first two generations of squarks are degenerate and light, while the lower set corresponds to only one light accessible squark.

3. Searches for 3$^{\rm rd}$ generation squarks

Searches for direct production of 3$^{\rm rd}$ generation squarks are of particular interest because the naturalness arguments require sbottom and stop squarks to be light, approximately below one

TeV. Therefore, the cross section for a direct production of stop and sbottom squarks, as shown in Figures 3, 4 can be sufficiently large. Two exemplary searches for 3rd generation squarks pair production are discussed in the following.

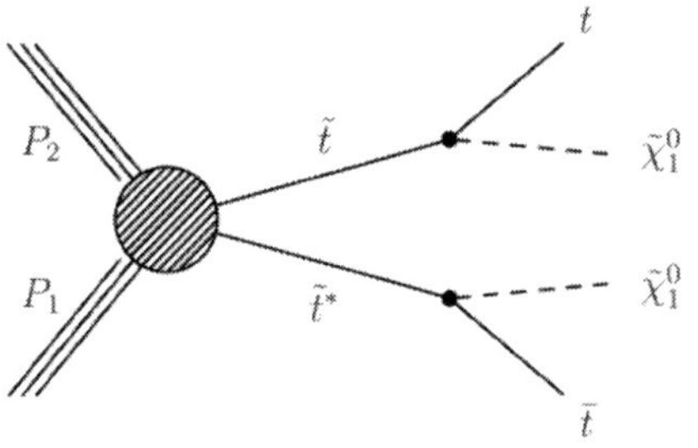

Figure 3. Simplified model for direct stop production

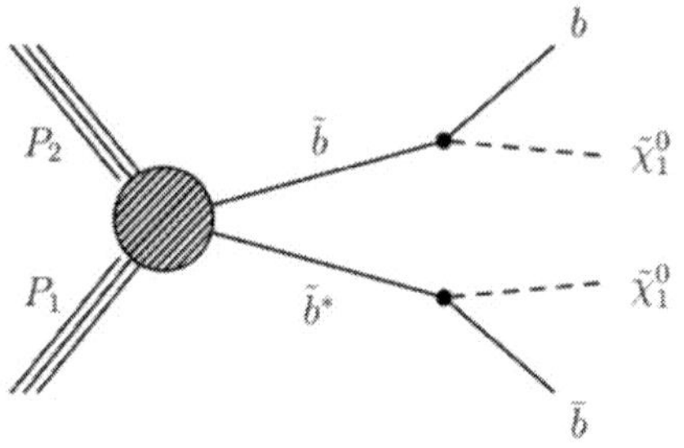

Figure 4. Simplified model for direct sbottom production

The search [7] for pairs of top squarks decaying to a top quark and a stable, weakly interacting, massive particle selects events containing multiple jets, with at least one identified as originating from a b-quark, and large missing transverse momentum. The search also employs a novel top quark tagging algorithm for identifying a top quark candidate decaying hadronically. The observed event yields are consistent with the SM expectations. Assuming both top squarks decay always into a top quark and a weakly interacting, neutral particle $\tilde{\chi}_1^0$, the production of top squarks with mass less than 535 GeV is excluded at 95% CL for $\tilde{\chi}_1^0$ masses less than 10 GeV.

The search [8] for direct production of bottom squark pairs is performed in a final state of two jets, where at least one is tagged as originating from a b-quark, accompanied by large missing transverse momentum, as well as large transverse mass and contransverse mass M_{CT} calculated from the jets and missing transverse momentum [9]. The signal region is defined by exclusive bins with respect to the b-tagged jet multiplicity and M_{CT}. The standard model backgrounds are modeled using μ + jets data samples, with and without a b-tag. The estimated standard model background is found to be in good agreement with the observed data. The results are interpreted in terms of a simplified model production of a pair of bottom squarks with each bottom squark decaying to a bottom quark and a weakly interacting stable neutral particle. The production of bottom squarks with mass up to 700 GeV is excluded at 95% CL for neutralino masses less than 50 GeV.

Figure 5 shows a summary of the CMS direct stop search results in the stop mass – neutralino mass plane. The exclusion contours reach up to stop masses of 750 GeV and neutralino masses of up to 250 GeV, covering a significant amount of the interesting phase-space. Nevertheless, white regions on the map, where a perfectly natural SUSY scenarios could hide at relative low masses, are observable. Here, future specialized analyses, as well as improved standard model $t\bar{t}$ and WW theory predictions that allow better distinction between signal and background, will be crucial to examine this interesting region further.

4. Searches for gauge mediated supersymmetry breaking scenarios

A different flavour of supersymmetry are models with gauge mediated supersymmetry breaking (GMSB), where the Gravitino $\tilde{G}$ with negligible mass $\lesssim$ 1 GeV is the LSP and the next-to-lightest supersymmetric particles (NLSP) determines the final state. In the following the NLSP is assumed to be the neutralino $\tilde{\chi}_1^0$.

Two searches for gauge mediated supersymmetry breaking scenarios are presented that are sensitive to a bino- or wino-like neutralino mixing. Here, decays like $\tilde{\chi}_1^0 \to \tilde{G}\gamma$ and $\tilde{\chi}_1^0 \to \tilde{G}Z$

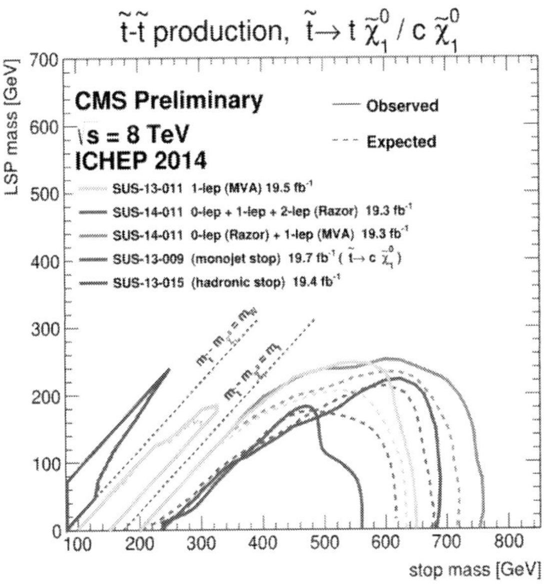

Figure 5. Summary of limits for direct stop searches in the stop mass – neutralino mass plane. The present searches are insensitive in regions where the mass difference beween the stop and the neutralino is similar to the standard model top or the W-boson mass. In these cases the analyses cannot distinguish the two stop decays from standard model $t\bar{t}$ or WW production. Similarly, near the diagonal where the stop and neutralino are nearly mass-degenerate the final state consists only of $E_{\rm T}^{\rm miss}$ and only very soft decay products. Specialized analyses can gain sensitivity in these regions.

are possible, leading to final states with $E_{\rm T}^{\rm miss}$, jets, and possibly photons. The analyses pursue different inclusive search strategies. One analysis [10] requires at least one photon and a large amount of missing transverse energy, while the other analysis [11] selects events with at least two photons and uses hemisphere algorithms to calculate kinematical "razor" variables $\rm M_R$ and $\rm R^2$ to search for the signal. The standard model background expectation is dominated by QCD-multijet and direct photon+jet production, as well as processes with initial state photon radiation. The background is mainly evaluated directly from the data.

While the above discussed searches assume strong production of squarks and gluinos in the context of a GMSB scenario with bino- and wino-like neutralino mixings, also more specialized searches were developed. The stop-higgsino search [12] assumes that only the supersymmetric partners of the top-quark and the Higgs boson (higgsino-like neutralino) are accessible, leading to direct stop pair production. In the stop decays standard model b-quarks and Higgs bosons h are created. Events with two photons forming a Higgs boson candidate and at least two b-quark jets are selected by the analysis.

The results are interpreted in the context of General Gauge Mediated (GGM) [13] supersymmetry and within simplified model scenarios. No evidence for SUSY production is found, and 95% CL upper limits are set on the production cross section. Lower limits from the single-photon analysis for wino-like neutralino mixings range beyond 0.8 TeV in gluino mass and 1 TeV in squark mass in the same GGM phase space, extending previous limits by about 100 GeV and 200 GeV, respectively. The results and limits are shown in Fig. 6. Limits from the diphoton razor analysis range beyond 1.3 TeV in gluino mass and beyond 1.5 TeV in squark mass for bino-like neutralino mixings in the studied GGM phase space as shown in Fig. 7, extending previous limits by up to 300 GeV and 500 GeV, respectively. In the case of higgsino-like neutralino mixing stop mass below 360 GeV to 410 GeV can be excluded, depending on the higgsino mass.

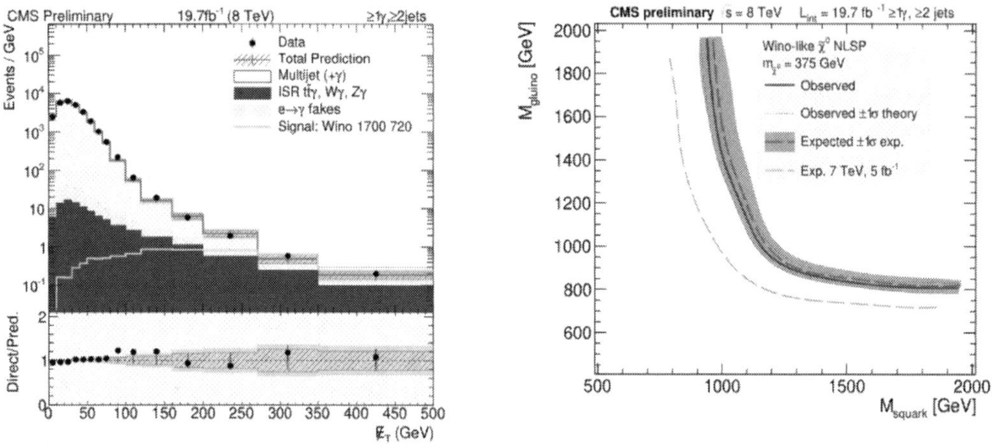

Figure 6. Observed $E_{\mathrm{T}}^{\mathrm{miss}}$ distribution of the single-photon analysis [10] compared to the standard model background estimated from data control regions (left). The $E_{\mathrm{T}}^{\mathrm{miss}} > 100$ GeV bins are combined, and interpreted in the GGM m(squark) - m(gluino) mass plane for wino-like neutralino mixing (right).

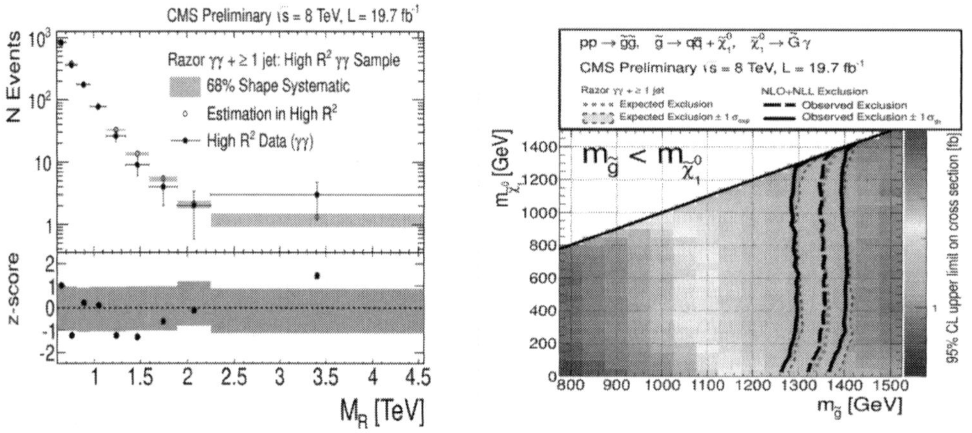

Figure 7. Observed distribution of the "razor" variable M_R [11] compared to the standard model background estimation from a data sideband region (left). Cross section limits and exclusion contours for a $\tilde{g}\tilde{g} \to q\bar{q}q\bar{q}\tilde{\chi}_1^0\tilde{\chi}_1^0$ simplified model are shown in the m(gluino) - m($\tilde{\chi}_1^0$) mass plane (right).

5. Searches for electroweak production of supersymmetry

Electroweak production of supersymmetry has significantly lower cross section compared to strong production. In addition, the particles in the final state have less energy compared to events of strong production, which makes a discovery of new physics in these channels extremely challenging. On the other hand, gaugino masses might be rather light in contrast to squark and gluinos. Here, a search [14] for electroweak pair production of neutralinos and charginos, leading to decay channels with Higgs, Z, and W bosons and undetected lightest SUSY particles (gravitinos $\tilde{G}$ in the GMSB framework or neutralinos $\tilde{\chi}_1^0$) is discussed. The produced neutralinos may decay either to a Higgs boson (h) and an LSP, or to a Z boson and an LSP, leading to hh, hZ, and ZZ states states with missing transverse energy $E_{\mathrm{T}}^{\mathrm{miss}}$. Charginos may directly decay to W bosons and an LSP, leading to hW states with $E_{\mathrm{T}}^{\mathrm{miss}}$. The decays of a Higgs boson to a bottom-quark pair, to a photon pair, and to final states with leptons are considered in conjunction with hadronic and leptonic decay modes of the Z and W bosons.

The signal sensitive variables utilized to discriminate signal against the standard model background depend on the analysis channel and include large values of either missing transverse energy $E_{\mathrm{T}}^{\mathrm{miss}}$, transverse mass M_{T}, or the scalar sum S_{T} of the two boson transverse momenta. In Fig. 8 the sensitivity of the individual analysis channels is shown in the plane of the higgsino-like neutralino mass versus the branching fraction of the neutralino decay into a standard model Higgs boson and the gravitino LSP.

No evidence is found for supersymmetric particles, and 95% CL upper limits are evaluated for the respective pair production cross sections and for neutralino and chargino mass values. Higgsinos with a mass value below 380 GeV are excluded. For small values of the LSP mass, chargino mass values up to 210 GeV can be excluded, where the $\tilde{\chi}_1^{\pm}$ and $\tilde{\chi}_2^0$ masses are taken to be equal. Figure 8 shows the obtained exclusion contours in the plane of the higgsino-like neutralino mass versus the branching fraction of the neutralino decay into a standard model Higgs boson and the gravitino LSP.

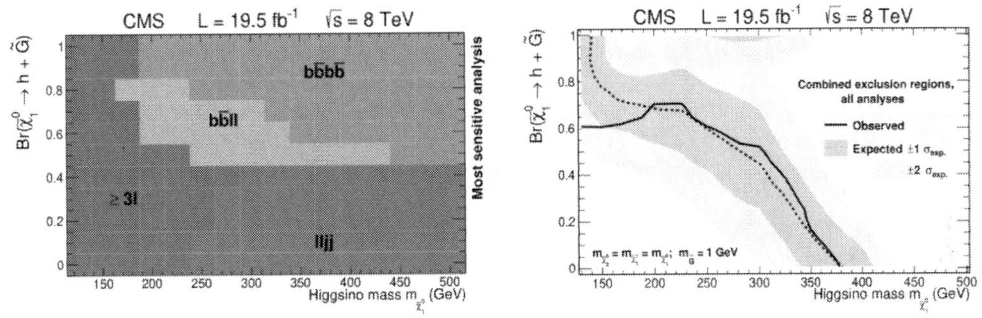

Figure 8. Sensitivity (left) of the individual analysis channels and (right) obtained exclusion contours of the search for electroweak pair production of gauginos [14] shown in the plane of the higgsino-like neutralino mass versus the branching fraction of the neutralino decay into a standard model Higgs boson and the gravitino LSP.

6. Searches for stealth supersymmetry

The previously discussed searches for supersymmetry assume R-parity conservation and are therefore characterized by large missing transverse momentum due to the undetectable LSPs. This, however, is not a necessary property of supersymmetry. The LSP might be charged and thus detectable, or it can be unstable if the R-parity is violated. As a benchmark scenario for supersymmetry with only little or no $E_{\mathrm{T}}^{\mathrm{miss}}$ a search [15] for stealth supersymmetry [16] is discussed. Stealth SUSY models naturally produce low-$E_{\mathrm{T}}^{\mathrm{miss}}$ signatures without violating R-parity. Low-scale SUSY breaking is assumed and a new hidden sector of particles at the weak scale is introduced, analogous to the SUSY-breaking sector, through which the typical non-stealth LSP, the neutralino $\tilde{\chi}_1^0$, can decay into a lighter hidden-sector SUSY particle $\tilde{S}$. The gaugino becomes the lightest visible sector SUSY particle (LVSP), the LSP $\tilde{G}$ in the stealth-model is produced from the decay of the hidden-sector SUSY particle to its standard model partner S. Since the superpartners are nearly mass degenerate, the produced LSP gains only little momentum, so that only little $E_{\mathrm{T}}^{\mathrm{miss}}$ is observable in the detector.

The analysis is perfomed in a final state with two photons, sensitive to $\tilde{\chi}_1^0 \rightarrow \tilde{S}\gamma$ decays or an oppositely charged electron and muon, sensitive to e.g. $\tilde{\chi}_1^{\mp} \rightarrow \tilde{S}W$ decays. The background is estimated from data side-band regions and found to be consistent with the observed data. The derived limits constrain the masses of squarks and neutralinos in the framework of stealth SUSY. In the photon-analysis 95% CL lower limits on squark masses between 700 GeV and 1050 GeV are derived, depending on the neutralino mass. In the lepton analysis squark masses below 550 GeV at the 95% CL can be excluded.

7. Searches for supersymmetry with a dilepton invariant mass edge

An example for a highly specialized supersymmetry analysis carried out at CMS is the search [17] for a kinematical edge in the invariant mass spectrum of two leptons of same flavor and opposite charge (e^+e^- or $\mu^+\mu^-$). The signal has a characteristic triangular shape, if the two leptons are created in a three body decay of a gaugino, e.g. in a supersymmetric cascade decay. In particular, a heavy neutralino can decay via a slepton into a lighter neutralino and two leptons, e.g. $\tilde{\chi}_2^0 \rightarrow \tilde{l}^+l^- \rightarrow \tilde{\chi}_1^0 l^+l^-$. The intermediate slepton can be on- or off-shell, which slightly influences the shape of the triangular invariant dilepton mass distribution. The upper edge is determined by the mass difference of the two involved neutralinos. A decay via an on-shell Z-boson is also possible $\tilde{\chi}_2^0 \rightarrow \tilde{\chi}_1^0 Z \rightarrow \tilde{\chi}_1^0 l^+l^-$.

The analysis is not particular sensitive to the details of the production process and requires jets and missing transverse momentum in addition to two opposite-sign same-flavor leptons. A feature of the analysis is, that the signal region is dominated ($\sim$ 99%) by $t\bar{t}$ background, which is flavor symmetric with respect to the two leptons, i.e. has as many $e^+e^- + \mu^+\mu^-$ events as it has $\mu^+e^- + \mu^-e^+$ events. Therefore, a data control-region with two opposite-charge but also opposite flavor leptons can be used to model the flavor-symmetric background with good precision. Systematic uncertainties arise only from differences in the lepton trigger, reconstruction, and isolation efficiencies and are of the same order as the statistical uncertainties of the data signal- or control-samples corresponding to 4%, if both leptons are reconstructed in the central detector.

Two analysis strategies are pursued: Counting experiments in three invariant dilepton mass m_{ll} regions below the Z-resonance, on-Z, and above the Z-resonance are performed, where the standard model background is estimated as explained above. The observed results are presented in Table 1. Remarkably, an excess of 2.6σ is observed in the central low-m_{ll} region. A second analysis approach is to perform an unbinned maximum likelihood fit simultaneously to all events in the dilepton same-flavor signal region and the opposite-flavor background control region. The fit determines the free parameters of the signal shape parametrization as shown in Figure 9; an edge position at $m_{ll} = 78.7 \pm 1.4$ GeV is found with a signal strength of 126 ± 41 events in the central detector region corresponding to a local significance of 2.4σ consistent with the result of the counting experiment.

The result has led to the publication of various interpretations, for example [18, 19]. Meanwhile, the ATLAS collaboration has performed a corresponding analysis [20] with slightly different selection criteria. Here, the observed data is well consistent with the standard model background expectations in the low-m_{ll} region beneath the Z-resonance. However, the same ATLAS analysis observes an excess of 3σ on the Z-resonance, where the CMS analysis observes good agreement, however in this case the CMS selection is significantly looser than the ATLAS selection with respect to $E_{\mathrm{T}}^{\mathrm{miss}}$ and the jet energy requirements. In summary, the best explanation for the observed excesses is at this point statistical fluctuations.

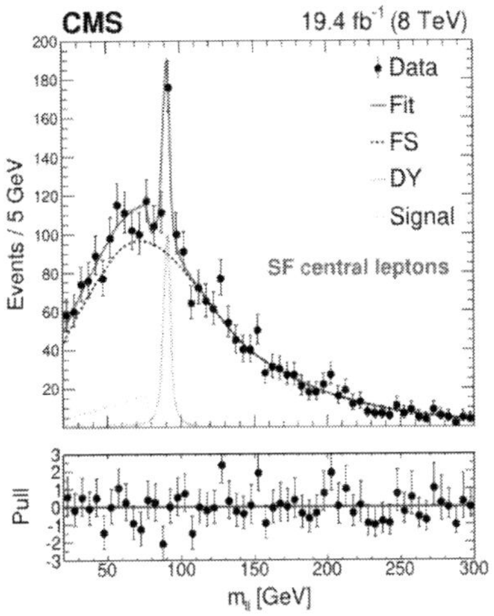

Figure 9. Result of the unbinned maximum-likelihood fit shown in the central dilepton same-flavor signal region of the edge analysis [17]. The fit was performed simultaneously to all events in the signal and in the opposite-flavor background control region. The background and signal shapes were parametrized and the free parameters were determined by the maximum-likelihood of the unbinned fit. The best fit finds a signal of 126 ± 41 events with an upper edge position of m_{ll} = 78.7 ± 1.4 GeV. This corresponds to a local significance of 2.4σ, consistent with the counting experiment.

Table 1. Results of the edge-search [17] counting experiment for event yields in the signal regions. Low-mass refers to $20 < m_{ll} < 70$ GeV, on-Z to $81 < m_{ll} < 101$ GeV, and high-mass to $m_{ll} > 120$ GeV.

	Low-mass		On-Z		High-mass	
	Central	Forward	Central	Forward	Central	Forward
Observed	860	163	487	170	818	368
Total estimated	730 ± 40	158 ± 16	471 ± 32	173 ± 17	771 ± 42	431 ± 35
Observed - estimated	130^{+48}_{-49}	5^{+20}_{-20}	16^{+37}_{-38}	-3^{+20}_{-21}	47^{+49}_{-50}	-62^{+37}_{-39}
Significance	2.6 σ	0.3 σ	0.4 σ	<0.1 σ	0.9 σ	<0.1 σ

8. Conclusion

CMS has searched for supersymmetry using the full dataset recorded until 2012 corresponding to about 20 fb^{-1} at a center-of-mass energy of 8 TeV. The observed data is in agreement with the expectations from the standard model, as demonstrated in Fig. 10. Exclusion limits for squark and gluino masses reach up to and beyond 1 TeV. Limits on 3$^{\rm rd}$ generation squark masses reach between 500 GeV and 800 GeV. The limits on electroweak gauginos and sleptons range typically between 200 GeV and 500 GeV. The reach with respect to all sparticle masses will be extended by at least a factor of two with the total expected data of the high luminosity LHC run. If the naturalness argument is a good guiding principle, than a discovery of supersymmetry might be just around the corner.

References

[1] S. Chatrchyan *et al.* [CMS Collaboration], "The CMS experiment at the CERN LHC," JINST **3** (2008) S08004.
[2] V. Khachatryan *et al.* [CMS and LHCb Collaborations], "Observation of the rare $B_s^0 \to \mu^+\mu^-$ decay from the combined analysis of CMS and LHCb data", arXiv:1411.4413 [hep-ex].
[3] G. Aad *et al.* [ATLAS and CMS Collaborations], "Combined Measurement of the Higgs Boson Mass in *pp* Collisions at $\sqrt{s} = 7$ and 8 TeV with the ATLAS and CMS Experiments," arXiv:1503.07589 [hep-ex].
[4] P. Bechtle, K. Desch, H. K. Dreiner, M. Hamer, M. Krämer, B. O'Leary, W. Porod and B. Sarrazin *et al.*, "How alive is constrained SUSY really?", arXiv:1410.6035 [hep-ph].

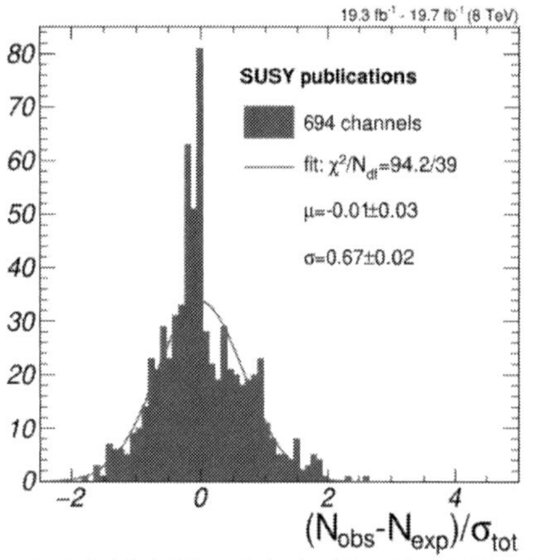

Figure 10. CMS supersymmetry search results per search channel. The distribution is a symmetric Gaussian with a mean at zero, indicating an overall good agreement with the standard model expectations. The peak at small values is caused by channels where only a small amount of background is expected and zero events are observed. The largest excess is from the dilepton analysis, discussed above. Correlation of channels are ignored and upward and downward systematic uncertainty fluctuations were symmetrized for this plot. The plot was prepared with the help of J. Lange.

[5] S. Chatrchyan *et al.* [CMS Collaboration], "Search for new physics in the multijet and missing transverse momentum final state in proton-proton collisions at $\sqrt{s}$= 8 TeV", JHEP **1406** (2014) 055 [arXiv:1402.4770 [hep-ex]].
[6] V. Khachatryan *et al.* [CMS Collaboration], "Searches for supersymmetry using the M_{T2} variable in hadronic events produced in pp collisions at 8 TeV", arXiv:1502.04358 [hep-ex].
[7] CMS Collaboration, "Search for top squarks in multijet events with large missing momentum in proton-proton collisions at 8 TeV", CMS-PAS-SUS-13-015.
[8] CMS Collaboration, "Search for direct production of bottom squark pairs", CMS-PAS-SUS-13-018.
[9] D. R. Tovey, "On measuring the masses of pair-produced semi-invisibly decaying particles at hadron colliders," JHEP **0804** (2008) 034 [arXiv:0802.2879 [hep-ph]].
[10] CMS Collaboration, "Search for supersymmetry in events with one photon, jets and missing transverse energy at sqrt(s) = 8 TeV", CMS-PAS-SUS-14-004.
[11] CMS Collaboration, "Search for supersymmetry in two-photons+jet events with razor variables in pp collisions at sqrt(s) = 8 TeV", CMS-PAS-SUS-14-008.
[12] S. Chatrchyan *et al.* [CMS Collaboration], "Search for top squark and higgsino production using diphoton Higgs boson decays", Phys. Rev. Lett. **112** (2014) 161802 [arXiv:1312.3310 [hep-ex]].
[13] P. Meade, N. Seiberg and D. Shih, "General Gauge Mediation," Prog. Theor. Phys. Suppl. **177** (2009) 143 [arXiv:0801.3278 [hep-ph]].
[14] V. Khachatryan *et al.* [CMS Collaboration], "Searches for electroweak neutralino and chargino production in channels with Higgs, Z, and W bosons in pp collisions at 8 TeV", Phys. Rev. D **90** (2014) 9, 092007 [arXiv:1409.3168 [hep-ex]].
[15] V. Khachatryan *et al.* [CMS Collaboration], "Search for stealth supersymmetry in events with jets, either photons or leptons, and low missing transverse momentum in pp collisions at 8 TeV", Phys. Lett. B **743** (2015) 503 [arXiv:1411.7255 [hep-ex]].
[16] J. Fan, M. Reece and J. T. Ruderman, "Stealth Supersymmetry," JHEP **1111** (2011) 012 [arXiv:1105.5135 [hep-ph]].
[17] V. Khachatryan *et al.* [CMS Collaboration], "Search for physics beyond the standard model in events with two leptons, jets, and missing transverse momentum in pp collisions at $\sqrt{s}$ = 8 TeV", arXiv:1502.06031 [hep-ex].
[18] B. Allanach, A. R. Raklev and A. Kvellestad, "Interpreting a CMS $lljjp_T^{\rm miss}$ Excess With the Golden Cascade of the MSSM," arXiv:1409.3532 [hep-ph].
[19] P. Huang and C. E. M. Wagner, "CMS kinematic edge from sbottoms," Phys. Rev. D **91** (2015) 1, 015014 [arXiv:1410.4998 [hep-ph]].
[20] G. Aad *et al.* [ATLAS Collaboration], "Search for supersymmetry in events containing a same-flavour opposite-sign dilepton pair, jets, and large missing transverse momentum in $\sqrt{s}$ = 8 TeV *pp* collisions with the ATLAS detector," arXiv:1503.03290 [hep-ex].

International Workshop on Discovery Physics at the LHC (Kruger2014) IOP Publishing
Journal of Physics: Conference Series **623** (2015) 012005 doi:10.1088/1742-6596/623/1/012005

The QGP dynamics in relativistic collisions

E. L. Bratkovskaya[1,2], **W. Cassing**[3], **V. P. Konchakovski**[3,4] **and V.D. Toneev**[5,2]

[1] Institute for Theoretical Physics, University of Frankfurt, Frankfurt, Germany
[2] Frankfurt Institute for Advanced Study, Frankfurt, Germany
[3] Institute for Theoretical Physics, University of Giessen, Giessen, Germany
[4] Bogolyubov Institute for Theoretical Physics, Kiev, Ukraine
[5] Joint Institute for Nuclear Research, Dubna, Russia

Abstract. The dynamics of partons and hadrons in ultra-relativistic nucleus-nucleus collisions is analyzed within the Parton-Hadron-String Dynamics (PHSD) transport approach, which is based on a dynamical quasiparticle model for the partonic phase (DQPM) including a dynamical hadronization scheme while reproducing lattice QCD results in thermodynamic equilibrium for the equation-of-state as well as transport coefficients like shear and bulk viscosities or the electric conductivity of the hot QCD medium. The PHSD model reproduces a large variety of observables from SPS to LHC energies, e.g. the quark-number scaling of elliptic flow, transverse mass and rapidity spectra of charged hadrons, dilepton spectra, open and hidden charm production, collective flow coefficients etc., which are associated with the observation of a strongly interacting QGP (sQGP). The 'highlights' of the latest results from LHC energies are presented with a focus on observable effects from the non-equilibrium initial dynamics.

1. Introduction

The dynamics of the early universe in terms of the 'Big Bang' may be studied experimentally by ultra-relativistic nucleus-nucleus collisions at Relativistic-Heavy-Ion-Collider (RHIC) or Large-Hadron-Collider (LHC) energies in terms of 'tiny bangs' in the laboratory. With sufficiently strong parton interactions, the medium in the collision zone can be expected to achieve local equilibrium after some initial delay and exhibit approximately hydrodynamic flow [1, 2, 3]. In these collisions a new state of strongly interacting matter is created, being characterized by a very low shear viscosity η to entropy density s ratio, η/s, close to a nearly perfect fluid [4, 5, 6]. Lattice QCD (lQCD) calculations [7, 8] indicate that a crossover region between hadron and quark-gluon matter should have been reached in these experiments.

In case of nuleus-nucleus (e.g. Pb+Pb) collisions the ideal or viscous hydro calculations the initial conditions – at some finite starting time of the order of 0.3 to 0.5 fm/c – have to be evaluated either in terms of the (standard) Glauber model or other initial state scenarios like in the IP-glasma model [9, 10] or the CGC approach [11, 12, 13, 14], respectively. Alternative initial state scenarios are a coherent or incoherent superposition of 'strings' [15] or just an incoherent superposition of hard nucleon-nucleon collisions as in the wounded nucleon model (WNM) [16]. Differences between the different initial state assumptions and dynamical evolutions thus have to be expected. The applicability of ideal or viscous hydrodynamic models to proton-nucleus reactions for low multiplicity events, however, is very much debated. This also holds for hybrid models [17, 18, 19] as long as they employ a hydro phase. To our knowledge only microscopic transport approaches [20, 21] allow to bridge the gap from p-p to p-A and A-A collisions in a unique way without introducing additional (and less controlled) parameters. We will here employ the microscopic Parton-Hadron-String Dynamics (PHSD) transport approach [22] to analyze early reaction dynamics in the context of collective flow coefficients to shed some light on the sensitivity of these observables on different assumptions on the initial state dynamics.

The complexity of heavy-ion collisions is reduced essentially in the case of proton-nucleus collisions owing to the expected dominance of the initial state effects. In 2013 the first ALICE measurement of the charged particle pseudorapidity density has been reported [23] for $|\eta| < 2$ in p-Pb collisions at a nucleon-nucleon center-of-mass energy $\sqrt{s_{NN}} = 5.02$ TeV. The measurement has been compared to two sets of particle production models that describe similar measurements for other collision systems: the saturation models employing coherence effects [12, 13, 14] and the two-component models combining perturbative QCD processes with soft interactions [24, 25]. A comparison of the model calculations with the data has shown that the results are model-dependent and predict the measured multiplicities only within 20%. Accordingly, the restrictions imposed by the measured minimal bias pseudorapidity spectra $dN_c/d\eta$ are not sufficient to disentangle different models for the very early interaction stage of ultra-relativistic collisions. Furthermore, a test of color coherence in proton-nucleus collisions at the LHC energy has been proposed in Ref. [16]. The idea of this proposal is based on the fact that the observed mean multiplicity of charged particles $\langle N_{ch} \rangle$ linearly depends on the number of participants N_{part} within the wounded nucleon model (WNM) of independent nucleon-nucleon scatterings, $\langle N_{ch} \rangle \sim N_{part}$, while in the CGC models this dependence is logarithmic, $\langle N_{ch} \rangle \sim \ln N_{part}$. For a small number of participants, $N_{part} \leq 10$, the mean multiplicities calculated in both approaches practically coincide (in agreement with experiment) but for $N_{part} \sim 25$ they differ by almost a factor of two [16]. Such large numbers of participant are possible at the LHC energy of 5.02 TeV in p-Pb collisions. Furthermore, as pointed out in Ref. [26] there should be a sizeable difference in the mean transverse momentum of particles versus the pseudorapidity $\langle p_T \rangle (\eta)$ with opposite slopes in η on the projectile side within the CGC framework relative to hydrodynamical calculations due the saturation scale Q_s in the CGC. We will report on these suggestions employing a study within the PHSD approach.

2. PHSD @ LHC

We briefly recall the ingredients of the PHSD which is a covariant dynamical approach for strongly interacting systems formulated on the basis of Kadanoff-Baym equations [27] or off-shell transport equations in phase-space representation, respectively. In the Kadanoff-Baym theory the field quanta are described in terms of dressed propagators with complex selfenergies. Whereas the real part of the selfenergies can be related to mean-field potentials (of Lorentz scalar, vector or tensor type), the imaginary parts provide information about the lifetime and/or reaction rates of time-like particles [28]. Once the proper (complex) selfenergies of the degrees-of-freedom are known, the time evolution of the system is fully governed by off-shell transport equations (as described in Refs. [27, 28]). The PHSD approach includes a dynamical hadronization scheme based on transition rates and incorporates elastic and inelastic interactions of hadrons in the final expansion stage as in the HSD model [29]. This approach allows for a simple and transparent interpretation of lattice QCD results for thermodynamic quantities as well as correlators in the partonic stage and leads to effective strongly interacting partonic quasiparticles with broad spectral functions. For a review on off-shell transport theory we refer the reader to Ref. [28]. Actual PHSD results and their comparison with experimental observables for heavy-ion collisions from the lower super-proton-synchrotron (SPS) to RHIC energies can be found in Refs. [22, 30, 31] including electromagnetic probes such as e^+e^- or $\mu^+\mu^-$ pairs [32, 33].

2.1. p-p reactions at the LHC

To extend the PHSD model to higher energies than $\sqrt{s_{NN}}$ = 200 GeV at RHIC, we have additionally implemented the PYTHIA 6.4 generator [34] for initial nucleon collisions at LHC energies. For the subsequent (lower energy) collisions the standard PHSD model [22] is applied (including PYTHIA v5.5 with JETSET v7.3 for the production and fragmentation of jets [35], i.e. for $\sqrt{s_{NN}} \leq 500$ GeV [35]). In this way all results from PHSD up to top RHIC energies are regained and a proper extension to LHC energies is achieved. At $\sim \sqrt{s_{NN}} = 500$ GeV both PYTHIA versions lead to very similar results. In PYTHIA 6.4 we use the Innsbruck pp tune (390) which allows to describe reasonably the p-p collisions at $\sqrt{s_{NN}} = 7$ TeV in the framework of the PHSD transport approach. The overall agreement with LHC experimental data for the distribution in the charged particle multiplicity N_{ch} (a), the charged particle pseudorapidity distribution (b), the transverse momentum p_T spectra (c) and the correlation of the average p_T with the number of charged particles (d) is satisfactory (cf. Fig. 1).

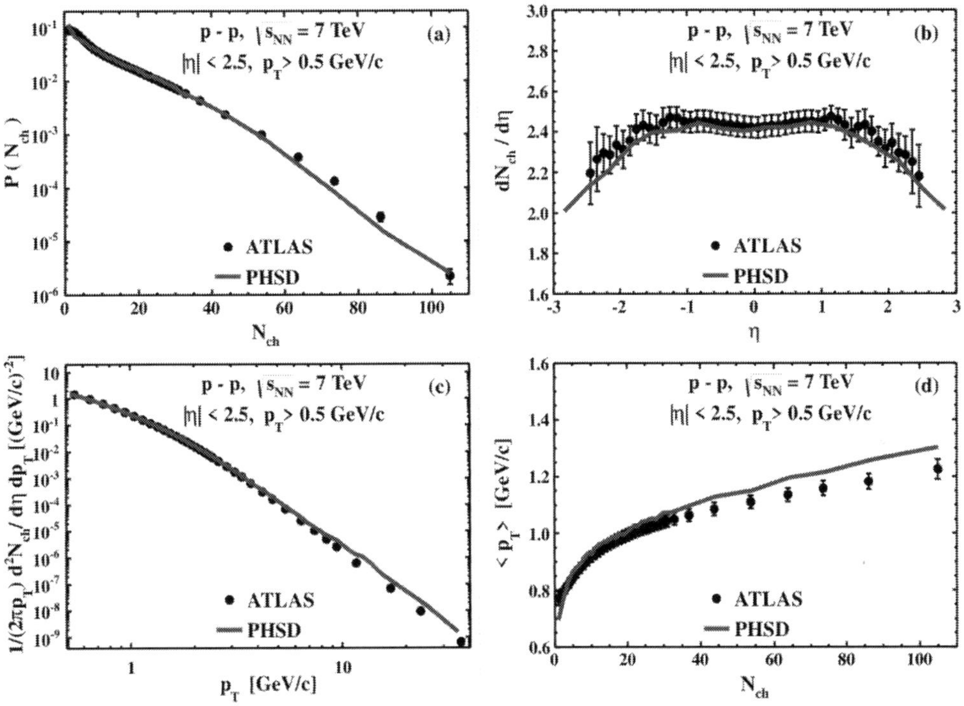

Figure 1. Comparison of the PHSD results (including PYTHIA 6.4) with LHC experimental data from the ATLAS Collaboration [36] for p-p collisions at $\sqrt{s_{NN}}$ = 7 TeV: (a) N_{ch} distribution, (b) $dN_{ch}/d\eta$ distribution, (c) p_T-spectra and (d) average p_T vs. N_{ch}.

2.2. Properties of p-Pb collisions at the LHC

With the elementary p-p collisions in the PHSD being adjusted at LHC energies via PYTHIA 6.4 (using the Innsbruck pp tune (390)) we proceed with observables and correlations from p-Pb collisions. In Fig. 2 (lhs) we present the probability distribution in the participant number N_{part} and the number of charged hadrons at midrapidity $N_{ch}(\eta = 0)$ as well as the ensemble average $< N_{ch} >$ vs. N_{part} (rhs) for p-Pb ($\sqrt{s_{NN}}$=5.02 TeV). As is seen from Fig. 2 (lhs), the number of charged particles at midrapidity correlates with the number of participants, $N_{ch}(\eta = 0) \sim N_{part}$, however, with a large dispersion in both quantities. When considering the ensemble average $< N_{ch}/d\eta > (\eta = 0)$ vs. N_{part} we obtain the solid (blue) line within PHSD while the dotted (red) line results from PHSD when including additionally fluctuations in the cross section (PHSD GG). For details we refer the reader to Ref. [37]. Both the standard Glauber and CGC results are presented and support the results of Ref. [16]. The two versions of the PHSD model, with (red dotted line) and without cross section fluctuations (blue solid line), predict that the multiplicity dependence turns out to be close to the CGC result but substantially differs from the wounded nucleon moden (WNM) for larger N_{part}. Thus, multiplicity distributions do not allow us to disentangle the different initial states under discussion from the CGC and a Glauber version as in PHSD. The reason of such a multiplicity suppression relative to the WNM is the energy-momentum conservation in PHSD which on average results in a decrease of particle multiplicity in subsequent scatterings as compared to the primary interaction.

A wide distribution is also observed in the number of participants or in the number of charged particles at midrapidity for a given impact parameter b, see Fig. 3. The solid lines in this figure show the mean values $\langle N_{part} \rangle$ and $\langle dN_{ch}/d\eta \rangle$, respectively. In contrast to nucleus-nucleus collisions, these mean quantities are almost independent of the impact parameter for central and semi-central collisions, $b <$ (4-5) fm. This fact should have been expected since the size of the projectile-proton is noticeably smaller than that of the target nucleus. From these results it follows that an event selection with respect to the number of participants or charged particles refers to a large range in impact parameter b.

The pseudorapidty distributions of charged particles from p-Pb minimum bias collisions at $\sqrt{s_{NN}} = 5.02$ TeV are compared with the experimental data [23] in Fig. 4. The data are displayed in the laboratory system which is shifted with respect to the nucleon-nucleon center-of-mass by $y_{cm} = -0.465$. The results of two versions of the parton-hadron string dynamics model (PHSD and PHSD-GG) differ only for backward-emitted particles and both versions are rather close to the measured data and the CGC result (open circles). Note that there are no modifications (or free parameters) in the PHSD except the extensions by PYTHIA 6.4 which implies that p-p, p-A

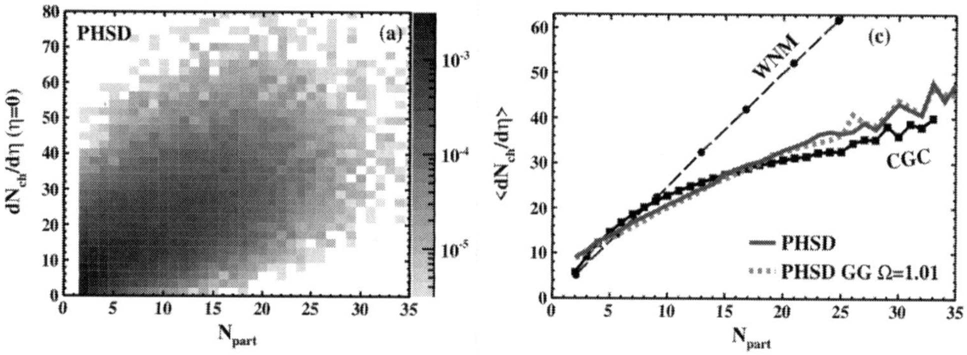

Figure 2. Probability distribution of the participant number and number of charged particles for Pb+Pb at $\sqrt{s_{NN}} = 5.02$ TeV at midrapidity (lhs) and its ensemle average in comparison to different models (rhs). The wounded nucleon model (WNM)(full dots) and color glass condensate (CGC) calculations (dotted line) are taken from Ref [16] while simulations in the Glauber-Gribov approximation (full squares and triangles) stem from Ref. [38]. The PHSD results are displayed in terms of the solid (blue) lines while the PHSD results including fluctuations in the cross section (PHSD GG) are shown in terms of the dotted (red) lines.

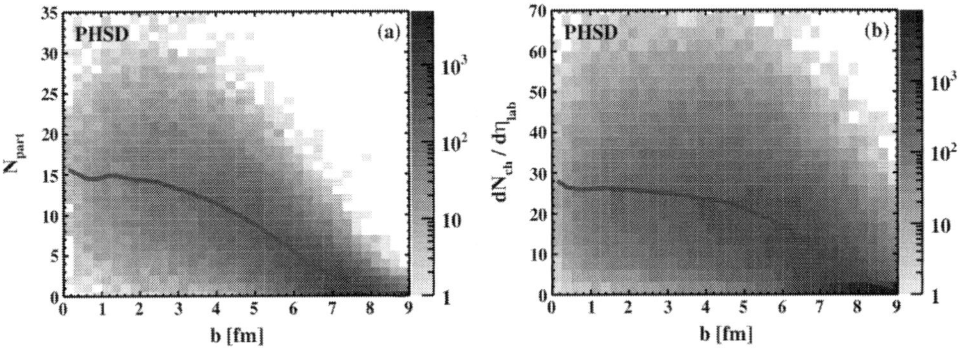

Figure 3. Event distributions of 2D-correlations for the participant number N_{part} (lhs) and charged particle multiplicity $dN_{ch}/d\eta$ (rhs) with the impact parameter b. The mean values of these distributions are shown by the solid (blue) lines.

and A-A collisions are consistently described from low SPS to LHC energies (within $\sim 10\%$).

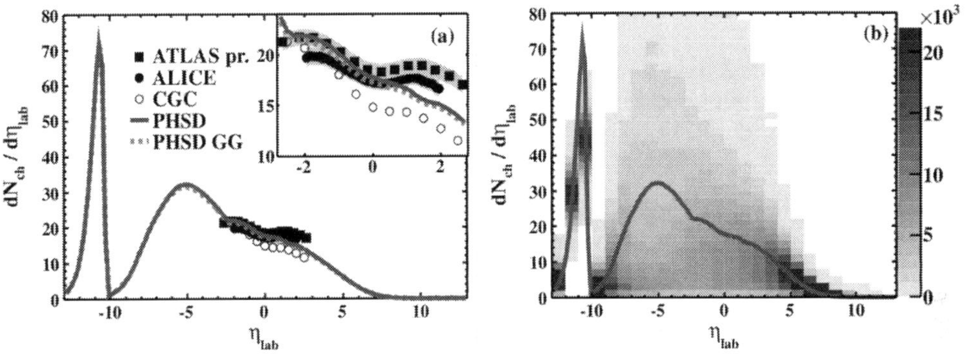

Figure 4. (lhs) Rapidity distribution of charged particles for minimum bias data from the ALICE [23] (full dots) and ATLAS [38] (full squares) collaborations for p-Pb collisions at $\sqrt{s_{NN}}$ = 5.02 TeV in comparison to the PHSD results (solid blue line) and the PHSD GG results including fluctuations in the cross section (dotted red line). The CGC results (open circles) have been taken from Ref. [14]. The zoomed results are displayed in the insertion. (rhs) Event-by-event fluctuations of the rapidity distribution. The blue solid line shows the average charged particle pseudorapidity distribution.

The CGC predictions from Ref. [14], performed earlier for the upcoming p-Pb run at the LHC, are plotted in the same figure (open circles). This result is based on the Balitsky-Kovchegov (BK) equation [39] which is the large-N_c limit of non-linear renormalization group equations such as the BK-JIMWLK hierarchy [40] which is tested with respect to $e+p$ data. An astonishing result is that the CGC and PHSD results almost coincide again. Note that this minimum-bias distribution corresponds to the mean charged particle multiplicity at the given value of pseudorapidity η. However, event fluctuations of $dN_{ch}/d\eta$ are very large as demonstrated in Fig. 4 (rhs). Thus, the study of minimum-bias $dN_{ch}/d\eta$ does not allow to disentangle the initial state concepts described within the PHSD and CGC approaches.

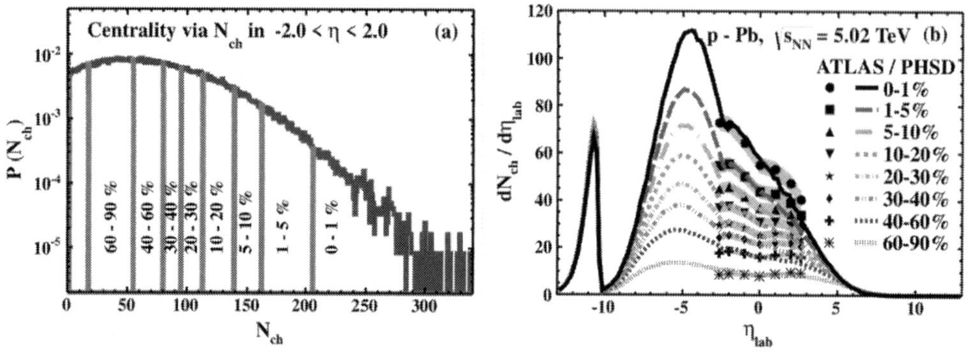

Figure 5. (lhs) Centrality bins for p-Pb collisions at $\sqrt{s_{NN}}$ = 5.02 TeV selected according to the charged particle multiplicity in the rapidity interval $|\eta| < 2$. (rhs) Comparison of the PHSD calculated rapidity distributions with ATLAS data [38] for charged particles in different centrality bins. The shaded bands show the experimental uncertainties.

Let us, furthermore, consider pseudorapidity distributions for fixed high-multiplicity events. Such distributions for different centrality bins have been measured by the ATLAS collaboration [38] for p-Pb collisions at $\sqrt{s_{NN}}$ = 5.02 TeV. Experimentally the centrality was defined according to selected bins in the transverse energy. We have defined corresponding bins in N_{ch} keeping the same percentage of the number of selected events as in [38] (the bin partition is shown in Fig. 5 (lhs) and the relative contribution of different centralities is given in the legend in Fig. 5 (rhs)). In this figure the PHSD results are based on 10^6 simulated events.

As is seen from Fig. 5 (rhs), the PHSD model quite well reproduces the shape of the $dN_{ch}/d\eta$ distributions and its variation with centrality, in particular the increase with centrality of the forward-backward asymmetry between the directions of the proton-beam and Pb-target. For the most central events the PHSD calculations very slightly overshoot this asymmetry, however, are in line with the data for the higher centralities within the experimental uncertainties (shaded areas in Fig. 5 (rhs)). We mention that the centrality sample of 40-60% with the maximal number $N_{ch} \sim 20$ roughly corresponds to the minimum-bias distribution. For events of the highest multiplicity which amount to (0-1)% – corresponding to $\sim 6.10^3$ simulated events – the number of charged particles at the maximum of the distribution is about 75. The agreement between calculations and data is not so bad taking into account the experimental error bands and the fact that PHSD has no free parameters once the p-p dynamics is fixed (by the PYTHIA tune).

The transverse momentum characteristics for charged particles from the PHSD for p-Pb collisions at $\sqrt{s_{NN}}$ = 5.02 TeV are compared with the ALICE data in Fig. 6. In the measured range $0.5 < p_T < 10$ GeV/c the yield changes by 7 orders of magnitude in a rough agreement with experiment [41]. Deviations by up to a factor of three are observed in the momentum range $p_T > 1.5$ GeV/c (see Fig. 6 (lhs)) and are presently not understood.

A remarkable difference in observables between the predictions of the saturated CGC and hydro models has been pointed out in Ref. [26]. Based on general arguments, it was shown that in the case of the CGC the mean transverse momentum slightly grows with increasing rapidity y on the proton side due to the increasing saturation momentum Q_s of the nucleus. On the contrary, the $\langle p_T\rangle_y / \langle p_T\rangle_{y=0}$ in the hydrodynamical framework (with Glauber initial conditions) decreases due to the decreasing number of particles with positive rapidity. This is due to the fact that the collective expansion scenario (in hydrodynamics) cannot lead in a simple way to

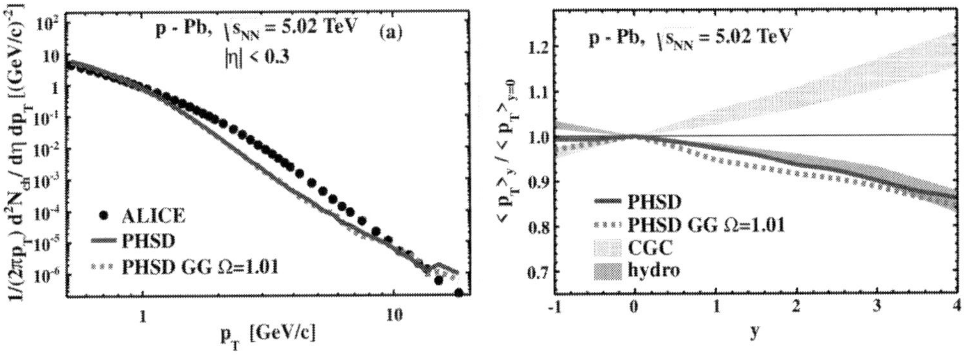

Figure 6. (lhs) Transverse momentum spectrum at midrapidity and (rhs) at $\sqrt{s_{NN}} = 5.02$ TeV in comparison to the ALICE data [41, 42].The average relative transverse momentum of produced charged particles as a function of rapidity for p-Pb collisions at $\sqrt{s_{NN}} = 5.02$ TeV. The solid (blue) line is the default PHSD prediction whereas the dotted (red) line shows the result including large fluctuations in the initial cross section. The CGC and hydrodynamic results are taken from Ref. [26]. The shaded areas correspond to the uncertainty in two selected CGC models and in the centrality selection in the hydro case, respectively.

an increase of the average transverse momentum on the proton side $y > 0$ [26] since there are less degrees-of-freedom to generate e.g. a transverse flow. The PHSD model predicts the $\langle p_T \rangle$ distribution (blue solid line) to be rather close to the hydrodynamic models since also in PHSD the collectivity is correlated with the density of the degrees-of-freedom which decreases with forward rapidity (cf. Fig. 5 (lhs)); cross section fluctuations have no essential influence on this result (dotted red line). It would be of great interest to check experimentally this clear difference in the $\langle p_T \rangle_y \,/\, \langle p_T \rangle_{y=0}$ distribution due to different initial state concepts.

2.3. Pb-Pb collisions at $\sqrt{s_{NN}}$=2.76 TeV

We recall that especially in case of heavy-ion reactions the flow harmonics v_n for the azimuthal angular distribution of hadrons have been found to be sensitive to the early stage of the nuclear interaction and in particular to their fluctuations. Indeed, the detailed heavy-ion analysis in Ref. [43] shows that Monte Carlo CGC approaches (MC-CGC) systematically give a larger initial eccentricity than Glauber models. However, it is unclear to what extent such properties of the CGC formalism are robust with respect to extended correlations. Also, studies of higher harmonics – as presented in [44] by the PHENIX or ALICE collaborations – do not clearly favor the CGC or Glauber assumptions for the initial state of the collision. The first LHC data on the bulk particle production in Pb-Pb collisions are in good agreement with improved CGC expectations but they are also compatible with Monte Carlo event generators [45]. We here compute bulk properties of Pb-Pb collisions at $\sqrt{s_{NN}}$=2.76 TeV and explore the sensitivity of the collective flow coefficients v_n on the size of initial state fluctuations within the (PHSD) transport approach [22] employed also in the previous subsection.

The first question to answer is whether the PHSD approach still works at LHC energies for nucleus-nucleus collisions at $\sqrt{s_{NN}}$=2.76 TeV although the invariant energy is higher by about a factor of 13.8 than at the top RHIC energy. In Fig. 7 we compare the average $< p_T >$ as a function of charged multiplicity N_{ch} in p-p reactions at $\sqrt{s_{NN}} = 7$ TeV, p-Pb collisions at $\sqrt{s_{NN}} = 5.02$ TeV and Pb-Pb collisions at $\sqrt{s_{NN}} = 2.76$ TeV from the PHSD to the experimental data from Ref. [42]. Note that for low multiplicities ($N_{ch} < 5$) the mean p_T

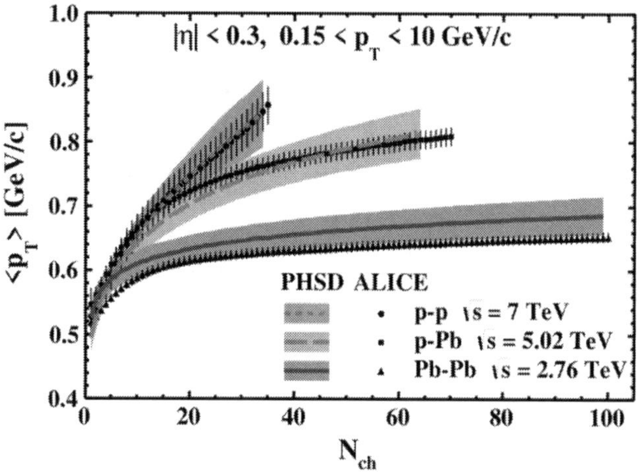

Figure 7. Average p_T results for p-p, p-Pb and Pb-Pb collisions from the PHSD transport approach in comparison to the ALICE experimental data from Ref. [42]. Note the different invariant energies for p-p, p-Pb and Pb-Pb collisions.

is almost independent on energy (see also Ref. [42]) which in PHSD can be traced back to the fact that (for the acceptance $|\eta| \le 0.3$, $0.15 \le p_T \le 10$ GeV/c) only events with one or two binary collisions N_{bin} are selected for all systems. Actually, the correlation $< p_T > (N_{ch})$ only weakly depends on $\sqrt{s_{NN}}$ for pp reactions at these LHC energies, however, when plotting $p_T(N_{ch})$ on an event-by-event basis, large fluctuations in p_T or N_{ch} are obtained within PHSD. The same holds true for p-Pb and Pb-Pb reactions where a fixed N_{ch} can be obtained by reactions with a varying number of binary collisions N_{bin}. Each of these binary reactions then has a low N_{ch} and $< p_T >$, respectively. The ensemble average finally leads to the average correlation shown in Fig. 7. Nevertheless, the agreement between data and calculations (within the statistical accuracy) is encouraging. Note again that only very peripheral Pb-Pb collisions are probed for $N_{ch} < 100$.

We continue with the transverse momentum spectra for central Pb-Pb reactions at $\sqrt{s_{NN}} = 2.76$ TeV (0-5% centrality) which are compared in Fig. 8 (lhs) with results of the ALICE Collaboration for all charged particles [46, 47] (PHSD: black solid line) as well as for charged pions [48] (PHSD: dashed blue line). Note that except for the upgrade in the PYTHIA version no additional parameters or changes have been introduced in the PHSD that had been employed before in Refs. [22, 30, 31, 32, 50, 51] from lower SPS up to top RHIC energies. In this respect the approximate reproduction of the midrapidity p_T spectra for central collisions over 7 orders of magnitude in Fig. 8 (lhs) is quite remarkable. A closer look at the low momentum spectra is offered in Fig. 8 (rhs) where the PHSD spectra for pions and kaons are compared to results of the ALICE Collaboration [46, 47, 48]. We recall, furthermore, that the charged particle pseudo-rapidity density $dN_{ch}/d\eta$ at midrapidity from PHSD matches well the experimental centrality dependence from ALICE when displayed as a function of the number of participants N_{part} (cf. Fig. 1 in Ref. [33]).

Whereas the transverse charged single-particle spectra compare quite well with the experimental observation the question remains for the collective behavior of the system. Anisotropic flow coefficients in both cases – i.e. experimental data and PHSD calculations – were obtained from the two-particle cumulant method [52] in the central pseudorapidity window

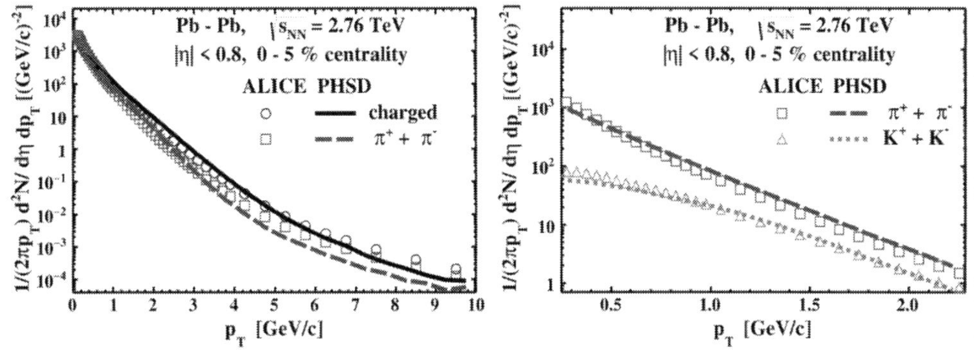

Figure 8. (lhs) Transverse momentum spectra from PHSD in comparison to the results of the ALICE Collaboration for all charged particles [46, 47] (solid line) as well as for charged pions [48] (dashed line). (rhs) Transverse momentum spectra from PHSD for $p_T \leq 2$ GeV/c in comparison to the results of the ALICE Collaboration [46, 47, 48] for pions and kaons.

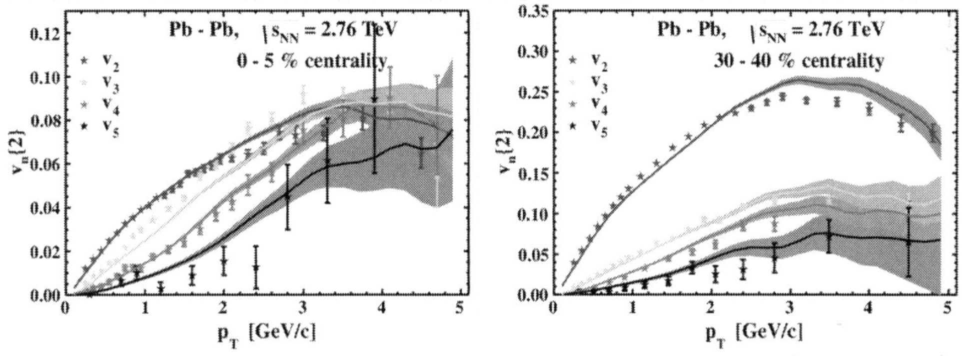

Figure 9. The flow coefficients v_2, v_3, v_4 and v_5 of all charged particles as a function of p_T for the centralities 0-5% (lhs) and 30-40% (rhs). The ALICE data have been taken from Ref. [49].

$|\eta| < 0.8$ and denoted as $v_n\{2\}$. The flow coefficients v_2, v_3, v_4 and v_5 of all charged particles from PHSD are shown in Fig. 9 as a function of p_T for the centralities 0-5% (lhs) and 30-40% (rhs) in comparison to the ALICE data from Ref. [49]. The PHSD results for $v_2(p_T), v_3(p_T)$ and $v_4(p_T)$ compare reasonably up to about 3.5 GeV/c whereas at higher transverse momenta the statistics is insufficient to draw robust conclusions. This also holds for the flow coefficient v_5 which still is in line with the data within error bars. It is quite remarkable that the collective behavior is reproduced not only for semi-central collisions (rhs) but also for 0-5% central collisions (lhs) that are more sensitive to the initial fluctuations [18]. These tests indicate that the 'soft' physics at LHC in central A-A reactions is very similar to the top RHIC energy regime although the invariant energy is higher by more than an order of magnitude. Furthermore, the PHSD approach seems to work from lower SPS energies up to LHC energies for p-p, p-A as well as A-A collisions, i.e. over a range of more than two orders in $\sqrt{s_{NN}}$.

A detailed study of the sensitivity of the initial state fluctuations in space on the collective flow coefficients v_2, v_3 and v_4 turned out to be negative even for very central (b=0) Pb-Pb collisions at the LHC (cf. Ref. [53]) which is essentially due to the fact that the global shape characteristics do not change much when increasing the local fluctuations. Thus our studies do not support the suggestion that the collective flow coefficients might be used to discriminate CGC from Glauber initial state conditions.

3. Conclusions

In this study the parton-hadron-string dynamics (PHSD) approach has been employed in the LHC energy range for Pb-Pb collisions at $\sqrt{s_{NN}} = 2.76$ TeV as well as p-Pb collisions at $\sqrt{s_{NN}} = 5.02$ TeV. We find that this approach works reasonably for both systems with respect to charged particle spectra as well as collective flow coefficients v_2, v_3, v_4 and v_5 for different centralities with a quality comparable to that achieved at RHIC energies before [22, 30, 31, 32, 50, 51]. Our finding implies that the 'soft' physics in Pb-Pb collisions at the LHC and Au-Au interactions at the top RHIC energies – despite a factor of ~ 14 in $\sqrt{s_{NN}}$ – is very similar and in line with the dynamical quasiparticle model (DQPM) that defines the parton properties for PHSD in equilibrium. This finding is common with earlier studies using viscous hydro approaches with varying initial conditions [10, 54].

The particular question addressed in this study has been the possibility to obtain information on the very early non-equilibrium dynamics of the reactions. We find for p-Pb reactions that the PHSD approach provides correlations between the charged particle multiplicity at midrapidity and the number of participant nucleons close to results from the CGC and differs substantially from results calculated with independent Glauber initial conditions (WNM). However, a sizeable difference is found between the PHSD approach and CGC models with respect to the rapidity dependence of the average transverse momentum. In Pb-Pb collisions the PHSD calculations have shown no sensitivity on the initial size of spatial fluctuations for the flow harmonics v_2 to v_4 which on one hand can be traced back to the low interaction rate in the initial nonequilibrium stage in PHSD (~ 0.3 fm/c) where effects from different granularities are already washed out to some extent. This is different from hydro calculations with varying granularity that instantly start to convert fluctuations in coordinate space to collective modes in momentum space. On the other hand our method for changing the size of initial fluctuations keeps the event shape in coordinate space approximately invariant (cf. [53]) which - in line with hydrodynamics - leads to very similar flow coefficients v_n. We mention that the low interaction rate in this very early phase in PHSD is common with the CGC concept and thus does not allow to disentangle or determine the effective degrees-of-freedom in this 'pre-hydro' phase.

Acknowledgments

This work in part was supported by the LOEWE center HIC for FAIR as well as by BMBF. V. D. was also partly supported by the Heisenberg-Landau grant of JINR.

4. References

[1] Ollitrault J Y 1992 *Phys. Rev.* D **46** 229
[2] Heinz U and Kolb P 2002 *Nucl. Phys.* A **702** 269
[3] Shuryak E V 2009 *Prog. Part. Nucl. Phys.* **62** 48
[4] Shuryak E V 2005 *Nucl. Phys.* A **750** 64
[5] Gyulassy M and McLerran L 2005 *Nucl. Phys.* A **750** 30
[6] Peshier A and Cassing W 2005 *Phys. Rev. Lett.* **94** 172301
[7] Cheng M *et al.* 2008 *Phys. Rev.* D **77** 014511
[8] Aoki Y *et al.* 2009 *JHEP* **0906** 088
[9] Schenke B Tribedy P and Venugopalan R 2012 *Phys. Rev.* **C** **86** 034908
[10] Gale C Jeon S Schenke B Tribedy P and Venugopalan R 2013 *Phys. Rev. Lett.* **110** 012302
[11] McLerran L 2007 *Nucl. Phys.* A **787** 1
McLerran L 2007 *Int. J. Mod. Phys.* A **21** 694
[12] Dumitru A Kharzeev D E Levin E M and Nara Y 2012 *Phys. Rev.* C **85** 044920
[13] Tribedy P and Venugopalan R 2012 *Phys. Lett.* B **710** 125

[14] Albacete J L Dumitru A Fujii A and Nara Y 2013 *Nucl. Phys.* A **897** 1
[15] Andrs C Moscoso A and Pajares C 2014 *Phys. Rev. C* **90** 054902
[16] Bzdak A and Skokov V 2013 *Phys. Rev. Lett.* **111** 182301
[17] Nonaka C and Bass S A 2007 *Phys. Rev.* C **75** 014902
[18] Petersen H and Bleicher M 2010 *Phys. Rev.* C **81** 044906
[19] Qin G-Y Petersen H Bass S A and Müller B 2010 *Phys. Rev.* C **82** 064903
[20] Plumari S Puglisi A Scardina F and Greco V 2012 *Phys. Rev.* C **86** 054902
[21] Xu Z and Greiner C 2005 *Phys. Rev.* C **71** 064901
[22] Cassing W and Bratkovskaya E L 2009 *Nucl. Phys.* A **831** 215
Bratkovskaya E L *et al.* 2011 *Nucl. Phys.* A **856** 162
[23] Abelev B *et al.* [ALICE Collaboration] 2013 *Phys. Rev. Lett.* **110** 032301
[24] Barnafoldi G G Barrette J Gyulassy M Levai P and Topor Pop V 2012 *Phys. Rev.* C **85** 024903
[25] Xu R Deng W-T and Wang X-N 2012 *Phys. Rev.* C **86** 051901
[26] Bozek P Bzdak A and Skokov V 2014 *Phys. Lett.* B **728** 662
[27] Juchem S Cassing W and Greiner C 2004 *Phys. Rev.* D **69** 025006
Juchem S Cassing W and Greiner C 2004 *Nucl. Phys.* A **743** 92
[28] Cassing W 2009 *E. Phys. J. ST* **168** 3
[29] Cassing W and Bratkovskaya E L 1999 *Phys. Rept.* **308** 65
Cassing W Bratkovskaya E L and Juchem S 2000 *Nucl. Phys.* A **674** 249
Bratkovskaya E L and Cassing W 1997 *Nucl. Phys.* A **619** 413
Geiss J Cassing W and Greiner C 1998 *Nucl. Phys.* A **644** 107
[30] Toneev V D *et al.* 2012 *Phys. Rev.* C **85** 034910
[31] Konchakovski V P Bratkovskaya E L Cassing W Toneev V D and Voronyuk V 2012 *Phys. Rev.* C **85** 011902
[32] Linnyk O Cassing W Manninen J Bratkovskaya E L and Ko C M 2012 *Phys. Rev.* C **85** 024910
Linnyk O Bratkovskaya E L Ozvenchuk V Cassing W and Ko C M 2012 *Phys. Rev.* C **84** 054917
[33] Linnyk O *et al.* 2013 *Phys. Rev.* **C 87** 014905
[34] Sjostrand T Mrenna S and Skands P Z 2006 *JHEP* **0605** 026
[35] Bengtsson H-U and Sjöstrand T 1987 *Comp. Phys. Commun.* **46** 43
[36] Aad G *et al.* [ATLAS Collaboration] 2011 *New J. Phys.* **13** 053033
[37] Konchakovski V P Cassing W and Toneev V D 2014 *J. Phys.* G **41** 105004
[38] ATLAS collaboration, ATLAS-CONF-2013-096
[39] Balitsky I 1996 *Nucl. Phys.* B **463** 99
Kovchegov Y V 1999 *Phys. Rev.* D **60** 034008
[40] Gelis F Iancu E Jalilian-Marian J and Venugopalan R, *Ann. Rev. Nucl. Part. Science* **60** 463
Iancu E and Venugopalan R 2003 *Quark Gluon Plasma*, edited by R. Hwa, X. N. Wang (World Scientific)
Weigert H 2005 *Prog. Part. Nucl. Phys.* **55** 461
[41] Abelev B B *et al.* [ALICE Collaboration] 2013 *Phys. Rev. Lett.* **110** 082302
[42] Abelev B B *et al.* [ALICE Collaboration] 2013 *Phys. Lett.* B **727** 371
[43] Albacete J L 2011 *J. Phys. G: Nucl. Part. Phys.* **38** 124006
[44] Schutz Y and Wiedemann U A (ed) 2011 Proc. of the 22nd Int. Conf. on Ultra-Relativistic Nucleus-Nucleus Collisions (Annecy, France, 23-28 May 2011) *J. Phys. G: Nucl. Part. Phys.* **38** 120301
[45] ATLAS Collaboration 2013 *JHEP* **11** 183
[46] Abelev B B *et al.* [ALICE Collaboration] 2013 *Phys. Lett.* B **720** 52
[47] Abelev B B *et al.* [ALICE Collaboration] 2013 *Eur. Phys. J.* C **73** 2662
[48] Abelev B B *et al.* [ALICE Collaboration] 2014 *Phys. Lett.* B **736** 196
[49] Aamodt K *et al.* [ALICE Collaboration] 2011 *Phys. Rev. Lett.* **107** 032301
[50] Linnyk O Cassing W and Bratkovskaya E L 2014 *Phys.Rev.* C **89** 034908
Linnyk O Konchakovski V P Cassing W and Bratkovskaya E L 2013 *Phys. Rev.* C **88** 034904
[51] Konchakovski V P *et al.* 2012 *Phys. Rev.* C **85** 044922
[52] Bilandzic A Snellings R and Voloshin S 2001 *Phys. Rev.* C **83** 044913
[53] Konchakovski V P Cassing W and Toneev V D 2015 *J. Phys. G* **42** arXiv:1411.5534
[54] Song H Bass S A and Heinz U 2011 *Phys. Rev.* **C 83** 054912
Song H Bass S A and Heinz U 2013 *Phys. Rev.* **C 87** 019902
Song H Bass S A and Heinz U 2014 *Phys. Rev.* **C 89** 034919

Journal of Physics: Conference Series **623** (2015) 012006 doi:10.1088/1742-6596/623/1/012006

The dynamics of Composite Higgses

Giacomo Cacciapaglia

Université de Lyon, France; Université Lyon 1, CNRS/IN2P3, UMR5822 IPNL, F-69622 Villeurbanne Cedex, France

E-mail: `g.cacciapaglia@ipnl.in2p3.fr`

Abstract. The nature of the 125 GeV Higgs recently discovered at CERN has not been established yet. One intriguing possibility is that it may arise as a light composite state from a confining dynamics at the TeV scale. I will review the recent progress in understanding the dynamics that may be behind this mechanism, focusing on what we can learn by knowing its details. The masses of the spin-1 resonances can in fact be extracted from lattice calculations thus providing an estimate of the mass scale of new states. Furthermore, a lot can be learned about the physics of eventual top partners.

1. Introduction

The discovery of the 125 Higgs boson at the LHC has undoubtedly opened a new era in particle physics. By analyzing its decay products, mainly from the golden channels $h \to \gamma\gamma$ and $h \to ZZ^* \to$ leptons, experimentalists at CMS and ATLAS have been able to establish with good accuracy that it is a spin-0 and CP-even state, which couples to particles proportionally to their mass, thus fulfilling the properties of the Higgs boson in the Standard Model (SM). In this sense, this discovery completes the list of particles expected in the SM after 50 years from its proposal. However, the discovery of the Higgs boson has not improved our comprehension of the Brout-Englert-Higgs mechanism [1, 2] responsible for the breaking of the electroweak (EW) symmetry: the SM contains a mere description of the spontaneous symmetry breaking via an elementary weakly-coupled scalar field transforming as a doublet of the weak SU(2) gauge symmetry, which develops a vacuum expectation value thanks to a mass term tuned to have "wrong" sign in the Lagrangian. Only a precise determination of the couplings of the discovered Higgs boson would allow to confirm this description of the EW symmetry breaking sector of the SM. However, the most recent data at the end of Run I of the LHC has allowed us to determine the couplings with an accuracy of 15–20% in the best channels [3, 4]. Even though the central values are in very good agreement with the SM predictions, there is plenty of space for New Physics effects. For comparison, the accuracy in the test of the EW gauge sector of the SM is of the order of 0.1%.

Furthermore, the discovery of a light scalar associated to the Higgs has brought in the realm of reality the old questions of *naturalness*, and the hierarchy problem. The mass of a scalar field is in fact very sensitive to quantum corrections deriving from its couplings to other particles in the model, the underlying reason being the absence of spin. For particles with spin, like fermions and vector bosons, sending the mass to zero allows for the recovery of a symmetry: chiral symmetry in the case of fermions and gauge symmetry in the case of vectors. In such cases one always finds, therefore, that loop corrections to the masses are proportional to the bare masses of the particles, thus a small mass-parameter in the Lagrangian stays small after quantum corrections are included. This is not true for scalar fields: their mass receives corrections proportional to the mass of any physics they couple to. For a generic scalar field ϕ, therefore, we can write that

$$m_\phi^2 = m_{\text{bare}}^2 + \frac{g^2}{16\pi^2} M^2\,, \tag{1}$$

where M is a mass scale associated to a sector of the model that couples to ϕ with coupling strength g. In the SM, we know that any field will ultimately couple to gravity, which carries an intrinsic scale $M = M_{\text{Planck}} = 10^{19}$ GeV. Following quantum corrections, thus, the mass of the Higgs boson should be very close to the Planck mass, unless the bare mass in the Lagrangian is tuned with a high accuracy to cancel the quantum corrections.

So far, the only mechanisms we know that can keep the Higgs mass (and thus the EW scale) under control are either based on a symmetry that relates the scalars to states carrying spin, or one that sets the Higgs mass to zero:

A- boson-fermion symmetry, dubbed supersymmetry [5], is an extension of space-time symmetries that allows to associate scalar fields to fermions, thus their masses enjoy the same protection from chiral symmetry;

B- extra dimensions, another extension of space time symmetry allowing for additional space-like dimensions, allow to associate 4-dimensional scalars to the additional polarisation of vector bosons, so that their masses are also protected by the extended gauge symmetry [6];

C- compositeness, where the scalar field is replaced by a bound state of fermions, whose mass is associated to a non-perturbative dynamical mass (like the condensation scale in QCD);

D- Goldstone symmetry, associating the scalar with the massless Goldstone boson of a spontaneously broken global symmetry.

In this contribution, we will focus on the possibility that the Higgs boson arises as a composite state: this is a very rich scenario that offers many qualitatively different candidates for the Higgs:

- *composite Higgs*, arising as a massive bound state in models where the dynamics breaks the EW symmetry at the condensation scale. This scenario is realized in old-school QCD-like Technicolor models [7, 8].
- *radion/dilaton scenario*, where the Higgs [9] can be associated to a classical conformal invariance of the theory in 4 dimensions, or extra dimensional models. This scenario is similar to the Technicolor case if the underlying dynamics is near conformal.
- *pNGB scenario*, where the Higgs is associated to a pseudo-Nambu-Goldstone-boson (pNGB) of the spontaneously broken global symmetry of the dynamics [10]. The mass is then generated by small explicit breaking of the global symmetry, like the gauging of the EW and the couplings responsible for the generation of the top mass. Holography allows to associate this case to models in (warped) extra dimensions, where the Higgs arise as a gauge boson.
- *holographic Higgs*, where the Higgs arises as a particle-like state associated with an operator of an underlying conformal field theory which couples marginally to the SM [11]. The AdS/CFT correspondence allows to relate this scenario with models in extra dimensions.

It should be noted here that the name "composite Higgs" [12] is currently associated to models of holographic Higgses, and that any dynamical model of pNGB Higgs automatically enjoys a Technicolor-like limit [13] where the Higgs arises as a composite bound state. So, in a sense, all these scenarios, are related to each other and have similar phenomenology.

A new approach to understanding the physics of a composite Higgs is to rely on a well defined underlying dynamics based on confining gauge interactions with matter fermions (thus qualitatively similar to QCD). This allows to work in a well-defined scenario where the global symmetries and properties of the composite states are determined [13]. In the following we will review the latest development achieved thanks to this approach.

2. Fundamental dynamics and the Higgs boson

Let us postulate the existence of a dynamics that is invariant under a global symmetry group $\mathcal{G}$, which is spontaneously broken to a subgroup $\mathcal{H}$. No matter what the details of the dynamics are, the spectrum of the theory will contain a number of massless Goldstone bosons, equal to the dimension of the coset $\mathcal{G}/\mathcal{H}$. One can further imagine to couple the dynamics to the SM by partially gauging the global symmetries. As a first example, we will assume that the EW symmetry is broken by the dynamics, thus it is sufficient to have

$$\mathcal{G} \supset \mathrm{SU}(2)_L \times \mathrm{U}(1)_Y\,, \quad \mathcal{H} \supset \mathrm{U}(1)_{\mathrm{em}}\,. \tag{2}$$

This is exactly what happens in Technicolor models: for instance, in QCD with 2 quarks, $\mathcal{G} = \mathrm{SU}(2)_L \times \mathrm{SU}(2)_R$, and $\mathcal{H} = \mathrm{SU}(2)_D \supset \mathrm{U}(1)_{\mathrm{em}}$. In this case, $\dim \mathcal{G}/\mathcal{H} = 3$, and one obtains exactly the 3 Goldstone bosons needed to give mass to the W and Z.

We can however be more ambitious and imagine a situation where

$$\mathcal{G} \supset \mathcal{H} \supset \mathrm{SO}(4) \supset \mathrm{SU}(2)_L \times \mathrm{U}(1)_Y\,, \tag{3}$$

where we extended the electroweak symmetry to the custodial $\mathrm{SO}(4) \sim \mathrm{SU}(2)_L \times \mathrm{SU}(2)_R$ in order to avoid three level corrections to the ρ (T) parameter [14]. If we properly choose the global symmetries, the coset may contain 4 scalars which transform as a bi-doublet of the custodial SO(4), and can thus play the role of the Higgs field. Giving a vacuum expectation value to such states would therefore be equivalent to misaligning the vacuum of the dynamics that breaks $\mathcal{G} \to \mathcal{H}$ along the direction of the generators of $\mathcal{G}$ associated to the Higgs field [10]. The misaligned vacuum will now break the EW symmetry in a way similar to Technicolor. The degree of misalignment can be described by an angular variable θ, so that:

- $\theta = 0$ corresponds to the vacuum that does not break the EW symmetry, and the coset contains a full bi-doublet of SO(4);
- $\theta = \pi/2$ corresponds to the Technicolor alignment that breaks the EW symmetry at the scale of the dynamics $f \sim v$;
- for intermediate θ, the breaking along the Higgs direction is $f \sin\theta \sim v$, so that $v/f < 1$.

The advantage of this scenario is in the possibility to induce a small hierarchy between the mass scale of the dynamics, where the spontaneous breaking takes place, and the EW scale v. Furthermore, one would naturally obtain

$$m_{W,Z} \sim m_h \sim f \sin\theta \sim v\,. \tag{4}$$

For the Higgs, this relation is due to the fact that for $\theta \to 0$, the EW symmetry is recovered and the fourth component of the bi-doublet must become an exact Goldstone boson, like the 3 eaten by W and Z. Note that the mass of the Higgs, and the alignment of the vacuum, are determined by explicit breaking of the global symmetry $\mathcal{G}$, which must be small in order not to deform the dynamics: as examples, one should mention the gauging of the EW symmetry, and the coupling to the standard fermions (primarily the top) apt to generate their mass (analogous to the Yukawa couplings in the standard model). If ones wants to generate a largish hierarchy between v and f, a certain degree of fine tuning is necessary in the explicit breaking terms.

coset	dim(coset)	Higgs(es)	$G_{\rm TC}$	fermions	
SU(4)/Sp(4)	5	1	SU(2)	2 doublets	pseudo-real
SU(5)/SO(5)	14	1	SO(N)	5 vectorial	real
$SU(3)^2/SU(3)$	8	1	SU(N)	3 fund.	complex
SU(6)/Sp(6)	14	2	SU(2)	3 doublets	pseudo-real
$SU(4)^2/SU(4)$	15	2	SU(N)	4 fund.	complex

Table 1. List of the smallest models of composite Higgs based on a simple underlying dynamics.

For a review of the possible global symmetries and a discussion of the fine tuning, we refer to the review in Ref. [12]. The minimal coset, in terms of dimension, is SO(5)/SO(4) [11], which contains exactly 4 states transforming as a bi-doublet of SO(4), and has been extensively studied in the literature.

3. The minimal dynamical model: SU(4)→Sp(4)

In the following, we will focus on the minimal dynamical model, which features the coset SU(4)/Sp(4) [15]: the microscopic model can be based on a very simple construction based on a hypercolor group $G_{TC} = SU(2)_{TC}$ with two Dirac fermions transforming as doublets. This microscopical model has been first proposed as a UV completion of Little Higgs models in Ref. [16], then as a candidate for a minimal Technicolor in Ref. [17], and finally as a candidate for pNGB Higgs in Ref. [18]. A list of other non-minimal model is in Table 1. One of the greatest advantaged of this model is that, due to its simplicity, it can be put on the lattice [19]: in fact, the symmetry breaking SU(4)/Sp(4) has been observed in lattice results, thus giving us confidence that the dynamics does what we need it for. Also, the model can be pushed close to the conformal limit by adding one or two extra fermions transforming as an adjoint of $SU(2)_{TC}$ [20].

Embedding the electroweak interactions $SU(2)_L \times U(1)_Y$ can be done by assigning EW quantum numbers to the fundamental quarks ψ: a minimal and consistent choice is to take the left-handed components ψ_L to transform as a doublet of $SU(2)_L$, while the right-handed ones ψ_R as a doublet of $SU(2)_R$ (*i.e.* they have opposite hypercharge equal to $\pm 1/2$). With this assignments, no gauge anomalies are introduced. The global $\mathcal{G} = SU(4)$ symmetry becomes evident once one builds a 4-component object with $\psi^i = (\psi_L, \psi_R^c)$. The condensate $\psi^i\psi^j$ transforms as:

$$\langle \psi^i \psi^j \rangle = \mathbf{6}_{\rm SU(4)} = \mathbf{5}_{\rm Sp(4)} \oplus \mathbf{1}_{\rm Sp(4)} \,. \tag{5}$$

The five-plet of Sp(4) contains the Goldstone bosons of the SU(4)/Sp(4) breaking, and they decompose under the custodial SO(4) ⊂ Sp(4) group as

$$\mathbf{5}_{\rm Sp(4)} = (\mathbf{2},\mathbf{2})_{\rm SO(4)} \oplus (\mathbf{1},\mathbf{1})_{\rm SO(4)} \,. \tag{6}$$

The spectrum, therefore, contains a bi-doublet (associable to the Higgs field) and an extra singlet η. On the other hand, the Sp(4) singlet corresponds to a massive scalar, similar to the σ in QCD: this is the candidate for the observed Higgs boson in the Technicolor limit. The most general alignment for the condensate can be obtained starting from the vacuum that preserves the EW symmetry, and applying a rotation generated by the broken generators:

$$\Sigma_0 = e^{i\gamma} \begin{pmatrix} i\sigma_2 \cos\theta & \sin\theta \\ -\sin\theta & -i\sigma_2\cos\theta \end{pmatrix} , \tag{7}$$

where σ_2 is the second Pauli matrix, and θ is the angle determining the alignment ($\theta = \pi/2$ corresponding to the Technicolor limit). The phase γ is generated by an anomalous global U(1) acting on the fermions ψ, and it potentially induces CP violation (therefore we will set it to zero in the following). In principle, a rotation along the singlet direction can also introduce a new parameter, however this direction commutes with the EW gauge generators and therefore will not affect the physics in the EW sector. From this vacuum, one can obtain the masses for the W and Z, and a Lagrangian for the Goldstone bosons (we refer the reader to Ref. [13]):

$$m_W^2 = m_Z^2 c_W^2 = 2g^2 f^2 \sin^2\theta\,, \quad v = 2\sqrt{2} f \sin\theta\,. \tag{8}$$

In the Unitary gauge, the spectrum contains two light scalars: a pNGB Higgs h, and the extra singlet η. On top of this, we have the heavy scalar σ, being a singlet of Sp(4). The masses of these states depend on the explicit breaking of the global symmetry, and we refer the reader to Ref.s [18, 13] for more details. Here, we will simply recap the main features. The breaking terms introduced by the coupling to standard model particles are due to gauge interactions (*i.e.* the partial gauging of SU(4)), and the coupling to the top. The latter can be introduced in the form of a four-fermion interaction

$$\frac{y_t}{\Lambda^2}(q_L t_R^c)(\psi_L \psi_R^c)\,, \tag{9}$$

i.e. a coupling of the elementary fermions q_L and t_R to the combination of ψ's that transforms like the Higgs doublet. After bosonizing the theory, the above four-fermion interaction will generate a mass for the top in the form

$$m_t = y_t' f \sin\theta\,, \tag{10}$$

where y_t' is an effective Yukawa coupling.

3.1. *Potential and fine tuning*

This interaction in Eq.(9) is crucial for the model, as loops of the elementary top induce an effective potential for the vacuum. Another breaking term is due to a mass for the fermions ψ, allowed by gauge interactions. The contribution of the top and ψ-mass to the potential is given by (here we neglect the contribution of gauge loops, which are of the same form of the top ones but smaller):

$$V(\theta) = y_t'^2 C_t \cos^2\theta - 4 C_m \cos\theta\,, \tag{11}$$

where C_t and C_m are $\mathcal{O}(1)$ loop coefficients determined by the dynamics, and the second term is generated by the mass term. This potential fixes the value of θ: the minimum is

$$|\cos\theta|_{\rm min} = \frac{2C_m}{y_t'^2 C_t}\,, \qquad \text{for} \quad y_t'^2 C_t > 2|C_m|\,, \tag{12}$$

else $\theta_{\rm min} = 0$. We can see here that obtaining a small θ, corresponding to a pNGB Higgs limit, requires a certain level of fine tuning between two terms whose origin is of very different nature. For instance, $\sin\theta = 0.2$ requires $\cos\theta = 0.98$, thus the two terms must be equal up to a 2%. It is also interesting to study the contribution to the masses of the two pNGBs h and η: the top loop and ψ-mass contribute

$$\begin{aligned} m_h^2 &= \frac{f^2}{4}\left(2C_m\cos\theta - y_t'^2 C_t \cos(2\theta)\right) \\ &= \frac{y_t'^2 C_t f^2}{4}\sin^2\theta_{\rm min}\,. \end{aligned} \tag{13}$$

In the second expression we have used the minimum condition to eliminate C_m in favor of θ. This formula is remarkable as if shows the pNGB mechanism at work: the naïve expectation for the top loop contribution to the Higgs mass in this model should be

$$\delta m_h^2\Big|_{\text{naïve}}(top) \sim -\frac{y_t'^2 \Lambda_{TC}^2}{16\pi^2} = -y_t'^2 f^2\,, \tag{14}$$

using $\Lambda_{TC} \sim 4\pi f$ as a cut-off, which is thus proportional to the scale of the condensate f. However, the minimization, which requires a tuning between the top loop and the ψ-mass to achieve a small θ, automatically generates an extra factor $\sin^2\theta$ in the Higgs mass, thus lowering the scale to v. In this model, also, the Higgs mass can be rewritten as

$$m_h^2 = \frac{y_t'^2 C_t f^2 \sin^2\theta}{4} = \frac{C_t m_t^2}{4}\,, \tag{15}$$

thus $m_h = 125$ GeV requires $C_t \sim 2$, which is a well reasonable value.

For the singlet, the mass reads

$$\begin{aligned} m_\eta^2 &= \frac{f^2}{4}\left(2C_m\cos\theta + y_t'^2 C_t \sin^2\theta\right) \\ &= \frac{y_t'^2 C_t f^2}{4} = \frac{m_h^2}{\sin^2\theta}\,. \end{aligned} \tag{16}$$

It is interesting to notice that in models where this symmetry breaking is achieved in a renormalizable way, these estimates change drastically [21]: in fact, the quadratic divergences would be renormalized (even though the naturalness problem is not solved) and the value of the masses will be determined by higher order terms in the θ-expansion.

3.2. EWPTs and LHC signatures

At this point, the value of the minimum θ is one of the free parameters of the model. The main constraint on it comes from effects on electroweak precision measurements, mainly to the S parameter. The value of θ that can be attained is crucial for predicting the LHC phenomenology of this model, because it allows to determine the mass of the additional pNGB η, and of the heavier spin-1 resonances. The correction to S can be estimated by calculating the effect of the modified couplings of the 125 GeV Higgs, and of the heavier singlet σ, while the contribution of the dynamics can be approximated by a loop of ψ's. Such an estimate makes sense if the contribution of heavier resonances is dominated by the lightest vector ones. The numerical impact of electroweak precision tests (EWPTs) has been studied in [22], and it depends crucially on the couplings of the σ to the SM particles. Neglecting the contribution of σ, we find that EWPTs prefer values $\sin\theta < 0.24$.

With this range in mind, we can estimate the production cross section for the η at the Run II at the LHC [22], which will collide protons at a center of mass energy of 13 TeV. The results, for the two leading channels, are shown in Figure 1: in blue we show the pair production by vector boson fusion (VBF), $pp \to jj\eta\eta$, as a function of the mass, where j are jets produced by forward light quarks from the proton and the two singlets couple to EW gauge bosons at tree level. We see that in the whole range considered, the cross section is always smaller than 0.5 fb, which would correspond to about 50 events produced after 100/fb of collected data (3 years of Run II). The second best channel is single production by gluon fusion, triggered by a top loop. We should also stress that in the region preferred by EWPTs, $m_\eta > 528$ GeV and the cross sections drop to very small values: $\sigma_{\text{VBF}} < 5$ ab and $\sigma_{\text{ggfusion}} < 1.5$ ab. We can therefore conclude that the singlet η cannot be observed at the LHC due to the too feeble production rates.

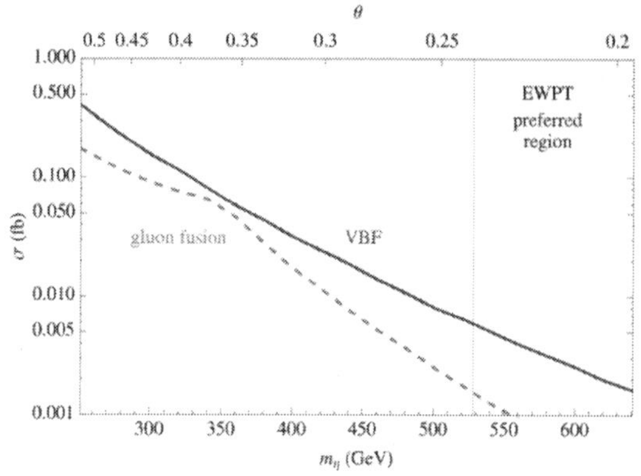

Figure 1. Production cross sections for the singlet pNGB η at the LHC Run II (13 TeV), as a function fo the η mass. In blue, pair production via vector boson fusion, in red single production via gluon fusion. On the upper axes we show the value of θ corresponding to the mass on the bottom axes: the vertical line corresponds to the limit preferred by EWPTs.

The model also contains spin-1 states, whose masses can be calculated on the Lattice [19]. By properly rescaling the Lattice results, the masses are predicted to lie in the range

$$m_a = \frac{3.3 \pm 0.7}{\sin\theta}\ \text{TeV} > 14 \pm 3\ \text{TeV}\,, \tag{17}$$

$$m_\rho = \frac{2.5 \pm 0.5}{\sin\theta}\ \text{TeV} > 10 \pm 2\ \text{TeV}\,, \tag{18}$$

where ρ and a label the vector and axial-vector resonances, and the lower limit corresponds to the EWPT preferred region. From these values we can immediately see that the LHC, even during Run II, does not have enough energy to produce on-shell resonances.

4. Conclusions

The possibility that the discovered 125 GeV Higgs is a composite state of an underlying strong dynamics is still viable, especially if the Higgs arises as a pseudo-Nambu-Goldstone boson of the breaking of a large global symmetry of the dynamics. In this paper we follow a new approach relying on the existence of a simple underlying dynamics based on a simple confining gauge group and fermionic matter. This approach allows some calculability, in the sense that the symmetry breaking pattern can be predicted uniquely and that Lattice results can be used to constrain the masses of some of the resonances.

We showed that the minimal case, based on a confining SU(2) gauge symmetry with 2 Dirac fermions in the fundamental representation, can lead to a realistic model for the discovered Higgs boson, which arises as a pNGB of the global symmetry breaking SU(4)/Sp(4). Electroweak precision tests can be evaded with a mild fine tuning in the potential, of the order of a few %. We also showed that the model predicts the absence of signals of new physics at the LHC, both Run I and Run II. In fact, the only new light state is a singlet η which can be pair and singly produced at the LHC with feeble sub-fb cross-sections. The event rates are therefore too small to yield detectable signals. At the same time, Lattice results show that heavier spin-1 states are too heavy to be produced, as their mass range is expected above 10 TeV. Therefore, this model is a good benchmark for a higher energy proton collider, following the recent proposals of 100 TeV center of mass energies.

We conclude by stressing that other possible symmetry patterns, and underlying dynamical models, are possible, which lead to additional light pNGBs. Thus, we can still hope that new scalar states can be observed at the LHC in these extended scenarios, and that maybe one such state can play the role of the Dark Matter in the Universe.

Acknowledgments

The author acknowledges partial support from the Labex-LIO (Lyon Institute of Origins) under grant ANR-10-LABX- 66.

References

[1] Englert F and Brout R 1964 Broken symmetry and the mass of gauge vector mesons *Phys. Rev. Lett.* **13** 321
[2] Higgs P W 1964 Broken symmetries and the masses of gauge bosons *Phys. Rev. Lett.* **13** 508
[3] Khachatryan V *et al.* [CMS Collaboration] 2014 Precise determination of the mass of the Higgs boson and tests of compatibility of its couplings with the standard model predictions using proton collisions at 7 and 8 TeV *Preprint arXiv:1412.8662 [hep-ex]*
[4] The ATLAS collaboration 2015 Measurements of the Higgs boson production and decay rates and coupling strengths using pp collision data at $\sqrt{s} = 7$ and 8 TeV in the ATLAS experiment *Preprint ATLAS-CONF-2015-007 and ATLAS-COM-CONF-2015-011*
[5] Sohnius M F (1985) Introducing supersymmetry *Phys. Rept.* **128** 39
[6] Hosotani Y 1983 Dynamical gauge symmetry breaking as the Casimir effect *Phys. Lett.* B **129** 193
[7] Dimopoulos S and Susskind L 1979 Mass without scalars *Nucl. Phys.* B **155** 237
[8] Eichten E and Lane K D 1980 Dynamical breaking of weak interaction symmetries *Phys. Lett.* B **90** 125
[9] Chacko Z, Franceschini R and Mishra R K 2013 Resonance at 125 GeV: Higgs or Dilaton/Radion? *J. High Energy Phys.* JHEP1304(2013)015 (*Preprint arXiv:1209.3259 [hep-ph]*)
[10] Kaplan D B and Georgi H 1984 SU(2) x U(1) breaking by vacuum misalignment *Phys. Lett.* B **136** 183
[11] Agashe K, Contino R and Pomarol A 2005 The minimal composite Higgs model *Nucl. Phys.* B **719** 165 (*Preprint hep-ph/0412089*)
[12] Bellazzini B, Csáki C and Serra J 2014 Composite Higgses *Eur. Phys. J.* C **74** 5, 2766 (*Preprint arXiv:1401.2457 [hep-ph]*)
[13] Cacciapaglia G and Sannino F 2014 Fundamental composite (Goldstone) Higgs dynamics *J. High Energy Phys.* JHEP1404(2014)111 (*Preprint arXiv:1402.0233 [hep-ph]*)
[14] Georgi H and Kaplan D B 1984 Composite Higgs and custodial SU(2) *Phys. Lett.* B **145** 216
[15] Katz E, Nelson A E and Walker D G E 2005 The intermediate Higgs *J. High Energy Phys.* JHEP0508(2005)074 (*Preprint hep-ph/0504252*)
[16] Batra P and Chacko Z 2008 Symmetry breaking patterns for the Little Higgs from strong dynamics *Phys. Rev.* D **77** 055015 (*Preprint arXiv:0710.0333 [hep-ph]*)
[17] Ryttov T A and Sannino F 2008 Ultra minimal technicolor and its dark matter TIMP *Phys. Rev.* D **78** 115010 (*Preprint arXiv:0809.0713 [hep-ph]*)
[18] Galloway J, Evans J A, Luty M A and Tacchi R A 2010 Minimal conformal Technicolor and precision electroweak tests *J. High Energy Phys.* JHEP1010(2010)086 (*Preprint arXiv:1001.1361 [hep-ph]*)
[19] Hietanen A, Lewis R, Pica C and Sannino F 2014 Fundamental composite Higgs dynamics on the Lattice: SU(2) with two flavors *J. High Energy Phys.* JHEP1407(2014)116 (*Preprint arXiv:1404.2794 [hep-lat]*)
[20] Ryttov T A and Sannino F 2010 Conformal house *Int. J. Mod. Phys.* A **25** 4603 (*Preprint arXiv:0906.0307 [hep-ph]*)
[21] Alanne T, Gertov H, Sannino F and Tuominen K 2014 Elementary Goldstone Higgs and dark matter *Preprint arXiv:1411.6132 [hep-ph]*
[22] Arbey A, Cacciapaglia G, Cai H, Deandrea A, Le Corre S and Sannino F 2015 Fundamental composite electroweak dynamics: status at the LHC *Preprint arXiv:1502.04718 [hep-ph]*

Search for heavy resonances at CMS

Otman CHARAF [1]

Department of Physics and Astronomy, University of Alabama, Tuscaloosa, AL 35487-0324, USA

E-mail: `Otman.Charaf@cern.ch`

Abstract. The search for new physics at the TeV scale is one of the major aspects of the CMS experimental program. This includes a myriad of theoretical models involving resonances that can decay to massive bosons, leptons or jets. An overview of such analyses is presented in this paper as well as the novel dedicated techniques related to the specificity of the event topologies. All analyses presented here use the full 2012 dataset, corresponding to an integrated luminosity of $\sim 20\ fb^{-1}$ recorded in proton-proton collisions at $\sqrt{s} = 8$ TeV with the CMS detector at the CERN LHC.

1. Introduction

The recent discovery by the ATLAS and CMS Collaborations [1, 2, 3] of a particle compatible with the Standard Model (SM) predictions for the Higgs boson enhanced our understanding of the electroweak symmetry (EW) breaking mechanism. But the large difference between the EW scale and the Planck scale, i.e. the hierarchy mass problem, stills remains unsolved. Many theoretical extensions of the SM (BSM theories) predict the existence of new gauge bosons or KK excitations of the graviton that could manifest at the TeV scale. Such resonances would appear as a peak over a continous background coming from the SM processes. The search for BSM resonances is one of the major aspects of the CMS [4] experimental program and includes dedicated techniques to demonstrate the background predictions are well under control. Almost all analyses presented here use the particle-flow (PF) algorithm [5] to reconstruct and identify all stable particles in the event (electrons, muons, photons, neutral and charged hadrons) by using a thorough recombination of all sub-detectors.

2. Dilepton resonances [6]

Dilepton resonances include the Sequential Standard Model Z' (Z'_{SSM}), the Z'_{ψ} predicted by E_6 gauge group grand unified theories, or the Randall-Sundrum graviton (G_{RS}), all of which can decay either to a dielectron or a dimuon pair. The main background is the irreducible Drell-Yan background and the selection is mainly based on shower shape variables as well as isolation variables. In order for the analysis to be robust, different data-driven background estimation techniques and cross-check methods have been derived. For the Drell-Yan contriubtion, the mass spectrum shape is derived from Monte-Carlo simulations and normalized to the number of events in data in the region $60 < M_{ll} < 120$ GeV. The $t\bar{t}$, tW, WW, WZ and $\tau\tau$ contributions are derived from Monte-Carlo simulations and cross-checked using the data-driven $e\mu$ method

[1] on behalf of the Compact Muon Solenoid (CMS) collaboration

which exploits the flavour-symmetric nature of such processes. Finally, the QCD multi-jet component is estimated using the data-driven fake rate method which calculates the probability for a jet, having been reconstructed as an electron, to pass the electron selection[2]. Figures 1 (left) and (middle) present the invariant mass spectra of the selected dielectron and dimuon pairs respectively. In each case, the spectrum from data is compared with the sum of the different background components. Excellent agreement is observed along the whole mass range and no significant deviation from the Standard Model processes is observed.

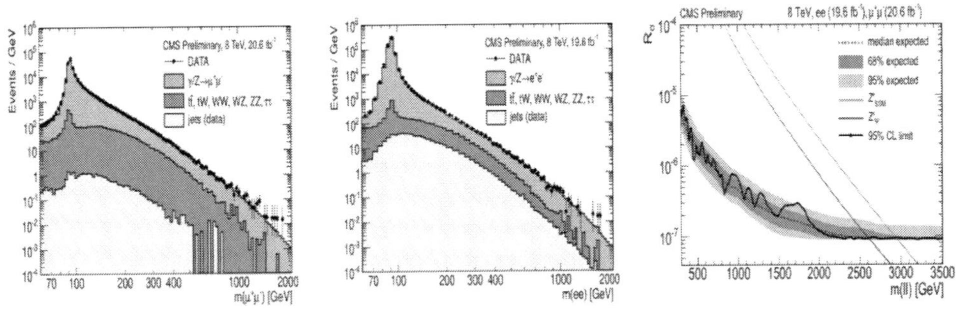

Figure 1. The invariant mass spectrum of $\mu^+\mu^-$ (left) and ee (middle) events. The points with error bars represent the data. The histograms represent the expectations from standard model processes (see text). Upper limits (right) as a function of resonance mass M on the production ratio R_σ of cross section times branching fraction into lepton pairs for Z'_{SSM} and Z'_ψ boson production to the same quantity for Z bosons.

To extract upper limits on the various model cross sections, a statistical interpretation tool is built based on an unbinned likelihood bayesian technique. The statistical method has the specificity of (a) combining both channels and (b) setting limits on the cross section ratio R_σ = $\sigma_{Z'}/\sigma_Z$ allowing therefore to cancel some systematic uncertainties. Figure 1 (right) gives the 95% CL limits as a function of the dilepton invariant mass for the Z'_ψ and Z'_{SSM} models. The median, 68% and 95% quantiles for the expected limits are also shown together with the theoretical predictions. A Z' with standard model-like couplings can be excluded below 2900 GeV and the superstring-inspired Z'_ψ below 2570 GeV. RS Kaluza-Klein gravitons are excluded below 2730 (2350) GeV for couplings of 0.10 (0.05), all at 95% CL.

3. Dijet resonances [7]

Dijet heavy resonances are predicted by the same models covering the dilepton resonances but, in addition, many other models such as string resonances, scalar diquarks, excited quarks, axigluons, colorons or color-octet scalar (s8) resonances can enter the scope of this analysis. The main background is the QCD multi jet background. All jets in the event are required to have a transverse momentum greater than 30 GeV and an absolute value of the pseudo rapidity less than 2.5. In order to reduce sensitivity to gluon radiation, this analysis uses the "wide jets" technique [8]. The method consists in considering the two leading jets in the event and merge all other jets to the closest leading jet if they are within a distance ΔR ¡ R_{wide} where $\Delta R = \sqrt{(\Delta\eta)^2 + (\Delta\phi)^2}$ and $R_{wide} = 1.1$. The invariant mass of the two wide jets is required to be higher than 890 GeV and their pseudo rapidity separation $|\Delta\eta_{j_1 j_2}|$ be less than 1.3. In addition the scalar sum of the transverse momenta of all jets in the event has to be higher than

[2] This source of background is negligible in the dimuon channel.

650 GeV. Figure 2 (left) shows the invariant mass spectrum of the dijet pair. The background is modeled by the so-called dijet function f(x) = $(p_0(1-x)^{p_1})/(x^{p_2+p_3 ln x})$ where x = $M_{jj}/\sqrt{s}$. The function describes very well the background spectrum as one sees from the residuals distribution. The main systematic uncertainties arise from the jet energy scale and resolution (1.25 % and 10 % on the invariant mass, respectively), and from the background parametrization. No significant deviation is observed from the known Standard Model processes. The statistical interpretation method is based on the bayesian formalism and is performed independently at each value of the resonance mass. Figure 2 (right) shows the upper limit on the cross section × branching ratio × acceptance as a function of the dijet pair invariant mass for the three decay modes (qq, gg, qg). The theoretical cross sections are also shown for the various models investigated. The upper limits on the cross sections are translated to lower limits on the resonances masses, summarized in Table 1.

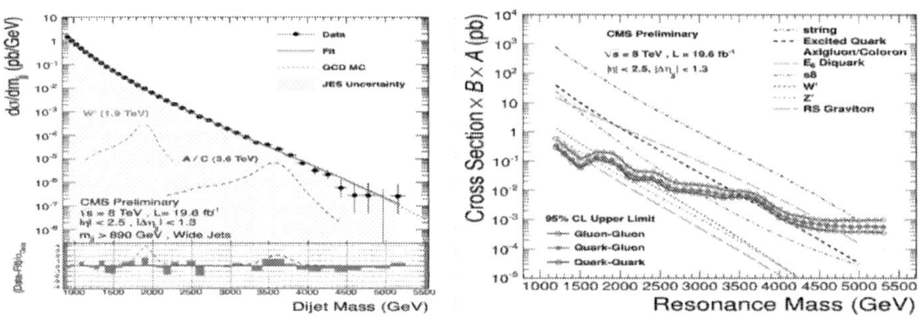

Figure 2. Left: Dijet mass spectrum from wide jets (points) compared to a smooth fit (solid) and to predictions including detector simulation of QCD and signal resonances. The QCD prediction has been normalized to the data. The error bars are statistical only. The bin-by-bin fit residuals, (data-fit)/σ_{data}, are shown at the bottom. Right: The observed 95% CL upper limits on σ × B × A for dijet resonances of the type gluon-gluon, quark-gluon, and quark-quark, compared to theoretical predictions for the various models investigated (see text).

Table 1. Observed and expected 95% CL exclusions on the mass of various resonances. Systematic uncertainties are included.

Model	Final State	Obs. Mass Excl. [TeV]	Exp. Mass Excl. [TeV]
String Resonance (S)	qg	[1.20,5.08]	[1.20,5.00]
Excited quark (q^*)	qg	[1.20,3.50]	[1.20,3.75]
E_6 Diquark (D)	qq	[1.20,4.75]	[1.20,4.50]
Axigluon (A)/Coloron (C)	$q\bar{q}$	[1.20,3.60] + [3.90,4.08]	[1.20,3.87]
Color Octet Scalar (s8)	gg	[1.20,2.79]	[1.20,2.74]
W' Boson (W')	$q\bar{q}$	[1.20,2.29]	[1.20,2.28]
Z' Boson (Z')	$q\bar{q}$	[1.20,1.68]	[1.20,1.87]
RS Graviton (G)	$q\bar{q}$+gg	[1.20,1.58]	[1.20,1.43]

4. VV resonances

Heavy resonances can also decay to a pair of boosted vector bosons which can then decay either leptonically or hadronically. In the leptonic decay, the isolation cones may overlap, requiring a modified isolation cone. In the hadronic decay, the vector boson can be tagged based on its invariant mass, using the so-called jet pruning algorithm [9] based on the Cambridge-Aachen (CA) algorithm [10]. It aims to reduce the soft radiation components in the jet and consists in reclustering each jet based on its original constituents using the CA algorithm while discarding, at each step, combinations considered to be soft. Often, an additional requirement on the jet invariant mass to be consistent with the W or Z mass is made (V-tagging). In addition, in the hadronic decay, the two jets resulting from the boosted vector boson decay may merge into one, which necessitates the use of so called jet substructure techniques [8, 11]. This method aims to identify the presence (or not) of subjets within the jet. To that regard, a metric is defined, called the N-subjettiness, to quantify the capability of finding N subjets within the considered jet.

$$\tau_N = \frac{1}{d_0}\sum_k p_{t,k} min(\Delta R_{1,k}, \Delta R_{2,k}, ..., \Delta R_{N,k}), d_0 = R_0 \sum_k p_{t,k},$$

where $p_{T,k}$ is the p_T of the particle constituent k of the original jet, and $\Delta R_{n,k}$ is its angular distance from the axis of the nth subjet (with n = 1, 2, . . . , N). d_0 is a normalization factor with R_0 set to the distance parameter R of the original CA jet (R = 0.8). The N-subjettiness is smaller for jets effectively formed from N subjets. To reduce as much as possible the QCD multi-jet contribution, the ratio $\tau_{21} = \tau_2/\tau_1$ is formed and used as a discriminating variable.

4.1. VV/qV resonances [12]

Such resonances can happen in the context of excited quarks ($q* \to qV$) or new heavy gauge bosons from GUT models ($W' \to WZ$) as well as KK excitations of the graviton in the context of the Randall-Sundrum or bulk scenarios. A focus is made here on the all-hadronic channel where the dominant source of background is the QCD multi-jet process. The selection makes use of the jet substructure technique, jet pruning algorithm and V-tagging method mentioned in Section 4. Events are selected that contain at least two jets with invariant mass higher than 890 GeV and pseudo rapidity separation $|\Delta\eta_{j_1 j_2}|$ less than 1.3. In addition, each jet is classified as a high-purity jet (HP-jet) or a low purity jet (LP-jet) depending if $\tau_{21} < 0.5$ or $0.5 < \tau_{21} < 0.75$, respectively. The remaining background is estimated, in each channel, using the fit function $f(x) = p_0 x^{p_1}/x^{p_2}$ with $x = M_{jj}/\sqrt{s}$. The main sources of systematic uncertainy arise from the V-tagging procedure, which amount to 7.5% (54%) in the HP (LP) category. Figure 3 shows the invariant mass spectra of the dijet pairs in the cases of single (top left) and double (top right) W/Z tag. In each case, the spectra are shown for the low and high purity categories and compared to the Monte-Carlo predictions from MADGRAPH/PYTHIA and HERWIG++. The agreement is very good in each case/category and no significant deviation is observed from the known Standard Model processes. The statistical interpretation is based on an asymptotic frequentist method (CLs method). Figure 3 (bottom row) shows the upper limits on the cross section × branching ratio for the models of excited quarks and Randall-Sundrum gravitons. Lower limits are set at the 95% confidence level on masses of excited quark resonances decaying into qW and qZ at 3.2 and 2.9 TeV, respectively. Randall-Sundrum gravitons G_{RS} decaying into WW are excluded up to 1.2 TeV, and W' bosons decaying into WZ, for masses less than 1.7 TeV. The corresponding expected limits amount to 3.0 and 2.6 TeV for excited quark resonances decaying into qW and qZ, respectively and to 1.3 and 1.6 TeV for Randall-Sundrum gravitons G_{RS} decaying into WW and W' bosons decaying into WZ, respectively. These mass limits are the most stringent up to date.

4.2. VV semileptonic resonances [13]

Such resonances are characterized by the presence, in the event, of at least 2 jets and either one lepton with missing transverse energy or two leptons ($X \to VV \to l\nu qq/llqq$). Two main

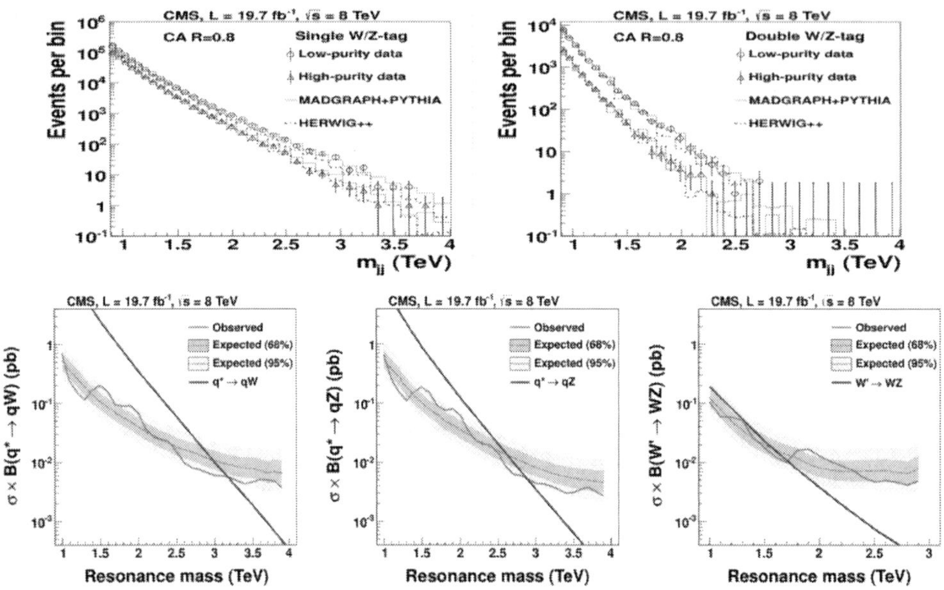

Figure 3. The m_{jj} distributions for (top left) singly and (top right) doubly tagged events in data, and for QCD multi-jet (MADGRAPH/PYTHIA and HERWIG++) simulations, normalized to data. Expected and observed 95% CL limits on the production cross section as a function of the resonance mass for (bottom left) qW resonances, (bottom middle) qZ resonances and (bottom right) WZ resonances, compared to their predicted cross sections for the corresponding benchmark models.

components arise in these searches: the $l\nu$ + V-jet and the ll + V-jet components depending whether the W or the Z decays leptonically. The lepton selection is based on shower shape and isolation in a way similar to the selection in the dilepton analysis (Section 2). The jet selection is subject to the jet pruning and jet substructure techniques in a similar way as in the VV/qV resonances analysis (Section 4.1). Each jet is classified as a high-purity jet (HP-jet) or a low purity jet (LP-jet) depending if $\tau_{21} < 0.5$ or $0.5 < \tau_{21} < 0.75$. In addition the transverse momentum of each jet is required to be greater than 30 GeV and its absolute pseudo rapidity be less than 2.4. The V-tagging requirement is also applied and each V-jet is tagged as a W-jet or a Z-jet depending on the pruned jet mass. Events are furthermore selected based on the V-boson transverse momentum and the presence of missing transverse energy. Additional criteria to ensure the back-to-back topology are required. After the full selection, the dominant source of background comes from the V+jet process. The overall normalization of this background contribution is estimated through a fit in sidebands of the pruned jet mass (M_j) while the shape is taken from the low M_j sideband and extrapolated to the signal region using an extrapolation function derived from Monte-Carlo predictions. Figure 4 (left) shows the pruned jet mass spectrum from data together with the different background predictions, in the case of the $e\nu$ channel for the HP category. Sources of background other than the V+jet are extracted from Monte-Carlo predictions. The residual distribution in the bottom pad shows a good level of agreement between the fit and the data. The statistical interpretation is based on an asymptotic frequentist method (CLs method) with an unbinned shape analysis. The categories are all combined and the results are further combined with the results from the corresponding analysis in the all hadronic channel (Section 4.1). Figure 4 (right) shows the 95% CL limit on the bulk graviton production cross section as a function of the graviton mass. In region I, only the ll+V-jet channel contributes. In region II, both ll+V-jet and $l\nu$+V-jet channels contribute. In region III, both the semi-leptonic and all-hadronic channels contribute. Upper limits at 95% CL are set on the bulk graviton production cross section in the range from 700 to 10 fb for resonance masses between 600 and 2500 GeV, respectively.

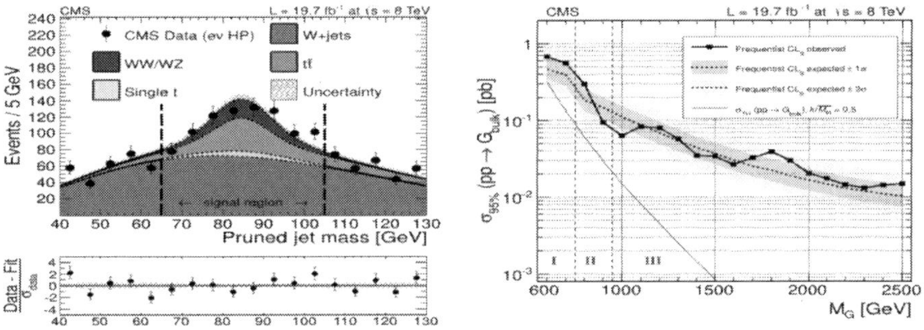

Figure 4. Left: Distribution of the pruned jet mass, mjet, in the $l\nu$+V-jet analysis in the electron channel for the HP category. Data are shown as black markers. At the bottom of each plot, the bin-by-bin fit residuals, (data-fit)/σ_{data}, are shown together with the uncertainty band of the fit normalized by σ_{data}. Right: observed (solid) and expected (dashed) 95% CL upper limit on the graviton production cross section obtained with this analysis and the analysis of the all-hadronic channel 4.1. The cross section for the production of a bulk graviton with $k/M_{Pl} = 0.5$ is shown as a red solid curve.

5. WZ resonances in leptonic channel [14]

Such processes are characterized by a resonance, such as the W' from the SSM or the ρ from Technicolor, that decays to a W and a Z, both decaying leptonically ($X \rightarrow WZ \rightarrow l\nu ll$). The dominant background is the irreducible SM WZ process. The lepton selection is based on shower shape and isolation criteria similar to the dilepton analysis (Section 2) where, in this case however, a modified isolation cone is used (Section 4). Events are then selected containing at least three leptons with two opposite charge leptons with mass $M_{l^+l^-}$ between 71 and 111 GeV. The mass of the three leptons M_{3l} has to be higher than 120 GeV in order to remove events where M_{3l} is close to the Z boson mass and a minimum missing transverse energy of 30 GeV is required to account for the neutrino from the W decay. Also, the distance between the lepton from the Z and the lepton from the W $\Delta R(l_Z, l_W)$ is to be higher than 0.3. Additional criteria such as the sum of the lepton transverse momenta L_t and the mass of the WZ pair M_{WZ} are also exploited and optimized for the best expected limit. Figure 5 shows the distributions of these two variables (left and middle) for the various sources of background, W' bosons of 1 and 1.5 TeV, as well as for the data. One sees that good discrimination is possible between the signal and the different background contributions. The statistical interpretation is based on a counting experiment (bayesian formalism). Figure 5 (right) shows the 95% CL upper limit on the cross section × branching ratio as a function of the resonance (W' or ρ_{TC}) mass. These upper limits get translated to lower limits on the resonance mass of 1470 GeV for the W' and 1140 GeV for the ρ_{TC}.

Journal of Physics: Conference Series **623** (2015) 012007 doi:10.1088/1742-6596/623/1/012007

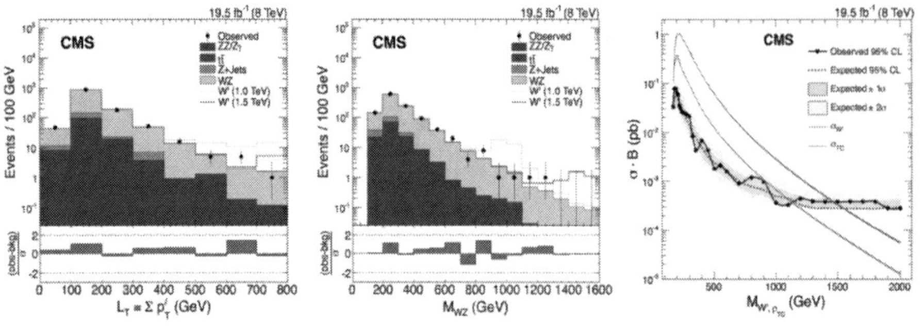

Figure 5. The WZ invariant mass (left) and L_T (middle) distributions for the background, signal, and observed events after the WZ candidate selection. The last bin includes overflow events. The (obs-bkg)/σ in the lower panel is defined as the difference between the number of observed events and the number of expected background events divided by the total statistical uncertainty. Limits (right) at 95% CL on $\sigma \times B(W' \rightarrow 3l\nu)$ as a function of the mass of the W' (blue) and ρ_{TC} (red), along with the 1 σ and 2 σ combined statistical and systematic uncertainties indicated by the green (dark) and yellow (light) band, respectively. The theoretical cross sections include a mass-dependent NNLO K-factor. The thickness of the theory lines represents the PDF uncertainty associated with the signal cross sections.

6. Light/heavy flavour three jet resonances [15]

Such resonances are predicted by scenarios like heavy coloured fermions or R-parity violating (RPV) gluinos where the pair-produced resonances can decay, individually to 3 jets ($XX \rightarrow qqqqqq$). The topology of the event is therefore characterized by the presence of at least 6 jets. In order to correctly reconstruct the event, this search uses the novel jet ensemble technique [16] that recombines jets to find correct triplets and is the first of its kind to include heavy flavour jets. It is divided into two categories: the light flavour inclusive mass search covering the mass range 400-1500 GeV and the heavy flavour low and high mass searches covering the mass ranges 200-600 GeV and 600-1500 GeV, respectively. The dominant source of background is the QCD multi-jet process which reduction is based on a selection that uses dedicated discrimination variables. First, a criterion on the transverse momentum of the fourth highest transverse momentum jet (sixth highest transverse momentum jet) is optimized for the low mass heavy search (inclusive and high mass heavy search). Second, a correlation is observed, in Monte-Carlo simulations, between the mass of the jet triplet M_{jjj} and the sum of their transverse momenta ($\sum_{i=1}^{3} p_{t,j}^{i}$). A criterion is therefore imposed on the triplet mass where: $M_{jjj} < \sum_{i=1}^{3} p_{t,j}^{i} - \Delta$. The value of Δ is determined from Monte-Carlo simulations and chosen so as to cover as broad a range as possible for values of the resonance mass. Finally, the sphericity of the signal events is exploited as they tend to be more spherical than background events. The sphericity tensor is defined as $S^{\alpha\beta} = \sum p^{\alpha} p^{\beta} / \sum |p|^2$ together with its eigenvalues λ_i and the sphericity defined as $S = (3/2) \times (\lambda_2 + \lambda_3)$. The background estimation proceeds twofold. In the light and high mass heavy flavour, the background is estimated by fitting the data with a dijet function. In the low mass heavy flavour, the $t\bar{t}$ background is determined from Monte-Carlo predictions while the SM multi-jet spectrum shape is extracted from the data in a b-jet control region. This estimation technique is further validated by considering the $t\bar{t}$ process as the signal, extracting its cross section and finding a good agreement with the theoretical value. No significant evidence of a new resonance is observed and 95 % CL upper limits on the resonance production cross section $\times$ branching ratio are derived using the asymptotic frequentist approach (CLs method) based on a profile likelihood. Figure 6 shows those limits as a function of the triplet invariant mass in the case of the light (left) and heavy (right) flavour search. The red dashed lines correspond to the $+1\sigma$ and -1σ theoretical predictions on the cross section times branching ratio, σ being the uncertainty on that prediction. This searches adopts a conservative approach and uses the -1σ theoretical curve to translate upper limits on cross section $\times$ branching ratio to lower limits on the resonance mass. For RPV gluinos, masses are excluded below 650 GeV in the light flavour search and in the region between 200 and 835 GeV in the heavy flavour search.

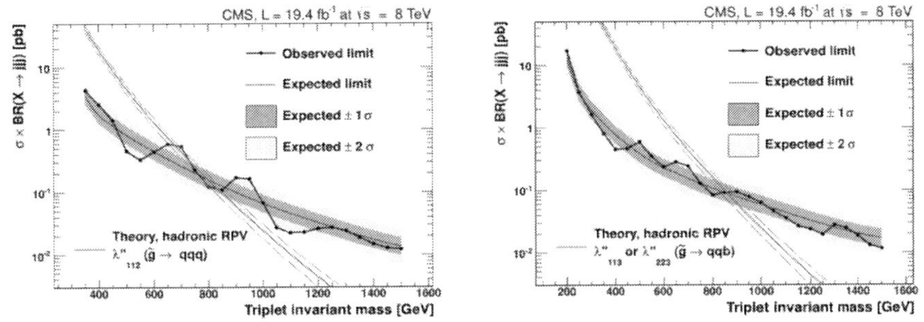

Figure 6. Observed and expected 95% CL cross section limits as a function of mass for the inclusive (left) and heavy-flavour searches (right). The limits for the heavy-flavour search cover two mass ranges, one for low-mass gluinos ranging from 200 to 600 GeV, and one for high-mass gluinos covering the remainder of the mass range up to 1500 GeV. The solid red lines show the NLO+NLL predictions, and the dashed red lines give the corresponding one-standard deviation uncertainty bands.

7. Conclusion

Results from searches for new massive resonances in CMS were presented, including a variety of theoretical models as well as many decay channels. No significant deviation from the corresponding SM processes was observed. Upper limits on the resonance production cross section were set at 95% CL and, when possible, translated to lower limits on the resonance mass. In many of the searches presented here, those results are the most stringent up to date. Those will be extended by the results from LHC Run II foreseen in mid 2015. In view of the higher instantaneous luminosity and centre of mass energy expected, most of those results will be superseded after $\sim 1\ fb^{-1}$ of data has been collected, which is likely to happen in a few weeks.

[1] Chatrchyan S *et al.* (CMS) 2013 *JHEP* **1306** 081 (*Preprint* 1303.4571)
[2] Chatrchyan S *et al.* (CMS) 2012 *Phys.Lett. B* **716** 30–61 (*Preprint* 1207.7235)
[3] Aad G *et al.* (ATLAS) 2012 *Phys.Lett. B* **716** 1–29 (*Preprint* 1207.7214)
[4] Chatrchyan S *et al.* (CMS) 2008 *JINST* **3** S08004
[5] Commissioning of the Particle-flow Event Reconstruction with the first LHC collisions recorded in the CMS detector Tech. Rep. CMS-PAS-PFT-10-001 2010
[6] 2013 Search for Resonances in the Dilepton Mass Distribution in pp Collisions at $\sqrt{s} = 8$ TeV Tech. Rep. CMS-PAS-EXO-12-061 CERN Geneva
[7] 2013 Search for Narrow Resonances using the Dijet Mass Spectrum with 19.6 fb^{-1} of pp Collisions at $\sqrt{s} =$ 8 TeV Tech. Rep. CMS-PAS-EXO-12-059 CERN Geneva
[8] 2011 Jet Substructure Algorithms Tech. Rep. CMS-PAS-JME-10-013 CERN Geneva
[9] Ellis S D, Vermilion C K and Walsh J R 2010 *Phys.Rev.* **D81** 094023 (*Preprint* 0912.0033)
[10] 2009 A Cambridge-Aachen (C-A) based Jet Algorithm for boosted top-jet tagging Tech. Rep. CMS-PAS-JME-09-001 CERN 2009. Geneva

[11] 2013 Identifying Hadronically Decaying Vector Bosons Merged into a Single Jet Tech. Rep. CMS-PAS-JME-13-006 CERN Geneva
[12] Khachatryan V *et al.* (CMS) 2014 *JHEP* **1408** 173 (*Preprint* 1405.1994)
[13] Khachatryan V *et al.* (CMS) 2014 *JHEP* **1408** 174 (*Preprint* 1405.3447)
[14] Khachatryan V *et al.* (CMS) 2015 *Phys.Lett.* **B740** 83–104 (*Preprint* 1407.3476)
[15] Chatrchyan S *et al.* (CMS) 2014 *Phys.Lett.* **B730** 193–214 (*Preprint* 1311.1799)
[16] Chatrchyan S *et al.* (CMS) 2011 *Phys.Rev.Lett.* **107** 101801 (*Preprint* 1107.3084)

Journal of Physics: Conference Series **623** (2015) 012008 doi:10.1088/1742-6596/623/1/012008

Vector-like Multiplets, Mixings and the LHC

Aldo Deandrea

Université de Lyon, France; Université Lyon 1, CNRS/IN2P3, UMR5822 IPNL, F-69622 Villeurbanne Cedex, France.

E-mail: `deandrea@ipnl.in2p3.fr`

Abstract. Vector-like quarks are an essential ingredient of many extensions of the Standard Model of particle physics. Moreover the presence of more than one vector-like multiplet is a common situation in many models. The interplay of these vector-like multiplet with precision electroweak bounds, flavour and collider phenomenology, especially at the Large Hadron Collider is a important point for the discovery of physics beyond the standard model. I consider the presence of two vector-like multiplets in order to show the constraints on such scenarios from tree-level data and oblique corrections in the case of a completely general flavour mixings with all the three standard model families of quarks. I also describe a framework for studying the interplay of any number of vector-like multiplets.

1. Introduction

The Large Hadron Collider (LHC) in the recent years has confirmed the effective description of the electroweak sector given by the Standard Model (SM) Lagrangian with the discovery of the Higgs boson and the analysis of its properties. It has also allowed to obtain exclusion limits on new particles, which are in some cases quite constraining. However such constrains are often model dependent or rely on simplified assumptions. With the higher energy runs of the LHC it will be possible to refine these analyses and in many cases to tackle more realistic situations concerning the models of physics beyond the standard model. New vector-like (VL) fermions are often present in many of the extensions of the SM, especially in relation with the top sector, extra-dimensional models, little Higgs models, gauge-Higgs models, gauge coupling unification and models with an extended custodial symmetry. Both CMS [1] and ATLAS [2] have recently devoted a considerable effort in the analyses establishing bounds on this type of new particles. Initially, simplifying assumptions were considered (mixing only with the third generation of SM quark family or specific decay modes) [3, 4, 5, 6, 7, 8, 22, 9, 10, 11]. However the most recent analyses, due to larger data samples, allow exploring more general situations with mixing of VL quarks with the first two generation of SM quarks [12, 13, 14, 15].

Considering the presence of a complete multiplet of the symmetries of the Standard Model is however not enough in some realistic scenarios: in fact, theoretically justified models often contain multiplet of larger global symmetries which can be described in terms of several multiplets which are close in mass. The various multiplet are then mixed with each other via the Higgs interactions. The presence of general mixing structures and the interplay of different multiplets typically affects the tree-level and loop-level bounds, thereby modifying the results expected by performing simplified analyses based on a single particle or a single multiplet [16]. I will describe in the following the general structures and mixing of more than one VL quark

Journal of Physics: Conference Series **623** (2015) 012008 doi:10.1088/1742-6596/623/1/012008

multiplet and, specifically, the implications of the presence of two VL quark multiplets mixings with the 3 SM quark generation. I will also focus on a specific sub-set of scenarios where both VL multiplets contain a top partner and where eventual bottom partners (i.e. with electric charge $-1/3\ e$) do not mix with the SM down sector. This choice is done to minimise the constraints from flavour, which are very severe on mixing in the down sector only. Typically larger mixing are allowed in this case, providing larger single production cross sections at the LHC. These scenarios are also theoretically justified in models where the new physics couples dominantly to the top quark. Depending upon the multiplet considered, non-SM quarks, *i.e.* quarks having non-SM electric charge, may be present in the considered multiplets. I will estimate the constraints on such scenarios from electroweak precision (EWP) data (oblique and non-oblique) and current LHC data.

2. Vector-like multiplets

Minimal sets of VL multiplets, which mix with SM quarks and a SM (or SM-like) Higgs boson have been extensively studied in literature [17, 3, 10, 7, 12, 13]. In the following we shall distinguish three groups of multiplets as follows:

- top-type multiplets: multiplets containing one VL top partner but no bottom partners (*i.e.* no VL quark with electric charge $-1/3\ e$). In addition to a top partner these multiplets may contain quarks with exotic charges $5/3\ e$ and $8/3\ e$.
- bottom-type multiplets: multiplets containing one VL bottom partner but no top partners (*i.e.* no VL quark with charge $2/3\ e$). In addition to a bottom partner these multiplets may contain quarks with exotic charges $-4/3\ e$ and $-7/3\ e$.
- mixed multiplets: multiplets containing both VL top and bottom partners. In addition these multiplets may contain all of the exotic charged VL quarks.

In the following we shall focus on the top-type multiplets as phenomenologically interesting in many different models. The first few multiplets of this type are given in Table 1.

Multiplet	ψ	$(SU(2)_L, U(1)_Y)$	T_3	Q_{EM}	Yukawa to SM
Singlet 2/3	U	$(\mathbf{1}, 2/3)$	0	+2/3	Yes
Doublet 7/6	$\begin{pmatrix} X^{5/3} \\ U \end{pmatrix}$	$(\mathbf{2}, 7/6)$	+1/2 −1/2	+5/3 +2/3	Yes
Triplet 5/3	$\begin{pmatrix} X^{8/3} \\ X^{5/3} \\ U \end{pmatrix}$	$(\mathbf{3}, 5/3)$	+2 +1 0	+8/3 +5/3 +2/3	No

Table 1. Quantum numbers for the top-type VL multiplets (up to triplets), explicitly indicating weak isospin, hypercharge, electric charge (Q_{EM}) and if a direct Yukawa coupling to SM quarks is allowed.

3. Vector-like quarks have chiral couplings to SM quarks

The structure of the couplings of the VL quarks to SM quarks and gauge bosons, in the case of an arbitrary number of VL representations, can be studied in a general way. Even in presence of mixing between various VL representations, the couplings are dominantly chiral like in the case of a single VL multiplet. Such couplings can be traced back to Yukawa couplings connecting the VL multiplets with a SM chiral fermion via the Higgs boson. This is true also for models with an extended Higgs sector: a singlet which acquires a VEV will generate masses in the form of VL masses as it does not break the gauge symmetries; additional doublets will generate the same structures as the SM Higgs, and only the coupling of the Higgs boson may be affected; for larger representation, like a triplet, the VEV will generally induce large corrections to the ρ parameter, thus it is bound to be very small and therefore generates small mixing terms. Following the above arguments, we can therefore state that our assumption of mixing mainly via the SM Higgs is solid. The Higgs in the SM is a doublet of SU(2), thus a field with weak isospin-1/2: a convenient way to classify the VL quarks is to use their weak isospin. VL multiplets with integer isospin (singlets, triplets, ...) can only couple via the Higgs to a left-handed doublet; on the other hand, VL multiplets with semi-integer isospin (doublets, quadruplets, ...) can only couple to a SM right-handed singlet. Also, one can potentially write a Yukawa coupling of a Higgs field with two VL quarks only if one of them has integer isospin (thus belonging to the first class) and the other semi-integer isospin. Of course, not all couplings are allowed as one needs to take into account the specific representation and the hypercharge. The FCNCs and the couplings between VL quarks and the standard ones are all proportional to the elements of the mixing matrices $V^{\alpha i}_{L/R}$, where α spans over the VL quarks, and $i = 1, 2, 3$ on the SM quarks.

Let us consider the most general case with $N-3$ VL quarks that mix via Yukawa interactions to the SM quarks, and to each other. In the starting basis, we consider that the SM Yukawa matrices are already diagonal (for simplicity), while the VL masses are also diagonal. We consider n_d semi-integer isospin states (doublets, ...) with potential mixing with the SM right-handed singlets, and $n_s = N - 3 - n_d$ integer isospin states (singlets, triplets, ...) with potential mixing with the SM left-handed doublets. The most general mass matrix, therefore, will have the following block form:

$$\mathcal{L}_{\rm mass} = \bar{q}_L \cdot \left(\begin{array}{ccc|ccc|ccc} \mu_1 & 0 & 0 & 0 & \dots & 0 & x_{1,n_d+4} & \dots & x_{1,N} \\ 0 & \mu_2 & 0 & 0 & \dots & 0 & x_{2,n_d+4} & \dots & x_{2,N} \\ 0 & 0 & \mu_3 & 0 & \dots & 0 & x_{3,n_d+4} & \dots & x_{3,N} \\ \hline y_{4,1} & y_{4,2} & y_{4,3} & M_4 & 0 & 0 & & & \\ \vdots & \vdots & \vdots & 0 & \ddots & 0 & & \omega_{\alpha\beta} & \\ y_{n_d+3,1} & y_{n_d+3,2} & y_{n_d+3,3} & 0 & 0 & M_{n_d+3} & & & \\ \hline 0 & 0 & 0 & & & & M_{n_d+4} & 0 & 0 \\ \vdots & \vdots & \vdots & & \omega'_{\alpha\beta} & & 0 & \ddots & 0 \\ 0 & 0 & 0 & & & & 0 & 0 & M_N \end{array} \right) \cdot q_R + h.c. \quad (1)$$

In this basis, the SM Yukawa masses μ_i are presented in a diagonalised form. We recognise the $3 \times n_s$ matrix x_{i,β_s} of the Yukawa couplings of the VL singlets/triplets (integer isospin), and the $n_d \times 3$ matrix $y_{\alpha_d,j}$ of the Yukawa couplings of the VL doublets (semi-integer isospin). M_α represent the VL masses of all the new representations, while the $n_d \times n_s$ matrix $\omega_{\alpha_d,\beta_s}$ and $n_s \times n_d$ matrix $\omega'_{\alpha_s,\beta_d}$ contain the eventual Yukawa couplings among VL representations. In general the Yukawa couplings between VL quarks distinguish between the chiral components of the VL quarks, therefore in general $\omega' \neq \omega^T$; furthermore, ω' corresponds to the "wrong" Yukawa couplings, in the sense that it connects left-handed singlets (integer isospin) with right-handed doublets (semi-integer isospin), which is the opposite chirality configuration of SM Yukawa couplings. A detailed proof is given in the appendix of [13]. For the semi-integer isospin (doublet) VL quarks, the dominant mixing angle is right-handed, while the left-handed one is suppressed by an extra factor of ω'/M_{VL}; the chiralities are exchanged for the integer isospin (singlets and triplets) VL quarks. It is interesting to stress that the subleading mixing angle is suppressed either by the "wrong" Yukawa couplings between VL quarks, ω', or by the light quark masses.

The leading mixing angle may be small in some cases, either because the relevant Yukawa coupling is absent or numerically small. However, in such a case, the couplings will receive an extra suppression and the single production will become subdominant with respect to the pair production: for the pair production studies, only the branching ratios are relevant, independently on the chirality of the couplings.

4. Numerical results

In the following we shall consider few examples of the interplay among different VL multiplets and the SM (for a complete analysis see [16]). We have considered the mass parameters M_1 and M_2 of the VL quarks to be same, *i.e.* $M_1 = M_2 = M$. As the models have too many parameters to make a meaningful scan, we will show results in two limiting cases, when possible:

- the VL quarks can mix with a single SM generation, but not with each other (i.e. $\omega \sim \omega' \ll M$);
- the VL quarks mix with each other (wherever possible), but the mixing with SM quarks is very small.

The results in these simple limits can give a general idea on the allowed value of the mixing parameters, even though the case where all of them are non-zero is more realistic. We will focus on the benchmark value for the VL mass of 800 GeV as a recent CMS analysis [23] sets a bound of 788 GeV under the assumption of strong pair production of VL quarks and 100% branching fractions to qW. In the following we also consider cases in which the VL quark does not decay to qW. For these cases the bound does not apply directly, but these VL quarks are in doublets containing also other VL quarks for which the bound applies. As it is reasonable to assume that mass splittings inside multiplets are not large compared to the mass scale of the multiplet, we shall apply this 800 GeV benchmark value to all cases.

4.1. *Singlet* $Y = 2/3$ *and Doublet* $Y = 7/6$

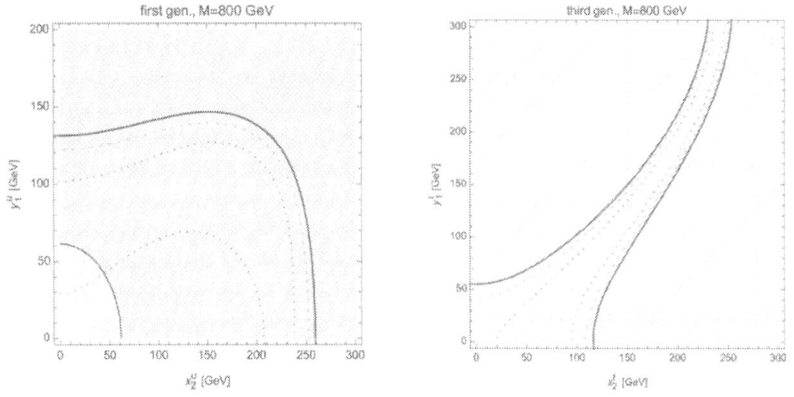

Figure 1. Singlet $Y = 2/3$ and Doublet $Y = 7/6$: EWP bounds at 1 σ (red-dashed), 2 σ (green-dashed) and 3 σ (blue) for VL quarks coupling with the first (left panel) and third (right panel) SM generations, compared with the region excluded at 3σ by tree-level bounds (yellow region). Here, $M = 800$ GeV, and $\omega = \omega' = 0$. Only the first quadrant is shown as the figures are symmetric with respect to a sign change in the coordinates in the other 3 quadrants. Similar considerations apply to all the other figures of the same type.

This scenario contains – besides the SM particle spectrum – two VL top quarks and one exotic quark with charge 5/3. The additional parameters (apart from the SM ones) are: x_2^k, y_1^k,

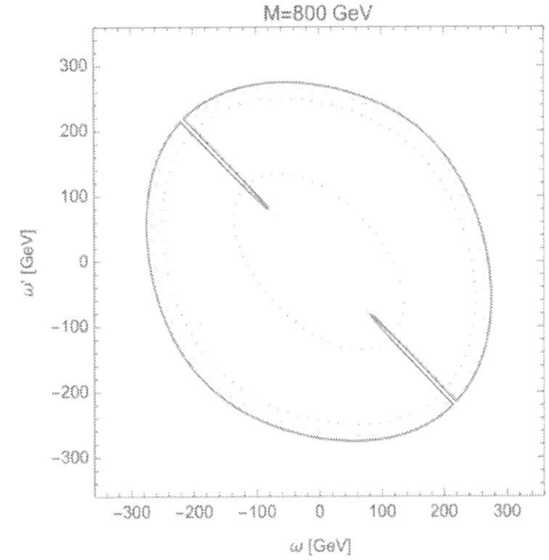

Figure 2. **Singlet $Y = 2/3$ and Doublet $Y = 7/6$** EWP bounds at 1σ (red-dashed), 2σ (green-dashed) and $3\,\sigma$ (blue) as a function of the new Yukawa couplings ω and ω' with $M = 800$ GeV. We have assumed that there is no mixing of VL quarks with the SM quark generations *i.e.* $x_2^k = y_1^k = 0$.

ω, ω' and M with k running on SM quark generations. We first study the case where $\omega' \sim \omega \sim 0$, and the VL quarks couple to a single generation: in this case, setting ω' to zero allows us to set both Yukawa couplings x_2 and y_1 to be real and positive. The allowed regions in the parameter space, given the constraints from tree-level and EWP tests, are presented in Figure 1. We see that for couplings to the light generations, the tree level bounds always dominate, and require the mixing of VL quarks to be rather small. The case of the third generation is very different: the tree level bounds are very weak as they only come from V_{tb}, while EWP tests allow for large mixings, especially via a compensation between the doublet and singlet (in particular, y_1^3 can assume very large values). This situation can be very interesting in the single-production channel, where for instance the top partner may be produced via couplings to the first generation (the smaller coupling is easily compensated by the valence quark in the initial state [13]) and then decay into a third generation quark [18, 19, 20].

In Figure 2 we show the EWP bounds in the plane of the Yukawa couplings between VL quarks, ω and ω', assuming that the other couplings are small. This plot gives a general idea on the allowed size of ω and ω': the bound is indeed not very strong, and values up to 300 GeV are allowed. The plot is clearly symmetric under change of sign of either ω or ω', reflecting the one arbitrary phase in this sector. As we are approximately decoupling the two t's from the SM quarks, the mixing is dominated by the 2×2 block of the VL quarks. The mixing angles vanish when $\omega = -\omega'$, thus explaining the sharp dents in the excluded region. This effect only appears in our limiting choice $M_1 = M_2$ and for negligible mixing to SM quarks.

4.2. Doublet $Y = 1/6$ and Doublet $Y = 7/6$

This scenario is of particular interest as it corresponds to a bi-doublet of the custodial SO(4) symmetry, which is a basic ingredient for top partial compositeness in models of composite Higgs (see [22]). It contains two VL top quarks, one VL bottom quark and one exotic quark with charge 5/3. The additional parameters are: y_1^k, y_2^k and M with k running on SM quark

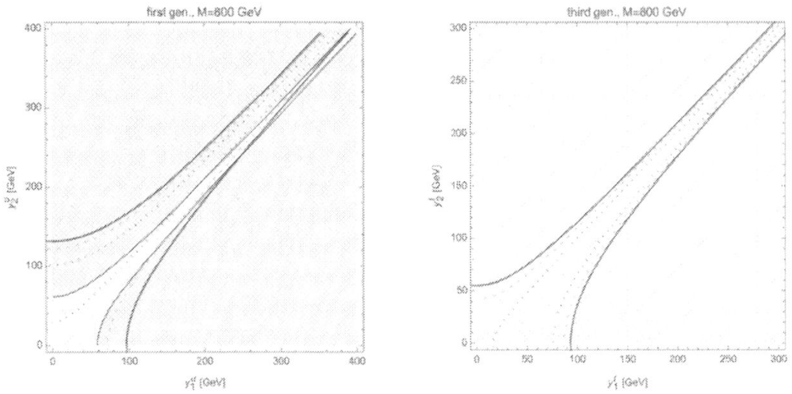

Figure 3. **Doublet $Y = 1/6$ and Doublet $Y = 7/6$** EWP bounds at 1σ (red-dashed), 2σ (green-dashed) and $3\ \sigma$ (blue) for VL quarks coupling with the first (left panel) and third (right panel) SM generations, compared with the region excluded at 3σ by tree-level bounds (yellow region in the left panel). $M = 800$ GeV, $\omega = \omega' = 0$.

generations. Note that for this case, mixing between VL quarks (*i.e.* ω and ω') is not allowed. This model also introduces an additional mixing in the bottom sector, which is independent from the the mixing in the top sector. It is therefore possible to impose the condition $y_{1d}^k = 0$ without affecting the top sector.

The results for the combined tree-level and EWP bounds are given in Figure 3. For the first generation, there is an interesting cancellation in the tree-level bounds for $|y_1^k| = |y_2^k|$: this is a consequence of an enhanced custodial symmetry, and this fact has been used in the literature to justify $\mathcal{O}(1)$ mixings of VL quarks with light generations [21]. EWP bounds show a similar cancellation, however along an axes which is a bit off compared to $|y_1^k| = |y_2^k|$, therefore a tension between the two allowed regions develops for large mixings. A similar behaviour in the EWP bounds can be seen in the case of mixing to the third generation only.

4.3. Single production cross sections

Tree-level, loop-level and the bounds from single production processes at the LHC can be compared for the cases discussed above. The relevance of single production is given by the fact that its cross-section depends on both the masses of the VL quarks and their couplings to the SM quarks; moreover, it is well known that single production becomes the dominant channel at the LHC, overcoming QCD pair production, when quark masses are higher than a certain (model-dependent) value. For typical scenarios where VL quarks mix predominantly with third generation and mixing parameters are not too constrained by flavour physics and EWP tests, the mass bounds from QCD pair production are already in the region where the single production channel is relevant or even dominant [12]. So far, few experimental searches for single production of VL quarks have been performed. To be specific, in the following we will consider the single production of a VL top partner in association with a light jet, and the mass of the VL quark will be fixed to 800 GeV. We consider exclusive coupling to each of the three SM generations for each scenario. In Figure 4, the tree-level bounds are the most stringent ones if VL quarks mix to the light generations, and they are stronger than the current bound form the ATLAS search. In both cases, the largest $Tjet$ cross section allowed is between 0.5 and 1 pb. In the case of mixing to the third generation only EWP and tree level bounds conspire to

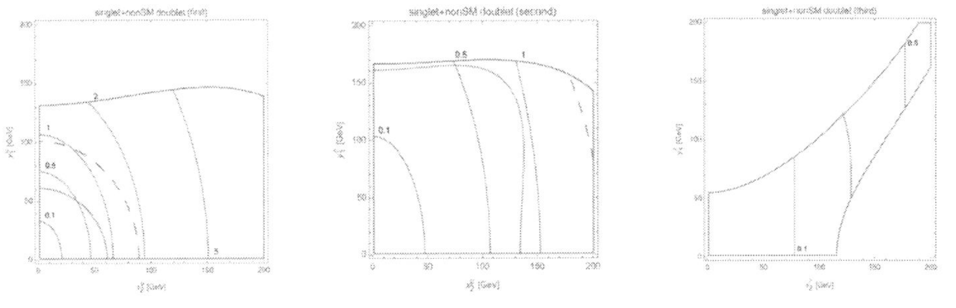

Figure 4. Singlet $Y = 2/3$ and Doublet $Y = 7/6$ for mixing with first generation only (left), second generation only (middle), third generation (right), and for a mass of the VL quarks of 800 GeV. The channel is T+jet. The grey contour lines correspond to cross-section values in picobarns at 14 TeV. The region inside the red line is allowed by the S and T parameters. The region inside the blue line is allowed by the tree-level bounds. The dashed black lines are the bounds from the ATLAS search.

select small mixing, and the single production at 14 TeV is limited to small values around 100 fb. in Figure 5 the results for the "Doublet ($Y = 1/6$) and Doublet ($Y = 7/6$)" scenario are presented. Again, though the presence of an exotic quark with charge 5/3 and of right-handed charged currents which contribute to the corrections to the oblique parameters, the tree-level constraints are stronger for the most part of the parameter space.

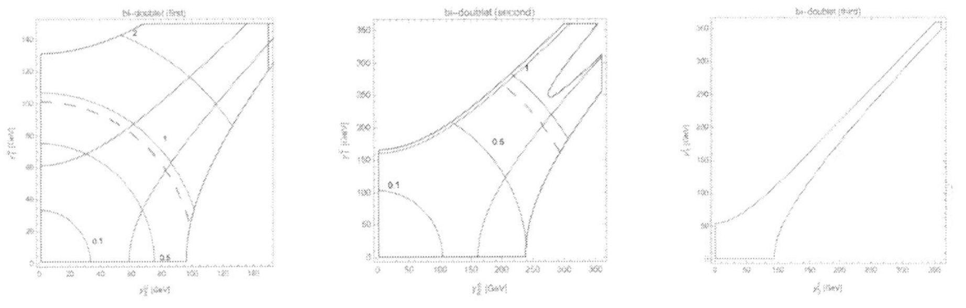

Figure 5. Doublet $Y = 1/6$ and Doublet $Y = 7/6$ for mixing with first generation only (left), second generation only (middle), third generation (right), and for a mass of the VL quarks of 800 GeV. The channel is T+jet. The grey contour lines correspond to cross-section values in picobarns at 14 TeV (this channel is not allowed in the case of the plot on the right). The region inside the red line is allowed by the S and T parameters. The region inside the blue line is allowed by the tree-level bounds. The dashed black lines are the bounds from the ATLAS search.

5. Conclusion

In most models beyond the SM, VL quarks appear in complete multiplets and usually more than one multiplet is present. In this analysis we have considered scenarios with multiple VL quarks both from the point of view of the general mixing structure with the three Standard Model generations and considering the mixing pattern of these multiplets for the determination of mixing effects and precision electroweak observables both at tree-level and at loop-level. The specific case of two different vector-like quark multiplets has been discussed in detail, with a special focus on an example containing top partners. The main result is that tree-level and loop-level constraints provide complementary information. Moreover the interplay of the vector-like multiplets among themselves and with the Standard model quarks have important consequences for phenomenology as in some cases large single production cross-sections are possible and coupling with light generations is not necessarily suppressed. These results have phenomenological implications for LHC searches as the bounds we have extracted pinpoint particular regions of the parameter space and suggest that in realistic cases containing multiple multiplets of vector-like quarks, cancellations are possible from tree-level bounds which allow large values of the mixing parameters. Direct searches by the LHC experiments in the near future will play a mayor role in studying physics beyond the Standard Model with VL multiplets.

Acknowledgements
I wish to thank my collaborators on this subject in the recent years and in particular G.Cacciapaglia, N. Gaur, D. Harada, Y.Okada, L.Panizzi for the many fruitful discussions and common work on vector-like quarks. AD is partially supported by Institut Universitaire de France, the Labex-LIO (Lyon Institute of Origins) under grant ANR-10-LABX-66, FRAMA (FR3127, Fédération de Recherche "André Marie Ampère") and IN2P3 Theory-LHC France funding.

References

[1] *https://twiki.cern.ch/twiki/bin/view/CMSPublic/PhysicsResultsB2G*
[2] *https://twiki.cern.ch/twiki/bin/view/AtlasPublic/ExoticsPublicResults*
[3] F. del Aguila, M. Perez-Victoria and J. Santiago, JHEP **0009** (2000) 011 [hep-ph/0007316].
[4] J. A. Aguilar-Saavedra, Phys. Lett. B **625** (2005) 234 [Erratum-ibid. B **633** (2006) 792] [hep-ph/0506187].
[5] C. Anastasiou, E. Furlan and J. Santiago, Phys. Rev. D **79** (2009) 075003 [arXiv:0901.2117 [hep-ph]].
[6] J. A. Aguilar-Saavedra, JHEP **0911** (2009) 030 [arXiv:0907.3155 [hep-ph]].
[7] G. Cacciapaglia, A. Deandrea, D. Harada and Y. Okada, JHEP **1011** (2010) 159 [arXiv:1007.2933 [hep-ph]].
[8] D. Marzocca, M. Serone and J. Shu, JHEP **1208** (2012) 013 [arXiv:1205.0770 [hep-ph]].
[9] A. Falkowski, D. M. Straub and A. Vicente, JHEP **1405** (2014) 092 [arXiv:1312.5329 [hep-ph]].
[10] J. A. Aguilar-Saavedra, EPJ Web Conf. **60** (2013) 16012 [arXiv:1306.4432 [hep-ph]].
[11] S. A. R. Ellis, R. M. Godbole, S. Gopalakrishna and J. D. Wells, JHEP **1409** (2014) 130 [arXiv:1404.4398 [hep-ph]].
[12] G. Cacciapaglia, A. Deandrea, L. Panizzi, N. Gaur, D. Harada and Y. Okada, JHEP **1203** (2012) 070 [arXiv:1108.6329 [hep-ph]].
[13] M. Buchkremer, G. Cacciapaglia, A. Deandrea and L. Panizzi, Nucl. Phys. B **876** (2013) 376 [arXiv:1305.4172 [hep-ph]].
[14] C. Delaunay, T. Flacke, J. Gonzalez-Fraile, S. J. Lee, G. Panico and G. Perez, JHEP **1402** (2014) 055 [arXiv:1311.2072 [hep-ph], arXiv:1311.2072].
[15] D. Barducci, A. Belyaev, M. Buchkremer, G. Cacciapaglia, A. Deandrea, S. De Curtis, J. Marrouche and S. Moretti *et al.*, JHEP **1412** (2014) 080 [arXiv:1405.0737 [hep-ph]].
[16] G. Cacciapaglia, A. Deandrea, N. Gaur, D. Harada, Y. Okada and L. Panizzi, arXiv:1502.00370 [hep-ph].
[17] F. del Aguila and M. J. Bowick, Nucl. Phys. B **224** (1983) 107.
[18] G. Brooijmans, R. Contino, B. Fuks, F. Moortgat, P. Richardson, S. Sekmen, A. Weiler and A. Alloul *et al.*, arXiv:1405.1617 [hep-ph].
[19] S. Beauceron, G. Cacciapaglia, A. Deandrea and J. D. Ruiz-Alvarez, Phys. Rev. D **90** (2014) 11, 115008 [arXiv:1401.5979 [hep-ph]].
[20] L. Basso and J. Andrea, JHEP **1502** (2015) 032 [arXiv:1411.7587 [hep-ph]].
[21] R. Barcelo, A. Carmona, M. Chala, M. Masip and J. Santiago, Nucl. Phys. B **857** (2012) 172 [arXiv:1110.5914 [hep-ph]].
[22] A. De Simone, O. Matsedonskyi, R. Rattazzi and A. Wulzer, JHEP **1304** (2013) 004 [arXiv:1211.5663 [hep-ph]].
[23] *https://twiki.cern.ch/twiki/bin/view/CMSPublic/PhysicsResultsB2G12017*

International Workshop on Discovery Physics at the LHC (Kruger2014) IOP Publishing
Journal of Physics: Conference Series **623** (2015) 012009 doi:10.1088/1742-6596/623/1/012009

Longitudinal Properties of High Energy Collisions

A. Deppman[1], **L. Marques**[1] **and J. Cleymans**[2]

[1] Instituto de Física, Universidade de São Paulo - IFUSP, Rua do Matão, Travessa R 187, 05508-900 São Paulo-SP, Brazil

[2] UCT-CERN Research Centre and Department of Physics, University of Cape Town, Rondebosch 7701, South Africa

Abstract. The rapidity distribution and longitudinal energy distribution are analysed within the thermodynamical formalism with the non-extensive Tsallis statistics. Our approach is based on a model for fireball rapidity distribution recently proposed where the fireball moves longitudinaly according to a q-Gaussian function rapidity distribution. We show that the rapidity distribution of charged hadrons are correctly obtained as compared to the experimental data for rapidities up to ~ 10.5. The energy density as a function of the rapidity is studied for the first time. We show that most of the energy carried by hadrons are in the high rapidity region. Our results show that the energy density is a good tool for studying the logitudinal aspects of high energy collisions.

1. Introduction

There is an increasing use of power-law distributions to describe high-energy experimental data [1–5], notably for p_T-distributions. Among the different formulas the one that can be consistently derived from the non-extensive ideal gas [6] presents some advantages, since its parameters are related to the known physical quantities through the usual equations from Therodynamics. An additional feature is that this non-extensive thermodynamics based on Tsallis statistics [7,8] allows the encompassment of the Hagedorn's self-consistent principle [9] giving as results a limiting temperature, T, which can be understood as a critical temperature for the transition between confined and deconfined regimes of the hadronic matter, a characteristic entropic index, q, that is a measure of the deviation from the usual Boltzmann-Gibbs statistics, and a new formula for the hadron-mass spectrum [10].

Several analyses [6, 11–13, 26, 27] of p_T-distributions have shown that the non-extensive thermodynamics can correctly reproduce the experimental data in a large range of p_T for different collision energies and particle especies [6, 11–13] with free parameters T and q resulting to be systematically around the values $T = 68$ MeV and $q = 1.146$, in accordance with the theoretical predictions [10], as shown in Fig. 1. Thus a complete thermodynamical description of the hot and dense system formed after high energy collisions, which we call fireballs, is obtained. Applications of this non-extensive thermodynamics to neutron-stars [14, 15], LQCD [16] and hydrodynamics have already been developed.

Recently the non-extensive approach was used to describe also rapidity distributions giving for the first time a complete description of the results of high energy in proton-proton(pp) collisions [17]. In this work we address the longitudinal distribution of the energy carried by hadrons of different masses.

2. Non-extensive Thermodynamics of fireballs

The starting point is the non-extensive entropy for an ideal quantum gas defined as [6]

$$S_q = -g\sum_i [n_i^q \ln_q n_i + (1-n_i)^q \ln_q(1-n_i)] \,. \tag{1}$$

From here one can derive all thermodynamical functions, particularly the occupation number, which is given by

$$n(\varepsilon) = \pm\{[1+(q-1)\beta(\varepsilon-\mu)] \mp 1\}^{-1} \tag{2}$$

where the + and − signs refer respectively to bosons and germions distributions.

More interesting for applications in High Energy Physics is the transverse mmentum distribution that can be obtained from Eq. 2 resulting [6]

$$\frac{dN}{dp_T} = \frac{p_T m_T}{T}\frac{dN}{dy}\bigg|_{y=0} \frac{(2-q)(3-2q)}{(2-q)m_0^2 + 2m_0 T + 2T^2}\left[1+(q-1)\frac{m_0}{T}\right]^{\frac{1}{q-1}}\left[1+(q-1)\frac{m_T}{T}\right]^{-\frac{q}{q-1}} \tag{3}$$

With the equation above it is possible to describe the p_T-distribution for any particle with mass m and relate it directly to the central rapidity density, $\frac{dN}{dy}|_{y=0}$ in terms of T and q.

For the description of rapidity distributions additional information is needed. Indeed the longitudinal distribution still carries some memory of the initial state which can be included in the theoretical formulations as a longitudinal expansion of the system described by a rapidity distribution of thermodynamically equilibrated fireballs [18, 19]. In Ref. [17] this distribution was

$$\nu(y_f) = G_{q'}(y_0, \sigma; y_f) + G_{q'}(-y_0, \sigma; y_f) \,, \tag{4}$$

where

$$G_q(y_0, \sigma; y_f) = \frac{1}{\sqrt{2\pi}\sigma} e_q\left(-\frac{(y_f - y_0)^2}{2\sigma^2}\right) \,, \tag{5}$$

and $e_q(x)$ is the q-exponential function defined as

$$e_q(x) \equiv [1-(q-1)x]^{-1/(q-1)} \,. \tag{6}$$

In Eq. 5 y_0 and σ are respectively the peak position and the width of the q-Gaussian function used as free parameters to be adjusted in analyses of experimental data.

The double differential yield turns out to be [17]

$$\frac{1}{N}\frac{d^2N}{dp_T\, dy} = \frac{1}{A}\int_{-\infty}^{\infty} \nu(y_f) \times \frac{p_T m_T \cosh(y-y_f)}{(2\pi)^2}\left[1+(q-1)\frac{m_T\cosh(y-y_f)}{T}\right]^{-\frac{q}{q-1}} dy_f \tag{7}$$

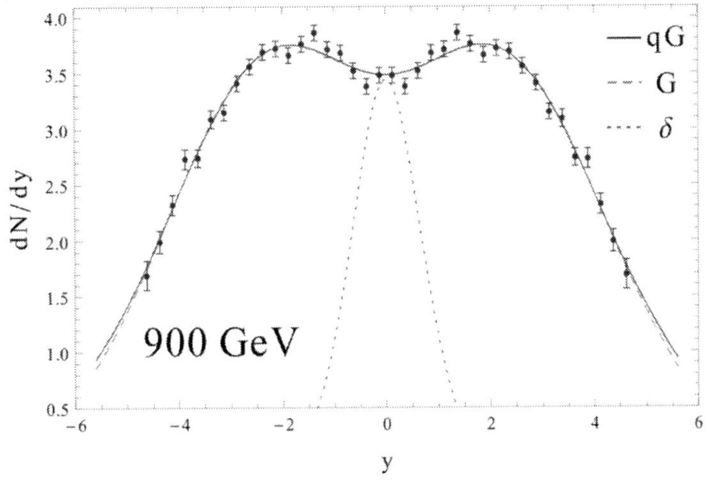

Figure 1. Eq. 3 fitted to experimental data for rapidity distribution from pp collision at 900 GeV from the UA5 collaboration [35]. Data for $y < 0$ are obtained symmetrically from $y > 0$. The full line represents the best fit using the q-Gaussian function ($q' = q$) and the dashed line corresponds to the best fit using the Gaussian function ($q' = 1$) in Eq. 4. The dotted line shows the results when a delta function is used for the fireball rapidity distribution, i.e., $\nu(y_f) = \delta(y_f)$. Figure taken from Ref. [17].

where N is the multiplicity and A a normalization constant. Integrating on p_T one obtains the rapitity distribution. In Fig. 1 we show the good agreement between the experimental data and theoretical calculation.

From the systematic analysis of rapidity distributions performed in Ref. [17] it turns out that the parameters y_o and σ in the q-Gaussian function show a simple pattern, as can be seen in Fig. 2. We note that the error bars are large particularly for energies above $\sqrt{s} = 1$ TeV because the available experimental data are limited to a small range of rapidity, however it is possible to observe that for energies above ~ 500 GeV both y_o and σ are approximately constant. While one would normally expect the peak position of the fireball rapidity distribution present a clear dependence on the beam rapidity, the fact that these parameters are approximately constant within a large range of collision energy could indicate that the expanding system detaches itself from the leading hadrons moving with rapidities close to the beam rapidity. Below ≈ 550 GeV the scaling observed above disappears, as can be seen in Fig. 2. It is possible that at these collision energies the chemical potential is not null hence more studies on the direction of verifying the role played by the chemical potential at low energies must be carried out, as performed in [20–22].

3. Yield and Energy Distributions

In what follows we assume that y_o and σ are constant even for energies above those analysed in the present work. Notice that this assumption is considered for identified hadrons but there are evidences that, at least for lower collision energies, different hadrons present slightly different rapidity

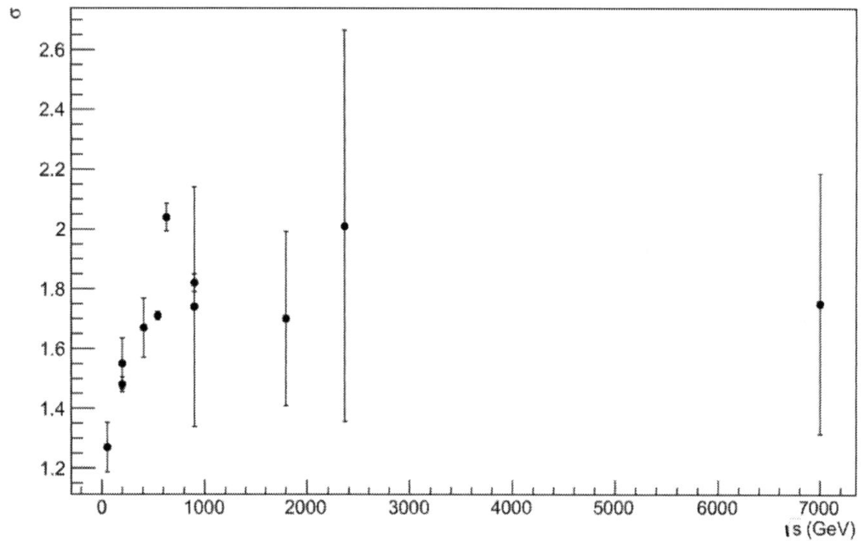

(a)

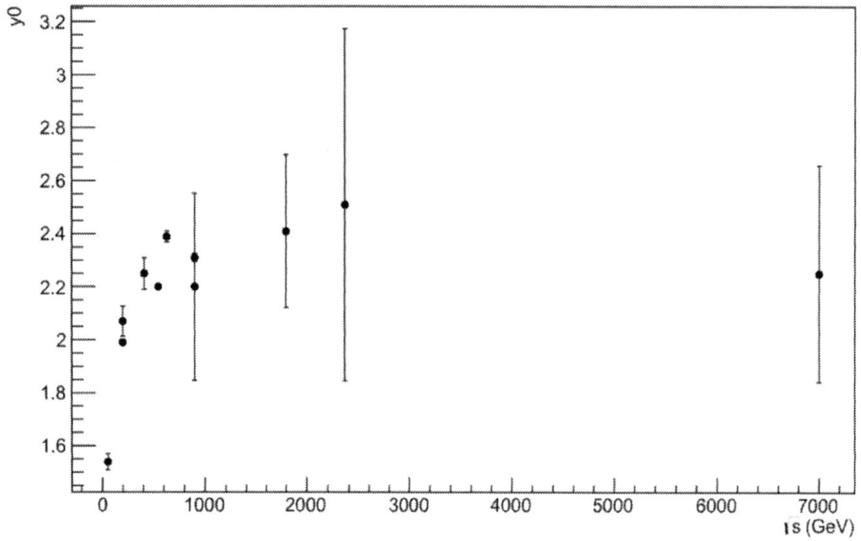

(b)

Figure 2. Values for the parameters σ and y_0 in Eq. 4 as a function of the collision energy. Figure taken from Ref. [17].

distributions [20]. However we will considere that for higher energies those differences can be negleted.

Within the hypothesis given above, the energy density carried by hadrons with mass m is given by

$$\frac{d^2E}{dp_T\,dy}(m) = \frac{gV}{(2\pi)^2}\rho(m)\varepsilon(p_T,y)\frac{d^2N}{dp_T\,dy}\,, \tag{8}$$

where

$$\varepsilon(p_T,y) = m_T\cosh(y)\,, \tag{9}$$

and the total energy carried by baryons is

$$E = \int_0^\infty \rho(m)E(m)dm \tag{10}$$

where $\rho(m)$ is the baryon mass spectrum. The function $\rho(m)$ was determined by the non-extensive self-consistent theory [10] and is given by

$$\rho(m) = \frac{\gamma}{m^{5/2}}\left[1+\frac{(q-1)m}{T}\right]^{\frac{1}{q-1}} \tag{11}$$

In Ref. [29] it was shown that this formula reproduces correctly the observed hadronic mass spectrum up to a mass of approximately 2 GeV.

For hot fireballs at temperatures close to the limiting temperature the calculation of the total energy with Eq. 10 is rather cumbersome because of the singularity at the critical point. An alternative way is to substitute the integration in that equation by a summation over the known hadrons, as

$$E = \sum_{i=1}^{m_h} E(m_i)\,. \tag{12}$$

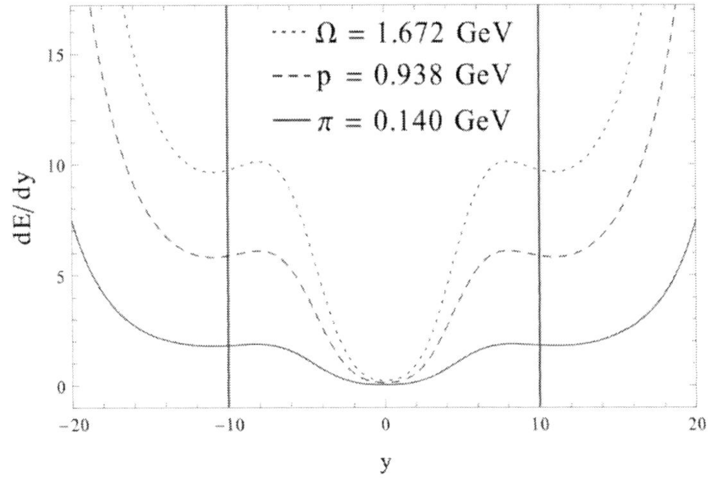

Figure 3. Energy density for different hadrons

Since according to our supposition in Eqs. 10 and 12 the parameters q, T, y_o and σ do not depend on the total energy, only the multiplicity N can be a function of the collision energy. Therefore one can write

$$N(E) = KE\,, \tag{13}$$

where

$$K = \left[\frac{1}{A}\sum_{i=1}^{n}\int_{-\infty}^{\infty}\int_{0}^{\infty}\varepsilon(p_T, y)\frac{d^2N}{dp_T\,dy}(m_i)dp_T\,dy\right]^{-1}. \tag{14}$$

In Ref. [17] considering charged hadrons with masses up to 5 GeV it was found that $K = 0.009\mathrm{GeV}^{-1}$, which falls a factor 2 above the value $K = 0.004\mathrm{GeV}^{-1}$ obtained by fitting the multiplicity for different energies. The difference was attributed to the fact that massive particles that were not taken into account in the calculation of K with Eq. 12 give a significant contribution to the total energy. It will be interesting, therefore, to analyse in more details the individual contributions of different hadrons.

Eq. 8 allows us to calculate the energy carried by any particle with mass m or even its energy density as a function of the rapidity, which is obtained by integrating that equation on p_T. In Fig. 3 the show the energy density calculated for a few hadrons where we observe that for all particles most of the carried energy is concentrated in the high rapidity region ($y \geq 5$) , while the minimum is located at the central region, $|y| \sim 0$. Also, for any fixed rapidity the energy increases as the particle mass increases. These results are straightforward consequences of the fact that the particle energy is given by

$$\varepsilon = m_T\,coshy \tag{15}$$

where $m_T = \sqrt{p_T^2 + m^2}$.

We also observe a peak, more pronounced for more massive particles, at intermediate rapidities ($y \sim 10$). This peak results from a combination of the decreasing q-exponential function from the thermodynamical factor (see Eq. 7) and the increasing hyperbolic cosine from the particle energy (Eq. 15). It is important to notice here that the q-exponential factor depends on $y - y_f$, therefore even high rapidity particles can be produced with high probablity for sufficiently high fireball rapidity. For $|y| > 10.5$ the energy density increases very fast showing a divergence as the particle rapidity increases. Of course this result evidences limitations in the model used here, namely, the fireball rapidity distribution cannot follow a simple q-Gaussian. as given by Eq. 4 function for rapidities close to the beam rapidity, but must decrease faster than the hyperbolic cosine of the rapidity.

The study of the energy density allow us to observe that this quantity is strongly dependent on the particle mass even for relatively low rapidities therefore it represents a good tool for investigation of logitudinal properties of identified particles. Measurements of energy density for hadrons would help us to test different models for rapidity distributions since it is more sensitive to small variations of yield density.

4. Conclusions

In this work we present a comprehesive study of the longitudinal properties of particles generated in high energy proton-proton collisions. The analysis is performed within the non-extensive thermodynamics approach using Tsallis statistics. The rapidity distribution is obtained with a model proposed recently, where it is supposed that the system formed after the high energy collision constitutes of two fireballs moving in the beam direction with opposite rapidities y_o. Each of these fireballs is formed by fireballs with rapidities distributed in the main fireball frame according to a q-gaussian function with width σ. It is assumed, based on a previous systematic study, that y_o and $\sigma$are independent of the collision energy.

We analize the the particle energy densityas a function of rapidity. The analysis evidences that the simple q-Gaussian distribution of the fireball rapidity, y_f, is not consistent for y_f close to the beam rapidity. although it leads to a good agreement of the particle rapidity distribution in a wide range of rapidities. We argue that for very high rapidities the fireball rapidity distributionmust decrease faster the the hyperbolic cosine function.

The results also show that most of the energy is carried by particles with rapidity around $y = 8$. In addition, it is discussed that the energy density is a good tool for investigating longitudinal properties of highenergy collisions since it is very sensitive to the particle mass and rapidity.

5. Acknowledgment

The authors are thankful to R. Martins and to S. F. Barros for their help with numerical calculations. This work received support from CNPq, under grant 305639/2010-2 (A.D.).

References

[1] B. I. Abelev et al. (STAR Collaboration), Phys. Rev. C **75**, 064901 (2007).
[2] A. Adare et al. (PHENIX Collaboration), Phys. Rev. C **83**, 052004, (2010); Phys. Rev. C **83**, 064903 (2011).
[3] K. Aamodt, et al. (ALICE Collaboration), Eur. Phys. J. C **71** (2011) 1655 .
[4] G. Aad, et al. (ATLAS Collaboration), New J. Phys. **13** (2011) 053033.
[5] V. Khachatryan, et al. (CMS Collaboration), JHEP **05** (2011) 064.
[6] J. Cleymans and D. Worku, J. Phys. G: Nucl. Part. Phys. **39** (2012) 025006.
[7] C. Tsallis, J. Stat. Phys. **52** (1988) 479.
[8] C. Tsallis, Braz. J. Phys. **29** (1999) 1.
[9] R. Hagedorn, Suppl. Nuovo Cimento **3** (1965) 147.
[10] A. Deppman, Physica A **391** (2012) 6380.
[11] M.D. Azmi and J. Cleymans, J. Phys. G: Nucl. Part. Phys. **41** (2014) 065001.
[12] I. Sena and A. Deppman, Eur. Phys. J. A **49** (2013) 17.
[13] I. Sena and A. Deppman, AIP Conf. Proc. **1520**, 172 (2013).
[14] E. Megías, D.P. Menezes and A. Deppman, Physica A **421** (2015) 15-24.
[15] E.P. Menezes, A. Deppman, E. Megías and PRD
[16] A. Deppman, Physica A (2014)
[17] L. Marques, J. Cleymans and A. Deppman, PRD
[18] J.D. Bjorken, Phys. Rev. D **27** (1983) 140.
[19] E.Schnerdermann, J. Sollfrank and U. Heinz, Phys. Rev.C (1993) 2462-2475.
[20] F. Becattini and J. Cleymans, J. Phys. G: Nucl. Part. Phys. **34** (2007) S959–S963.
[21] J. Cleymans, J. Phys. G: Nucl. Part. Phys. **35** (2008) 044017.
[22] L.A. Stiles and M. Murray,nucl-ex/0601039.
[23] T.S. Biro, "Is There a Temperature?" Fundamental Theories of Physics, Vol 171 (2011), Springer Verlag.
[24] C. Beck, Physica A **286** (2000) 164.

[25] R. Hagedorn, Ref.TH.3684-CERN, September (1983).
[26] C.-Y. Wong and G. Wilk, Acta Phys. Polon. **B 43**, 2047 (2012).
[27] C.-Y. Wong and G. Wilk, Phys. Rev. D **87**, 114007 (2013).
[28] M. Rybczyński, Z. Wlodarczyk, Eur. Phys. J. C **74** (2014) 2785.
[29] L. Marques, E. Andrade-II and A. Deppman, Phys. Rev. D **87** (2013) 114022.
[30] A. Adare, et al. Phys. Rev. C **83** (2011) 064903.
[31] B. Abelev, et al. Phys. Lett. B **717** (2012) 162.
[32] S. Chatrchyan, et al. Eur. Phys. J. C **72** (2012) 2164.
[33] B. Abelev *et al.* (ALICE Collaboration), Eur. Phys. J. C **72** (2012) 2183.
[34] A. Kumar, P.K. Srivastava, B.K. Singh and C.P. Singh, Adv. High En. Phys. **2013** (2013) 352180.
[35] G. J. Alner, Z. Phys. C - Particles and Fields **33** (1986) 1.
[36] B. Alver, et al. Phys. Rev. C **83** (2011) 024913.
[37] R. Harr, et al. Phys. Lett. B **401** (1997) 176.
[38] V. Khachatryan, et al. J. High Energy Phys. **2** (2010) 041.
[39] F. Abe, et al. Phys. Rev. D (R) **41** (1990) 2330.
[40] V. Khachatryan, et al. Phys. Rev. Lett. **105** (2010) 022002.

Recent Standard Model results from CMS

Simon de Visscher on behalf of the CMS collaboration
CERN

E-mail: sdevissc@cern.ch

March 2015

Abstract. The data recorded by the Compact Muon Solenoid experiment during the first LHC run (2010-2012) has allowed to refine our knowledge on the production of jets and vector boson using the proton-proton collisions at $\sqrt{s}$= 7 or 8 TeV. Representative studies concerning the PDF and α_s extractions, the study of inclusive vector boson production, of production vector bosons in association with jets, and some recent results on anomalous gauge couplings evaluation are discussed.

1. Introduction

These proceedings are devoted to a few representative Standard Model (SM) measurements, which were performed by the CMS experiment using data collected during the LHC Run I operation. In Sec. 2 the impact of the CMS measurements on the parton density function (PDF) constraints as well as on the strong coupling constant α_s extraction is discussed. In Sec 3, studies of the inclusive vector boson (V=W,Z,γ) production as well as V production in association with light or heavy jets (V+jets) are presented. In Sec. 4, the latest results on the extraction of anomalous triple/quartic gauge couplings (aTGC/aQGC) are reviewed. Full list of publications discussing SM-related results are accessible through [1].

2. PDF and α_s determination

The PDFs are essential ingredients to estimate the inclusive and differential cross-sections of all types of processes in proton-proton (p-p) collisions at the LHC. Several of them are needed to constrain the valence quarks, sea quarks and gluon PDFs: while QCD production of jets allows to refine the gluon PDF, the inclusive production of a W boson is more specifically sensitive to the valence- and sea-quarks PDFs, and that of the W boson in association with a charm quark has a large sensitivity to the strange-quark PDF. The data collected by the CMS collaboration in 2010 and 2011 was successfully used to refine our knowledge of these PDFs.

First, the inclusive jet production study [2] has allowed to add additional constraints especially on the gluon PDF. Final state is defined by the presence of at least two anti-k_T jets [3] with size parameter R=0.7, p_T >114 GeV and $|y|$ <2.5. The full integrated luminosity (5fb^{-1}) of the 7 TeV p-p collisions is used. The difference between the gluon PDF uncertainty with HERA-I DIS data only [4] and with HERA-I

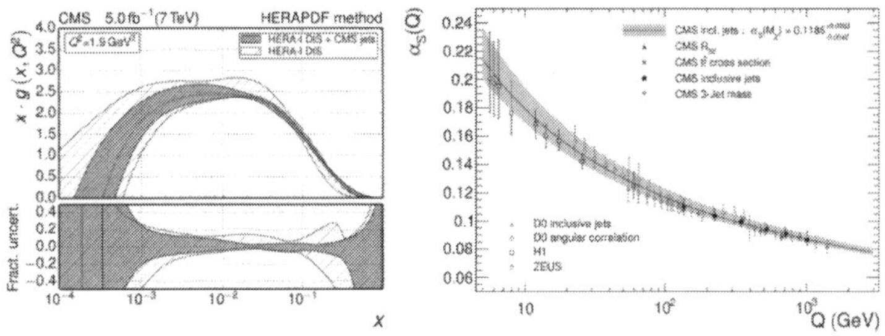

(a) (b)

Figure 1: (a) Comparison of the gluon PDF from HERA-I DIS data only, and after combining with CMS multijets analysis result. The error band represent the total (experimental, statistical, modeling) uncertainties. (b) Evolution of α_s with the scattering scale Q.

DIS and CMS measurement added is shown in Fig. 1a: the uncertainty on the gluon PDF is significantly reduced for fractional parton momenta x>0.01. More modest, although measurable, improvements on the u and d valence quarks PDFs are also present.The measurement of α_s was performed as well, found to be

$$\alpha_s(\mathrm{M_Z}) = 0.1185 \pm 0.0019(\mathrm{exp}) \pm 0.0028(\mathrm{PDF}) \pm 0.0004(\mathrm{NP})^{+0.0053}_{-0.0024}(\mathrm{scales}),$$

where NP stands for the uncertainties related to the non-perturbative corrections brought to the data to correct the measured quantities to the parton level, scales for the factorization and renormalization scale uncertainties. The running of α_s was also evaluated in several ranges of scattering scale Q^2 and is shown to be inline with previous measurements, as shown in Fig. 1b

The quarks PDFs were also extracted using the muon charge asymmetry and W+c measurements. The muon charge asymmetry measurement [5] especially impacts the quark PDFs in the Bjorken scaling variable from 0.001 to 0.1, while the W+c measurement allows to probe in particular the s-quark pdf [6] . Both measurements have been combined together, with the HERA e-p cross-section measurement, see [7] for the details and conclusions.

3. Production of a W or Z boson

The study of W and Z bosons production is essential for two reasons. First, these processes are large backgrounds for numerous searches and measurements. A precise estimate of the corresponding differential cross-sections is therefore essential in this context. Second, the understanding of their production mechanism is important to refine our knowledge of QCD and QED. The p-p collisions data taken during the Run I allowed to compare data with predictions from theory to a unprecedented level, both in terms of statistics and range of energy.

3.1. Differential W bosons and Z bosons production cross-section at $\sqrt{s}$= 8 TeV

Theoretical predictions for the production of a vector boson W or Z in hadron collisions are computed to the next-to-next-to-leading-order (NNLO) in perturbative quantum chromodynamics (pQCD). The uncertainties are essentially related to the PDFs, α_s and the scales choice. The complication in the prediction of vector boson spectra, independently of the number of jets from initial-state radiation (ISR), is the necessity to describe production mechanisms simultaneously at soft and hard scales, including the transition region. These studies therefore constitute stress tests of sophisticated event generation/calculation techniques. Here below the example of the study of the W and Z vector bosons transverse momenta with the full integrated luminosity ($\sim 20\text{fb}^{-1}$) of the 8 TeV p-p collisions are given. The measurement of the W boson transverse momentum (see Fig. 2a) is compared with different theoretical predictions: `POWHEG` [8] (NLO in QCD), `FEWZ` [9] (NNLO QCD), `RESBOS` [10] (NNLL/NLO QCD).

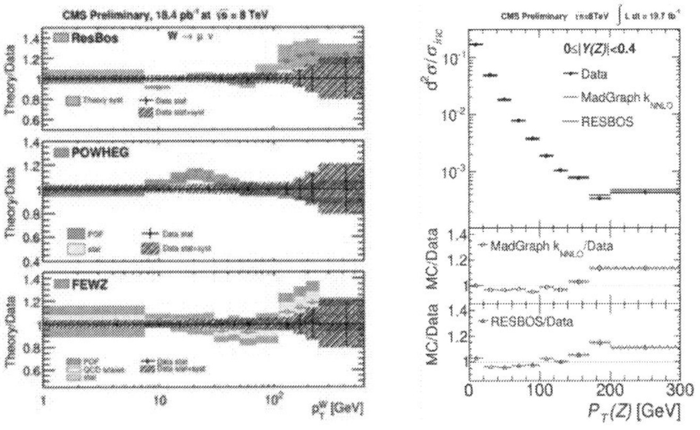

(a) (b)

Figure 2: (a) Transverse momentum of the W boson. The uncertainties in the data points are systematical and statistical errors added in quadrature. The uncertainties in the theory predictions represent the scale, PDF, and statistical errors, also added in quadrature. (b) Transverse momentum of the Z boson for $|y| < 0.4$. The error bars represent the statistical uncertainty of data and predictions, while the shaded band represent the systematic uncertainties from data only.

The comparison is done in a phase-space region defined by the presence of one pre-FSR muon (electron) with $p_T > 20$ (25) GeV, $|\eta| < 2.1$ (2.5). Correction of the detector effects is performed with the SVD technique [11]. The measurement of the Z boson $d\sigma/dp_T|dy|$ (see Fig. 2b) is compared with `RESBOS` and the tree-level multi-leg event generator `Madgraph` [12]. The comparison is done in a phase-space defined by the presence of two muons. The leading (second) with $p_T > 25$ (10) GeV, $|\eta| < 2.1$ (2.4). Correction of the detector effects is performed with the D'Agostini technique [13].

Both `Madgraph` and `RESBOS` predictions severely overshoot the data in the high

p_T^Z region (>100 GeV). A similar effect is observed with `RESBOS` and `FEWZ` for the transverse momentum of the W boson, while `POWHEG` exhibits an event excess in the region between 10 and roughly 30 GeV. More details concerning these studies can be found in [14, 15], and other CMS studies can be found elsewhere [1].

3.2. V + jets

The modeling of the production of V+jets processes is sensitive to several effects: the presence of higher order QCD or QED corrections, the choice of factorization and renormalization scales, the merging procedure used to get the leading jets from the matrix-element (ME) calculation (see e.g. [16–19]), and the characteristics of the parton shower (PS).

The CMS collaboration has published several results concerning the V+jets processes. Comparison between data and prediction from theory is shown for the production of a Z boson in association with jets [20, 21] in Fig. 3. The comparison is done in a phase-space region defined by the presence of two same-flavour and opposite-charge leptons (e or μ) with p_T >20 GeV, $|\eta|$ <2.4. Each lepton is dressed, i.e. is defined from the clustering of a final state lepton and FSR photons in a cone with 0.1 radius centered on this final state lepton. In addition, the lepton invariant mass M_{ll} must fulfill de condition $|M_Z - M_{ll}| < 20$ GeV. Jets are defined at particle level using anti-k_T (R=0.5) algorithm, p_T >30 GeV and $|\eta|$ <2.4. A minimal ΔR of 0.5 between the jets and the leptons is required. The correction of detector effects is performed using the D'Agostini method. The number of jets shown in Fig. 3a turns out to be reasonably described by both `Sherpa2` [22] and `Madgraph`. Deviations are observed essentially for N_{jet} >5, where the parton shower prediction is dominant. That means the ME calculation are performing well, both at tree-level and NLO. In Fig. 3b, the transverse momentum of the leading jet turns out to be better described by the NLO prediction, although a trend is visible, especially in the low p_T region. The scalar sum of jet p_T shown in Fig. 3c is very well described by `Sherpa2`, while a `Madgraph` undershoot the data in the low H_T region.

The comparison and ratio of momenta of Z boson and photon produced in association with jets [23] is shown in Fig. 3d and 3e, respectively. The Z+jets events selection is essentially the same as described above. For the γ+jets, a minimal transverse momentum of 100 GeV is required for the photon, as well as $|\eta|$ <1.4. It turns out that `Madgraph` predicts the good shape, but fails at estimating the normalization. One possible source for this deviation is the missing QCD and QED higher order terms. This is particularly important for searches for supersymmetric particles, where the $Z \to \nu\nu$ background processes can be deduced by estimating the normalization and shape of γ+jet process. Other results which concern the jets production in association with a W boson, the azimuthal and longitudinal correlation in Z+jets, can be found in [1].

3.3. W and Z + heavy flavour quark pairs

The production of a Z or W boson in association with a heavy flavour quark pair c or b constitute a subclass of the V+jets events copiously produced in the p-p collisions at the LHC. Due to their much smaller cross-section, the production mechanisms of the heavy quarks are only scarcely known and large uncertainties remain up to now. Dominant uncertainties are related to the choice of factorization and renormalization

Journal of Physics: Conference Series **623** (2015) 012010 doi:10.1088/1742-6596/623/1/012010

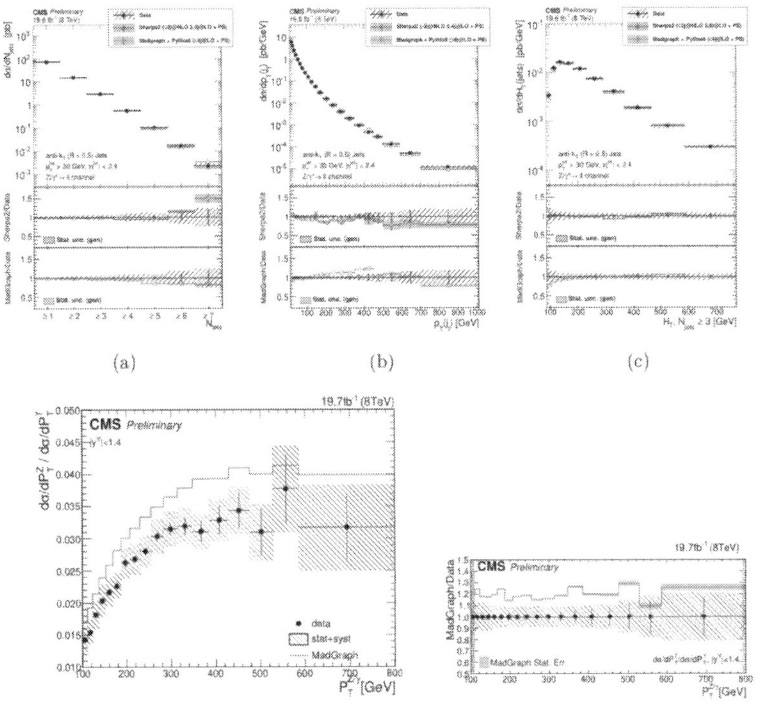

(a) (b) (c) (d) (e)

Figure 3: (a) Number of jets produced in association with the Z boson (b) p_T of the leading jet produced in association with the Z boson. (c) Scalar sum of the jets p_T, assuming the presence of at least three jets. The uncertainty band associated to the predictions from theory show the statistical uncertainty. (d) Differential cross-section ratio of leptonic Z/γ w.r.t. p_T^V for $N_{jet} > 0$. (e) Ratio of the Z and γ+jets predictions from theory and data. The error bars reflect the statistical uncertainty of the ratio, the hatched band is the total uncertainty.

scales, flavour numbering scheme and related ME-PS merging approach (which includes the treatment of FSR gluons radiated off the heavy quark states), and the b-quark mass. Several analyses were completed by the CMS collaboration, all related to the production of a Z or W boson in association with a b-jet pair [24, 25]. The cross-sections are given for the events with exactly one b-jet (Fig. 4a) and events with at least two b-jets (Fig. 4b and 4c). Comparisons with `aMC@NLO` [26] and `Madgraph` in both the four flavour (4F) and five-flavour (5F) numbering schemes, and with `MCFM` (5F) were done. For the Z+b-jet(s) analysis, the phase-space is defined by the leptons: $p_T > 20$ GeV, $|\eta| < 2.4$, $|M_Z - M_{ll}| < 15$ GeV, the presence of one or at least two b-jet:

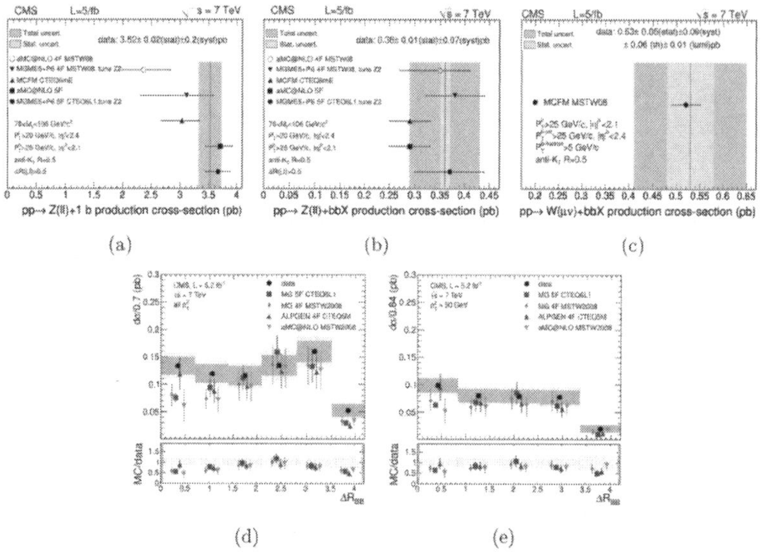

Figure 4: (a), (b) and (c): cross-sections for Z+1 b-jet, Z+>1 b-jet and W+2 b-jets, respectively. Both statistical and total uncertainties are represented. (d) and (e): differential cross-section of Z+2 B-hadrons w.r.t. the ΔR separation between the hadrons. The darker band is the statistical uncertainty, the brighter one is total uncertainy.

anti-k_T(R=0.5), p_T >25 GeV, $|\eta|$ <2.4, ΔR(j,l) >0.5 and the presence of a b-hadron inside the jet. For the W+2 b-jets analysis, the phase-space definition is the same apart that the lepton selection is defined from the presence of one muon with p_T >25 GeV, $|\eta|$ <2.1.

For Z + exactly one b-jet, the 4F predictions from both `Madgraph` and `aMC@NLO` tend to undershoot the data, while the 5F predictions are more inline with the data. For Z and W + at least two b-jets all predictions are in agreement (within the uncertainties) with the data.

Besides the inclusive measurements, the differential cross-section of the production of a Z boson in association with a pair of B-hadron (identified through the presence of secondary vertices) was measured [27]. Several variables are studied, including the ΔR between the B-hadrons, which is the most sensitive variable to test the modeling of the collinear production of B-hadrons and suffers from large theoretical uncertainties. In Fig. 4d (4e) the ΔR between the B-hadrons is shown for events with $p_T^Z > 0$ (50) GeV. There the phase-space is defined by the presence of two leptons with p_T >20 GeV, $|\eta|$ <2.4, $|M_Z - M_ll|$ <10 GeV, and two B-hadrons with p_T >15 GeV and $|\eta|$ <2.0.

These first studies of the production of a vector boson in association with b-quark pairs shows that as soon as the two heavy states are sufficiently separated, i.e. if the production mechanism of the b-quarks is not likely identified as from a gluon splitting, the prediction do behave well, while the collinear region is generally poorly described.

4. Anomalous gauge couplings

The study of anomalous triple (quartic) gauge couplings aTGC (aQGC) is of prime importance to test the dimension-4 SM hypothesis: besides the searches for the presence of new resonances, a seducing idea is to consider the SM as an effective field theory described by a higher dimensional (6 or 8) lagrangian with as consequence a change of gauge-couplings strengths. Recent results concerning the aTGC $ZZ\gamma$ and $Z\gamma\gamma$ [28, 29] as well as WWWW [30] aQCG are quoted here. In the first case. The experimental final state is a dilepton final state and a photon, while in the second case, the hypothesis of the aQCG is tested using the vector-boson-scattering process leading to a pair of same-sign W bosons.

Exclusion limits on the operator strength were established, as shown in Fig. 5a and 5b.

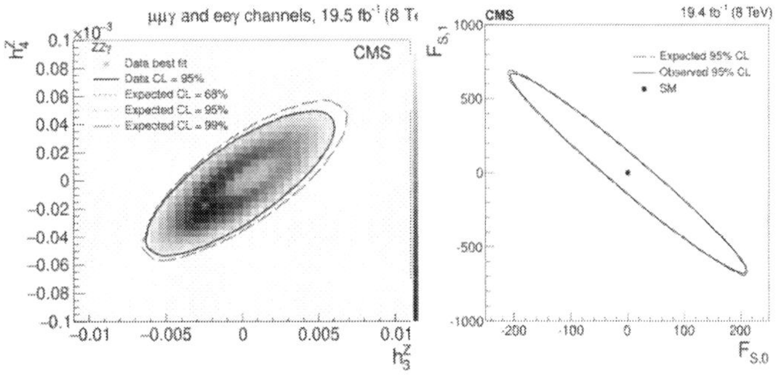

(a) (b)

Figure 5: (a) Best fit between data and model of anomalous ZZ? couplings [28]. The blue marker indicates the point of highest probability. The level of gray represents the exclusion probability and the black line corresponds to the 95% CL limit. Several expected contours from SM simulation are also shown.(b)Observed and expected 2D limit on FS_0 and FS_1 for the WWWW aQGC [30] .

The general conclusion from the diverse aTGC/QCG searches is that as for today there is no observed deviation from the dimension-4 SM hypothesis. All CMS results concerning searches for anomalous triple (WWZ, $WW\gamma$, $ZZ\gamma$, $Z\gamma\gamma$) and quartic ($\gamma\gamma$WW, γZWW) gauge couplings can be found elsewhere [1].

References

[1] Smp public twiki. `https://twiki.cern.ch/twiki/bin/view/CMSPublic/PhysicsResultsSMP`.
[2] Vardan Khachatryan et al. Constraints on parton distribution functions and extraction of the strong coupling constant from the inclusive jet cross section in pp collisions at $\sqrt{s} = 7$ TeV. 2014.
[3] Matteo Cacciari, Gavin P. Salam, and Gregory Soyez. The Anti-k(t) jet clustering algorithm. *JHEP*, 0804:063, 2008.

[4] F.D. Aaron et al. Combined Measurement and QCD Analysis of the Inclusive e+- p Scattering Cross Sections at HERA. *JHEP*, 1001:109, 2010.

[5] Serguei Chatrchyan et al. Measurement of the electron charge asymmetry in inclusive W production in pp collisions at $\sqrt{s} = 7$ TeV. *Phys.Rev.Lett.*, 109:111806, 2012.

[6] Serguei Chatrchyan et al. Measurement of associated W + charm production in pp collisions at $\sqrt{s} = 7$ TeV. *JHEP*, 1402:013, 2014.

[7] Serguei Chatrchyan et al. Measurement of the muon charge asymmetry in inclusive $pp \to W + X$ production at $\sqrt{s} = 7$ TeV and an improved determination of light parton distribution functions. *Phys.Rev.*, D90(3):032004, 2014.

[8] Simone Alioli, Paolo Nason, Carlo Oleari, and Emanuele Re. A general framework for implementing NLO calculations in shower Monte Carlo programs: the POWHEG BOX. *JHEP*, 1006:043, 2010.

[9] Ryan Gavin, Ye Li, Frank Petriello, and Seth Quackenbush. FEWZ 2.0: A code for hadronic Z production at next-to-next-to-leading order. *Comput.Phys.Commun.*, 182:2388–2403, 2011.

[10] G.A. Ladinsky and C.P. Yuan. The Nonperturbative regime in QCD resummation for gauge boson production at hadron colliders. *Phys.Rev.*, D50:4239, 1994.

[11] Andreas Hocker and Vakhtang Kartvelishvili. SVD approach to data unfolding. *Nucl.Instrum.Meth.*, A372:469–481, 1996.

[12] Johan Alwall, Michel Herquet, Fabio Maltoni, Olivier Mattelaer, and Tim Stelzer. MadGraph 5 : Going Beyond. *JHEP*, 1106:128, 2011.

[13] G. D'Agostini. A Multidimensional unfolding method based on Bayes' theorem. *Nucl.Instrum.Meth.*, A362:487–498, 1995.

[14] CMS Collaboration. Measurement of Z production as a function of transverse momentum and rapidity. 2014.

[15] CMS Collaboration. Measurement of the transverse momentum distribution of W bosons in pp collisions at $\sqrt{s} = 8$ TeV. 2014.

[16] Johan Alwall et al. MadGraph/MadEvent v4: The New Web Generation. *JHEP*, 09:028, 2007.

[17] Johan Alwall, Simon de Visscher, and Fabio Maltoni. QCD radiation in the production of heavy colored particles at the LHC. *JHEP*, 0902:017, 2009.

[18] Leif Lonnblad and Stefan Prestel. Matching Tree-Level Matrix Elements with Interleaved Showers. *JHEP*, 1203:019, 2012.

[19] Rikkert Frederix and Stefano Frixione. Merging meets matching in MC@NLO. *JHEP*, 1212:061, 2012.

[20] CMS Collaboration. Measurement of the differential production cross section of Z bosons in association with jets in pp collisions at sqrt(s) = 8 TeV. 2014.

[21] CMS Collaboration. Measurement of the double differential cross section of Z bosons produced in association with jets in pp collisions at $\sqrt{s} = 8$ TeV. 2014.

[22] T. Gleisberg, Stefan. Hoeche, F. Krauss, M. Schonherr, S. Schumann, et al. Event generation with SHERPA 1.1. *JHEP*, 0902:007, 2009.

[23] CMS Collaboration. Measurement of the Z/gamma*+jets/photon+jets cross section ratio in pp collisions at sqrt(s)=8 TeV. 2014.

[24] Serguei Chatrchyan et al. Measurement of the production cross sections for a Z boson and one or more b jets in pp collisions at sqrt(s) = 7 TeV. *JHEP*, 1406:120, 2014.

[25] Serguei Chatrchyan et al. Measurement of the production cross section for a W boson and two b jets in pp collisions at $\sqrt{s}$=7 TeV. *Phys.Lett.*, B735:204–225, 2014.

[26] J. Alwall, R. Frederix, S. Frixione, V. Hirschi, F. Maltoni, et al. The automated computation of tree-level and next-to-leading order differential cross sections, and their matching to parton shower simulations. *JHEP*, 1407:079, 2014.

[27] Serguei Chatrchyan et al. Measurement of the cross section and angular correlations for associated production of a Z boson with b hadrons in pp collisions at $\sqrt{s} = 7$ TeV. *JHEP*, 1312:039, 2013.

[28] Vardan Khachatryan et al. Measurement of the Z gamma production cross section in pp collisions at 8 TeV and search for anomalous triple gauge boson couplings. 2015.

[29] Vardan Khachatryan et al. Measurements of the ZZ production cross sections in the 2 l 2 nu channel in proton-proton collisions at sqrt(s) = 7 and 8 TeV and combined constraints on triple gauge couplings. 2015.

[30] Khachatryan et al. Study of vector boson scattering and search for new physics in events with two same-sign leptons and two jets. *Phys. Rev. Lett.*, 114(arXiv:1410.6315. CMS-SMP-13-015. CERN-PH-EP-2014-250):051801. 27 p, Oct 2014. Replaced with published version. Added journal reference and DOI.

Measurements of the top-quark mass and properties at CMS

Marc Dünser[1]
[1]**ETH Zürich** (on behalf of the CMS Collaboration)
Institute for Particle Physics (IPP)
HPK E 32
Otto-Stern-Weg 5
CH - 8093 Zürich

E-mail: marc.dunser@cern.ch

Abstract. Measurements of the top-quark mass and other top-quark properties are presented, obtained from the CMS data collected in 2011 and 2012 at centre-of-mass energies of 7 and 8 TeV. The mass of the top quark is measured using several methods and decay channels. The measurements of the top-quark properties include the W helicity in top-quark decays, the search for anomalous couplings, and the ratio of top-quarks decaying to bW over qW in order to gain information on $|V_{tb}|$ using both $t\bar{t}$ and single-top quark event samples. The results are compared with predictions from the standard model as well as new physics models. The cross section of $t\bar{t}$ events produced in association with a W, Z boson or a photon is also measured.

1. Introduction

The top-quark is the heaviest fundamental particle known in the Standard Model of particle physics (SM) and exhibits features which render it unique among quarks. Its large mass results in its decay before hadronization processes take place, hence the properties of the quark are directly accessible. After its discovery about 20 years ago at the Tevatron, studying the top-quark has been an active field of research [1, 2]. Knowledge of the precise value of its mass is of paramount importance to the understanding of the SM, as it can help constraining the SM and also plays an important role in physics beyond the standard model. Figure 1 shows an example in which a global fit on the free parameters of the SM is performed and compared in a plane of m_{top} and m_W, the mass of the W boson.

Besides measurements of its mass, some measurements of its properties are presented in the following which constitute some of the most precise measurements to date.

2. Measurements of m_{top}

Measuring the mass of the top-quark can be done in many different ways. The most common procedure is to construct templates from Monte-Carlo simulation and to extract the top-quark mass parameter of the simulation by interpolating between the different mass templates.

Such a technique is implemented in CMS [4] in the measurement via the so-called m_{lb} variable, the invariant mass between the lepton and the b-tagged quark in di-leptonic $e^{\pm}\mu^{\mp}$ events. The lepton is chosen as the one which yields minimal m_{lb} with the highest p_T b-tagged jet in the

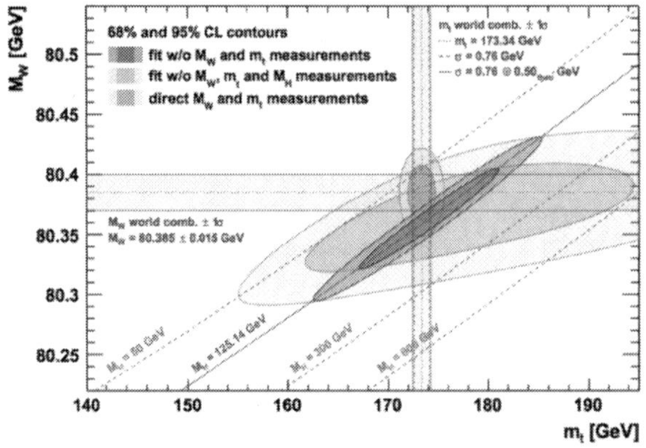

Figure 1. Comparison of a global fit to the SM with the values of the W boson versus the mass of the top-quark. Direct measurements with their errors are shown as green bands while the global fit yields the blue regions [3].

event [5]. The final distribution of the variable with MC templates at different top-quark masses is shown in Fig. 2. A final mass of 172.3±1.3 GeV is extracted.

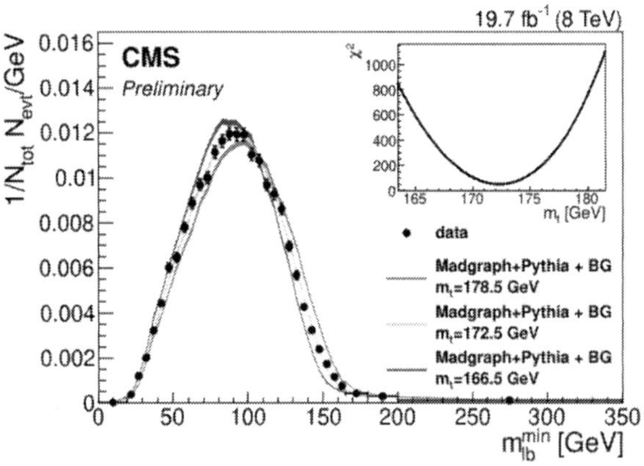

Figure 2. Distribution of m_{lb} overlaid with different simulated templates at masses of 178.5 (red), 172.5 (green), and 166.5 (blue) GeV. The inlay shows the interpolated χ^2 fit to the data, yielding a minimum at 172.3 GeV.

A similar approach has been taken for the most precise single measurement of m_{top}. Single lepton events are selected with at least four accompanying jets of which exactly two are required to be b-tagged. A kinematic fit to the $t\bar{t}$ hypothesis is performed along with the implementation of the ideogram method in order to extract additional information in cases where the physical objects are incorrectly assigned [6, 7]. The final measurement from fits to seven MC templates where not only m_{top}, but also an overall jet-energy scale factor is obtained yields a value of

172.04 ± 0.19 ± 0.75 GeV, where the first error is statistical and the second systematic. The overall jet scale factor is fit to 1.007 ± 0.002 ± 0.012 pertaining to a sub-percent correction to the jet energies. Both results are displayed in Fig. 3.

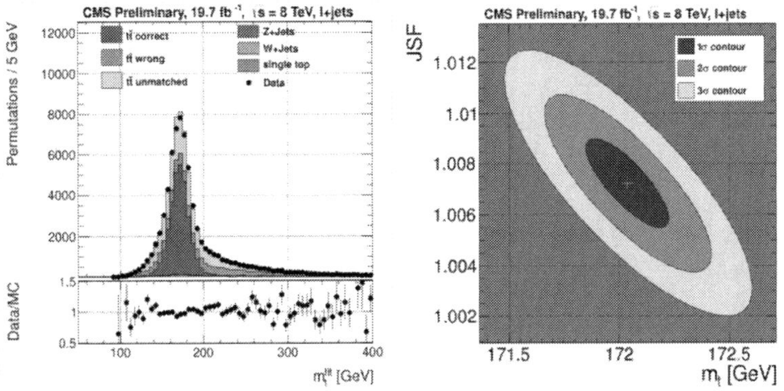

Figure 3. Fitted m_{top} (left) after event selection and a goodness-of-fit requirement as well as the final fit to the jet scale (right).

Two more measurements of the top-quark mass in CMS were presented. These are a measurement in the di-leptonic final state where a very similar technique to the one in the aforementioned single leptonic measurement was employed via the implementation of a kinematic fit with the added complication of an additional neutrino. The final presented measurement is a fully hadronic measurement in which a global jet scale factor is extracted simultaneously with m_{top}. The obtained values are m_{top} = 172.47 ± 0.17 ± 1.40 GeV for the high-statistics di-lepton measurement and m_{top} = 172.08 ± 0.36 ± 0.83 GeV along with a jet scale factor of 1.007 ± 0.003 ± 0.011 in splendid agreement with the previously presented measurements [8, 9]. Two pieces of these results are shown in Fig. 4, where the χ^2 fit to the seven templates is shown on the left for the di-leptonic analysis, and the fitted top mass of the hadronic analysis is displayed on the right.

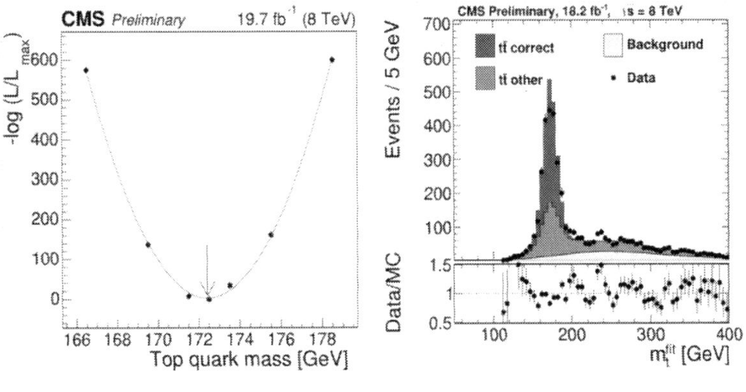

Figure 4. χ^2 fit to the seven MC templates with best fit value for the di-leptonic analysis (left) as well as the top mass as extracted from the kinematic fit of the fully hadronic analysis.

2.1. Combination of m_{top} measurements

Combinations of different measurements of m_{top} from CMS and some global combinations are presented as well. Many measurements from both the $\sqrt{s} = 7$ TeV and 8 TeV datasets collected by CMS enter in the combination procedure [10]. In total eight different analyses are combined by measure of the BLUE (Best Linear Unbiased Estimator) [11] where correlations among these measurements and their associated theoretical and experimental uncertainties are treated consistently. Of the order of 25 uncertainties are investigated and combined with the central values into a combined best estimate of the top-quark mass of CMS at $m_{top} = 172.38 \pm 0.10 \pm 0.65$ GeV. This represents the most accurate experimental measurement to date and improves upon the preceding world average published by the CDF, D0, ATLAS, and CMS collaborations in early 2014 [12].

Summaries of all the measurements going into the CMS combination as well as the averages of the Tevatron experiments and the world average are presented in Fig. 5.

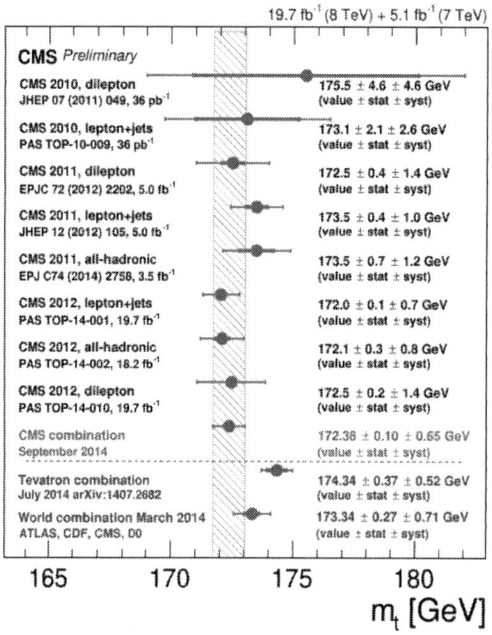

Figure 5. Summary of the eight input measurements from 2010 through 2012 from CMS and the resulting combined measurement. Markers indicate the central values with the statistical errors in thick red lines and the systematic uncertainties in thin red lines. Also given are the Tevatron combination and the world average.

3. Top-quark properties

As briefly stated before, the top-quark is the only quark which does not hadronize. Thus, its spin relation and many other quantities are relatively directly observable. A short selection of measurements of such properties is given, including the decay ratio $R = t \rightarrow \frac{bW}{qW}$ and the helicity of the W boson from single-top events leading to limits on anomalous couplings. Additionally, some rare process cross-sections are presented.

The measurement of the decay ratio $R = t \rightarrow \frac{bW}{qW}$ can be interpreted as a measurement of the $|V_{tb}|$ element of the CKM matrix with the ratio R being the square of V_{tb}. Previous measurements performed at D0 indicate some tension with the constraints derived from the unitarity of the CKM matrix of $|V_{tb}| = 0.999146^{+0.000021}_{-0.000046}$ [13, 14]. Di-leptonic (e or μ) events are selected and the spectra of the number of b-tagged jets is analyzed. Systematic uncertainties are dominated by the uncertainties of the experimental tagging efficiency [15]. By fitting the final distributions a ratio of $R = 1.015 \pm 0.003 \pm 0.031$ is extracted, corresponding to a central value of $|V_{tb}|$ of 1.007. This is very consistent with the expectations of unitarity of the CKM matrix and can not confirm the observed tensions observed by the D0 collaboration. The results of the fit as well as the distributions are shown in Fig. 6.

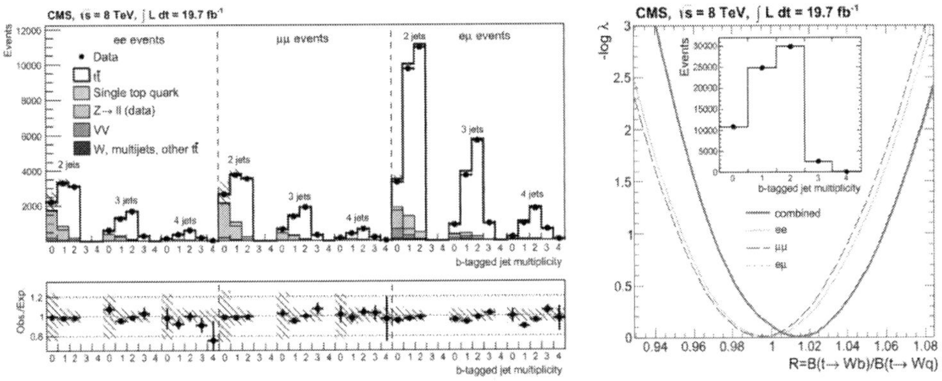

Figure 6. Spectrum of N_{b-jets} as a function of the number of jets and the flavor of the di-leptonic events from which the ratio R is extracted (left). The χ^2 of the flavor dependent fit as well as the combined fit is shown on the right.

The W-boson helicity can be measured directly through the angle between the W-boson in the top-quark rest frame and the momentum of the down-type decay fermion in the rest-frame of the W-boson. This angular distribution contains contributions from left-handed, right-handed, and longitudinal partial decay widths of the W-boson and can be put formally into a probability density function

$$\rho(\cos\theta_l^\star) = \frac{3}{8}(1 - \cos\theta_l^\star)^2 F_L + \frac{3}{4}\sin^2\theta_l^\star F_0 + \frac{3}{8}(1 + \cos\theta_l^\star)^2 F_R, \tag{1}$$

where F_i are the partial decay widths. In this measurement, the goal is to quantify the helicity from single-top events. A single lepton is selected with exactly two jets accompanying it of which one is required to be b-tagged. The final event selection is not very pure in the single-top process, but most of the contamination stems from $t\bar{t}$ events, which contain the same information to begin with. Therefore, this measurement represents an orthogonal analysis to the W-helicity measurement of $t\bar{t}$ events [16, 17]. The results for the F_i are 0.298(43), 0.720(54), and −0.018(22) for the left-handed, longitudinal, and right-handed partial decays widths, respectively. The negative value for the right-handed fraction originates from the constraint $\sum F_i = 1$ imposed in the fit on the angular distribution. This compares well with the theoretical prediction of 0.311(5), 0.687(5), and 0.0017(1). These results are then interpreted in terms of anomalous tWb couplings g_L and g_R, which are predicted to vanish at tree level in the SM. Both the angular distribution of $\cos\theta_l^\star$ and the limits on the anomalous couplings are shown in Fig. 7.

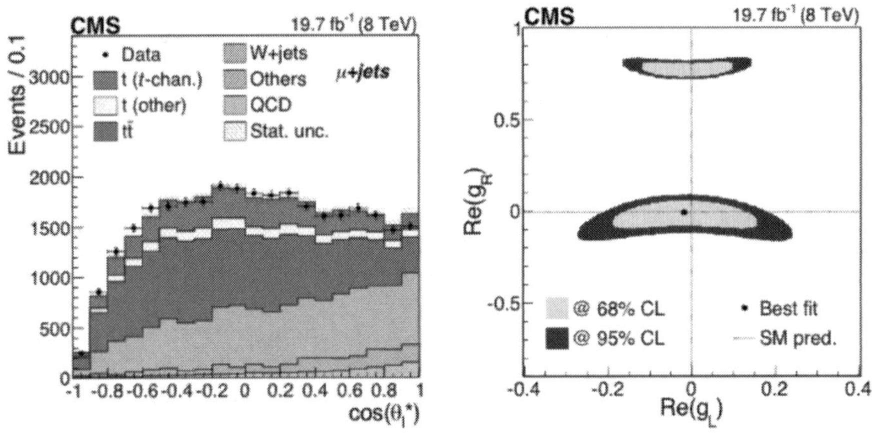

Figure 7. Observation versus prediction on the angular distribution of $\cos\theta_l^\star$ (left) and the limits on the real, anomalous coupling constants g_L and g_R (right). Both distributions agree well with the expectation from the SM.

Finally, results on the rare SM processes of $t\bar{t}$ $+W/Z/\gamma$ are presented. The cross-sections of the $t\bar{t}$ plus massive boson production is measured in leptonic final states combining the same-sign di-lepton, tri-lepton and quadri-lepton final states in an exclusive manner. A simultaneous fit to the observed data is then performed in the cross-sections of $t\bar{t}$ $+W$ and $t\bar{t}$ $+Z$ to extract the values for the cross section from a simple counting experiment. The results of $\sigma_{ttW} = 170^{+110}_{-100}$ fb and $\sigma_{ttZ} = 200^{+90}_{-90}$ fb agree well with the respective SM theoretical expectations of $\sigma_{ttW}^{theory} = 206^{+21}_{-23}$ fb and $\sigma_{ttZ}^{theory} = 197^{+22}_{-25}$ fb, albeit with large uncertainties due to the limited statistics of this measurement [18]. The results of the simultaneous fit is compared with the SM expectation in the left part of Fig. 8.

The cross-section of the $t\bar{t}$ $+\gamma$ process is investigated in the single muonic decay mode of $pp \to W^+W^-b\bar{b}\gamma$, where the γ is required to have at least 20 GeV of p_T. At least four jets of which one is required to be tagged as originating from a b-quark are selected in the final state to suppress non $t\bar{t}$ backgrounds. A maximum likelihood method is introduced to estimate the fraction of real, prompt emissions of photons and the ratio $R = \sigma_{tt+\gamma}/\sigma_{tt}$ is extracted with a $t\bar{t}$ cross-section measurement of CMS as an input [19]. It is evaluated to $(1.07 \pm 0.07 \pm 0.27)$ $\times 10^{-2}$ yielding a final estimate for the cross section of $\sigma_{tt+\gamma} = 2.4 \pm 0.2 \pm 0.6$ pb [20]. This is in good agreement with the SM expectation of $\sigma_{tt+\gamma}^{theory} = 1.8 \pm 0.5$ pb.

4. Concluding remarks

A number of different measurements on the top-quark mass were presented along with the CMS combination of said quantity. The precision of single measurements has arrived at the sub-percent level due to the large dataset and advances in both theoretical and experimental understanding. The relative error of the CMS combination of 0.38% stands as the most precise value to date.

Some select properties of the top-quark were presented as well, including the ratio bW/qW, the helicity of the W-boson in the top-quark decay as well as measurements on the $t\bar{t}$ +boson processes. These latter processes suffer from large statistical and systematic uncertainties, though greater precision is expected to be achieved at higher center-of-mass energy in Run

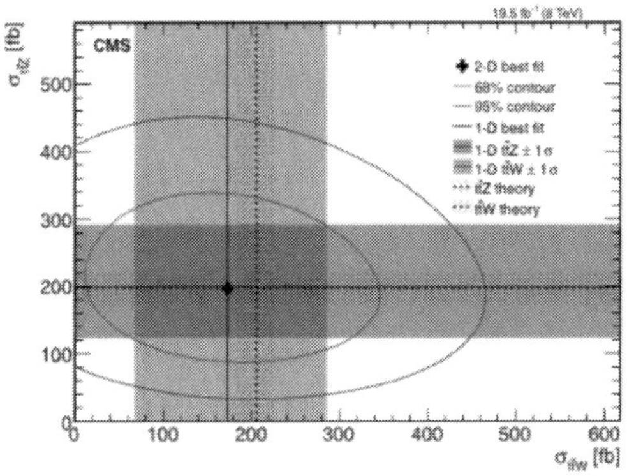

Figure 8. Simultaneous fit to the cross-sections of the $t\bar{t}$ $+W$ and $t\bar{t}$ $+Z$ processes as observed in leptonic final state.

2 of the LHC.

References

[1] CDF Collaboration. Observation of top quark production in $\bar{p}p$ collisions. *Phys.Rev.Lett.*, 74:2626–2631, 1995.

[2] D0 Collaboration. Search for high mass top quark production in $p\bar{p}$ collisions at $\sqrt{s}$ = 1.8 TeV. *Phys.Rev.Lett.*, 74:2422–2426, 1995.

[3] M. Baak et al. The global electroweak fit at NNLO and prospects for the LHC and ILC. *Eur.Phys.J.*, C74:3046, 2014.

[4] CMS Collaboration. The CMS experiment at the CERN LHC. *JINST*, 3:S08004, 2008.

[5] CMS Collaboration. Determination of the top-quark mass from the m_{lb} distribution in dileptonic ttbar events at $\sqrt{s} = 8$ TeV. Technical Report CMS-PAS-TOP-14-014, CERN, Geneva, 2014.

[6] CMS Collaboration. Measurement of the top-quark mass in $t\bar{t}$ events with lepton+jets final states in pp collisions at $\sqrt{s} = 7$ TeV. *JHEP*, 1212:105, 2012.

[7] CMS Collaboration. Measurement of the top-quark mass in $t\bar{t}$ events with lepton+jets final states in pp collisions at $\sqrt{s}$=8 TeV. Technical Report CMS-PAS-TOP-14-001, CERN, Geneva, 2014.

[8] CMS Collaboration. Measurement of the Top Quark Mass in Dilepton $t\bar{t}$ Decays at $\sqrt{s} = 8$ TeV. Technical Report CMS-PAS-TOP-14-010, CERN, Geneva, 2014.

[9] CMS Collaboration. Measurement of the top-quark mass in $t\bar{t}$ events with all-jets final states in pp collisions at $\sqrt{s}$=8 TeV. Technical Report CMS-PAS-TOP-14-002, CERN, Geneva, 2014.

[10] CMS Collaboration. Combination of the CMS top-quark mass measurements from Run 1 of the LHC. Technical Report CMS-PAS-TOP-14-015, CERN, Geneva, 2014.

[11] Louis Lyons, Duncan Gibaut, and Peter Clifford. How to Combine Correlated Estimates of a Single Physical Quantity. *Nucl.Instrum.Meth.*, A270:110, 1988.

[12] The ATLAS, CDF, CMS, and D0 Collaborations. First combination of Tevatron and LHC measurements of the top-quark mass. 2014.

[13] J. Beringer et al. Review of Particle Physics (RPP). *Phys.Rev.*, D86:010001, 2012.

[14] D0 Collaboration. Precision measurement of the ratio B($t \to Wb$)/B($t \to Wq$) and Extraction of V_{tb}. *Phys.Rev.Lett.*, 107:121802, 2011.

[15] CMS Collaboration. Measurement of the ratio $B(t \to Wb)/B(t \to Wq)$ in pp collisions at $\sqrt{s} = 8$ TeV. *Phys.Lett.*, B736:33–57, 2014.

[16] CMS Collaboration. Measurement of the W boson helicity in events with a single reconstructed top quark in pp collisions at $\sqrt{s} = 8$ TeV. *JHEP*, 1501:053, 2015.

[17] CMS Collaboration. Measurement of the W-boson helicity in top-quark decays from $t\bar{t}$ production in lepton+jets events in pp collisions at $\sqrt{s} = 7$ TeV. *JHEP*, 1310:167, 2013.

[18] CMS Collaboration. Measurement of top quark-antiquark pair production in association with a W or Z boson in pp collisions at $\sqrt{s} = 8$ TeV. *Eur.Phys.J.*, C74(9):3060, 2014.

[19] CMS Collaboration. Top pair cross section in dileptons. Technical Report CMS-PAS-TOP-12-007, CERN, Geneva, 2012.
[20] CMS Collaboration. Measurement of the inclusive top-quark pair + photon production cross section in the muon + jets channel in pp collisions at 8 TeV. Technical Report CMS-PAS-TOP-13-011, CERN, Geneva, 2014.

Journal of Physics: Conference Series **623** (2015) 012012 doi:10.1088/1742-6596/623/1/012012

Searching for P- and CP-odd effects in HIC

D Espriu

ICCUB, University of Barcelona, Martí i Franquès, 1, 08928 Barcelona, Catalonia,Spain

E-mail: espriu@icc.ub.edu

Abstract. In this talk we will summarize the main results from our recent work concerning the possibility that a new metastable phase occurs in some heavy ion collisions (HIC). This phase would be characterized by the breaking of two characteristic symmetries of strong interactions; namely P and CP. We investigate the experimental consequences of parity breaking in such a situation and propose suitable observables to elucidate the presence this phenomenon.

1. Why P and CP might not be good symmetries in HIC

Parity is one of the well established global symmetries of strong interactions. While there are arguments to think that P or CP cannot be broken in the usual QCD vacuum[1], this may not be the case under extreme conditions of temperature and density.

For a long time the possibility that P- and CP-odd condensates may exist as regular phase at finite density has been contemplated, going back all the way to the work of Migdal[2]. The debate was inconclusive by using simple nucleon-meson models and, if anything, the arguments suggested that a condensate of this type was not physically viable. However, more recent –and more complete– effective theory studies indicate that such a phase is a real possibility[3] at 'moderate' densities (3 to 8 $\times \rho_N$).

This possibility is surely relevant in astrophysical contexts but it may still be posible to ephemerally produce such a phase in HIC in conditions where the density is high (i.e. large baryonic chemical potential) and temperature relatively low. These conditions may be within reach of future experiments at FAIR and NICA[4]. However this is not the main subject of this presentation. Rather we would like to consider the possibility that in a violent collision it could be póssible to produce long-lived bubbles with a non-vanishing value of the topological charge, where parity is locally broken. The possibility that these topological fluctuations would take place in a finite volume and large T was first proposed by Kharzeev, Pisarski and Tytgat[5] and later considered in a number of works[6, 7]

Large fluctuations in the gauge field could indeed exist in a hot environment and could generate a local imbalance of topological charge. This picture is supported by lattice simulations[8] although results obtained in euclidean simulations are hard to connect with the time dependent dynamics that exist in the early stages of a HIC.

The possibility of domains where P is broken is also supported by the *glasma* picture[9]. However domains in this model are typically very small ($<< 1$ fm) in the initial QGP phase. Of course these domains should expand and grow at later stages. In our work we will be interested exclusively in what happens in the final hadronic phase and assume that some domains with a net topological charge and spatial extent > 1 fm exist in this hadronic phase, originally created via the glasma mechanism or any other.

1.1. *Generating an effective chiral chemical potential*

If a topological charge ΔT_5 arises in a finite volume due to quantum fluctuations and it is conserved for a sufficiently long time in a quasi-equilibrium situation, we can introduce a topological chemical potential μ_θ conjugate to it into the QCD lagrangian via $\Delta\mathcal{L}_{top} = \mu_\theta \Delta T_5$ that plays the role of an effective θ term.

For *peripheral* heavy ion collisions such a situation may lead to the Chiral Magnetic Effect[10] whereby a large θ term combined with the large magnetic field due to the colliding nuclei generates a large electric field and originates charge separation.

For *central* collisions (and light quarks) a large θ term may trigger an ephemeral phase with axial chemical potential $\mu_5 \neq 0$. This comes about because the PCAC equation predicts an induced axial charge to be conserved in the chiral ($m = 0$) limit:

$$\frac{d}{dt}(Q_5^q - 2N_f T_5) \simeq 0, \quad Q_5^q = \int_{\mathrm{vol.}} d^3x \bar{q}\gamma_0\gamma_5 q = \langle N_L - N_R \rangle \tag{1}$$

Neglecting the quark mass is a good approximation for the lightest u and d quarks only. The strange quark is already too heavy and erases the chiral charge in the time scales where this phenomenon could be relevant. In a quasi-equilibrium situation the appearence of a nearly conserved chiral charge can be incorporated with the help of a chiral chemical potential μ_5.

2. Effective meson theory in a medium with local parity breaking

In principle, from the discussion in the previous section we have two possible isospin structures for μ_5: (a) an isosinglet pseudoscalar or (b) an isotriplet Only the situation (a) will be considered in this talk.

Having in mind that although the parity breaking domain may have formed in the early stages of the HIC we are only concerned about its consequences in the later stage hadronic phase we will now proceed to build an effective theory for mesons in a P- and CP-odd environment. We will deal in turn with scalars and vector mesons. Note that in both cases a breaking of Lorentz symmetry will occur.

The $J = 0$ sector can be described by using the spurion technique in the Lagrangian

$$D_\mu \Longrightarrow D_\mu - i\{\mu_5\delta_{0\mu}, \cdot\} \tag{2}$$

and constructing a generalised sigma model[18] with the light scalar mesons $\sigma, \vec{\pi}, \eta, \eta', \vec{a}_0$. The most relevant operator is $D = 2$. All these hadronic states naturally have a well defined parity in 'normal' QCD and partly because of that only the σ and ρ interact strongly with the pion fireball.

The new eigenstates of the hamiltonian do not have a well defined parity. Now due to parity breaking there is mixing in the $\sigma - \eta - \eta'$ and $\vec{\pi} - \vec{a}_0$ channels. The resulting $J = 0$ eigenstates are *all* coupled and are expected to *thermalize* in the HIC fireball.

For vector mesons the most relevant operator has $D = 3$. P- and CP-odd effects will appear through the Chern-Simons term[7, 11]

$$\Delta\mathcal{L} \simeq \varepsilon^{\mu\nu\rho\sigma}\mathrm{Tr}\left[\hat{\zeta}_\mu V_\nu V_{\rho\sigma}\right] \tag{3}$$

with $\hat{\zeta}_\mu = \partial_\mu \hat{a}(\vec{x}, t) = \delta_{\mu 0}\partial_0 \hat{a}(t)$ for a spatially homogeneous, time dependent background. We shall assume here $\partial_0 \hat{a}(t) \sim \hat{\zeta} \propto \mu_5$, as follows from the discussion in the previous section.

Vector mesons will be introduced and treated in the conventional way using the Vector Meson Dominance model[12] and enter the above lagrangian via the corresponding current V_μ. Note that for vector mesons there will be no mixing at this order resulting from the previous lagrangian but rather a distortion of the spectrum.

The current V_μ formed by a combination of vector mesons and the photon couples to fermions

$$\mathcal{L}_{\rm int} = \bar{q}\gamma_\mu \hat{V}^\mu q; \quad \hat{V}_\mu \equiv -eA_\mu Q + \frac{1}{2}g_\omega \omega_\mu \mathbb{I} + \frac{1}{2}g_\rho \rho^0_\mu \tau_3, \tag{4}$$

where $Q = \frac{\tau_3}{2} + \frac{1}{6}$, $g_\omega \simeq g_\rho \equiv g \simeq 6$. In addition we have to include Maxwell and mass terms

$$\mathcal{L}_{\rm kin} = -\frac{1}{4}\left(F_{\mu\nu}F^{\mu\nu} + \omega_{\mu\nu}\omega^{\mu\nu} + \rho_{\mu\nu}\rho^{\mu\nu}\right) + \frac{1}{2}V_{\mu,a}(\hat{m}^2)_{a,b}V^\mu_b \tag{5}$$

$$\hat{m}^2 \simeq m_V^2 \begin{pmatrix} \frac{10e^2}{9g^2} & -\frac{e}{3g} & -\frac{e}{g} \\ -\frac{e}{3g} & 1 & 0 \\ -\frac{e}{g} & 0 & 1 \end{pmatrix}, \quad \det\left(\hat{m}^2\right) = 0. \tag{6}$$

Particularizing to the present situation we get the following P-odd interaction

$$\mathcal{L}_{P-odd}(k) = \frac{1}{2}\zeta\epsilon_{jkl}\,V_{j,a}\,N_{ab}\,\partial_k V_{l,b}, \tag{7}$$

where in the case of an isosinglet pseudoscalar background

$$N^\theta_{ab} \simeq \begin{pmatrix} \frac{10e^2}{9g^2} & -\frac{e}{3g} & -\frac{e}{g} \\ -\frac{e}{3g} & 1 & 0 \\ -\frac{e}{g} & 0 & 1 \end{pmatrix}, \quad \det\left(N^\theta\right) = 0. \tag{8}$$

Due to (7) different vector meson helicities get a different momentum-dependent correction to its mass[13]. Vector mesons exhibit the following dispersion relation:

$$m^2_{V,\epsilon} = m_V^2 - \epsilon\zeta|\vec{k}|, \tag{9}$$

where $\epsilon = 0, \pm 1$ is the meson polarization. The photon is unaffected. The position of the poles for $\pm$ polarized mesons is moving with wave vector $|\vec{k}|$. Massive vector mesons split into three polarizations with masses $m^2_{V,+} < m^2_{V,L} < m^2_{V,-}$. This splitting unambiguously signifies P breaking. Could this effect be measured?

3. Possible manifestations of P-odd effects in HIC

When trying to understand the nature of the fireball produced in a HIC it is natural to investigate electromagnetic probes such as photons and leptons. In the previous section we have shown that P breaking has substantial effects on the meson spectrum and this could eventually reflect in their leptonic decay products.

Let us then proceed to investigate possible 'anomalies' in dilepton production in various meson decays such as $\rho, \omega \to e^+e^-$. The total dilepton production also receives potential contribution from the pseudoscalar Dalitz decays $\eta, \eta' \to \gamma e^+e^-$ and the ω Dalitz decay $\omega \to \pi^0 e^+e^-$

In fact dilepton production shows a number of anomalies. Perhaps the most obvious one is a surprising enhancement in the low dilepton invariant mass region that has been observed in virtually all experiments for a long time [14, 15, 16]. This excess is to this date unexplained.

Another peculiarity is a large broadening of the ρ spectrum. This was very clearly observed by the NA60 collaboration[17] several years ago by measuring dimuon production around the combined $\rho - \omega$ peak and carefully subtracting contributions from Dalitz decays. Interpreting these results is not easy and it has remained a big puzzle for a long time. On the other hand, the NA60 measuremnt seems to give credence to the idea of 'broadening' for in-medium ρ mesons as opposed to the 'shifting mass' scenario. It is claimed that the broadening can be understood by 'conventional' mechanisms although this involves a certain amount of parameter fitting and it is certainly more of a post-diction than a prediction.

3.1. *Dalitz decays in a P-odd environment*

Having seen that there a number of features in dileptons produced in HIC it is legitimate to ask oneself if local parity breaking could be of some relevance to account for, perhaps in combination with more conventional explanations, some of these effects.

Let us first turn to the modifications induced in the $J = 0$ sector that contributes to dilepton production via a number of Dalitz decays. In what follows we will always compare to the PHENIX results and for that reason use the same set of experimental cuts: $|y_{ee}| < 0.35$, $|\vec{p}_t^{\,e}| > 200$ MeV and gaussian M_{ee} smearing (width=10 MeV), and the effective temperature $T = 220$ MeV.

Using the effective lagrangian briefly described before for $J = 0$ mesons we determine[18] the spectrum (that as described contains states without definite parity) and, given that they mix among themselves strongly, assume that they thermalize in the fireball that now consists mostly of the lightest of such states. After that we proceed to compute the corresponding Dalitz decays. Many more details can be found in X.Planells' Ph D thesis[19].

The results represent a net enhancement to dilepton production. Even though they are clearly insufficient to explain the dramatic enhancement in PHENIX in the region 200-500 MeV, they help vis a vis STAR data. However, the reader should be warned that except for very low values of μ_5 the effective lagrangian so obtained ceases to make sense very soon, a fact that may reflect the need to introduce more resonances. Thus this approach is of limited validity and we do not exclude that a more accurate treatment may actually provide a higher dilepton yield and be par of the solution of the dilepton puzzle in the 200-500 MeV region. We do regard this as a totally open issue at present.

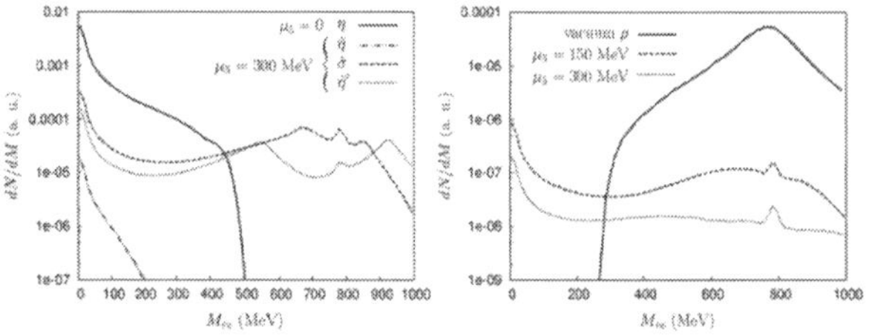

Figure 1. Left: the production from the three new eigenstates is compared to the one from η decays (that is the only relevant one if undistorted vacuum properties are assumed). The latter is depleted but there is a net enhancement at larger invariant masses. Right: the total dilepton production from the discussed mechanism is shown for two values of μ_5 and compared to the vacuum ρ peak as a refernce. However be warned that beyond $\mu_5 \sim 100 MeV$ the effective lagrangian technique shows clear signs of failure.

3.2. *ρ broadening*

P-odd effects introduce broadening in a totally natural manner. Because of the split in masses depending on the polarization, the original Breit-Wigner actually splits in three different peaks. There is a similar effect for the ω meson. It is of course very tempting to compare this 'automatic' broadening with the very precise experimental results obtained by the NA60 collaboration from dimuon production. And indeed the peculiar shape of the ρ spectrum measured by NA60 is grossly reproduced. We used our best guess from the available published data by the NA60 collaboration itself to implement the experimental cuts. As the reader can see, we have not attempted to superimpose the two plots beacuse the NA60 data is not properly acceptance corrected. We have tried hard to obtain more information to make a meaningful comparison but found the collaboration to be unresponsive.

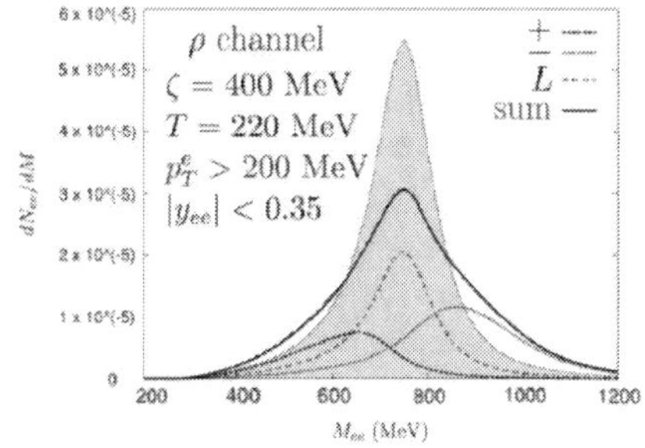

Figure 2. Polarization splitting in ρ spectral function for LPB $\zeta = 400$ MeV ($\mu_5 = 290$ MeV) compared with $\zeta = 0$ (shaded region). Note that the temperature is an effective or 'boosted' one so it is not surprising that it is actually larger than the deconfining temperature.

In any case the comparison is tantalizing. Within the assumption of local parity breaking only one parameter is fitted —the value of μ_5; the best fit is obtained for values of μ_5 close to the constituent mass. Incidentally these values are preferred in NJL type analysis for a stable parity breaking phase to exist[20]

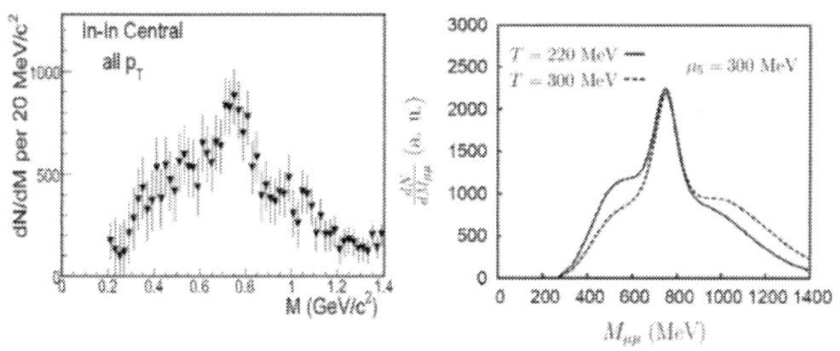

Figure 3. Left: NA60 results for the ρ spectral line-shape as a funtion of the dimuon invariant mass. Right: the analogous quantity obtained from assuming local parity breaking with a value of $\mu_5 = 300$ and two values for the effective temperature

3.3. In-medium $V \to \ell^+\ell^-$ decays

We shall be even briefer here, as this particular point has already been reported and discussed in several conferences. We will now include both the ρ and ω mesons and consider the corresponding distortions in their respective spectra for dilepton production.

The associated enhancement of the dilepton yield could (at least partly) explain the anomalous enhancement seen by PHENIX and STAR.

4. Observables sensitive to P-odd effects

One of the unambiguous signals of P-odd effects is the separation between polarizations. Is there any way to study these decays in order to separate the different polarizations and thus confirm or rule out the presence of local parity breaking in HIC?

It is well known that the angular distribution of leptons carries the information on the polarization of the decaying meson. However, current angular distribution studies are not thought to detect possible P-odd effects.

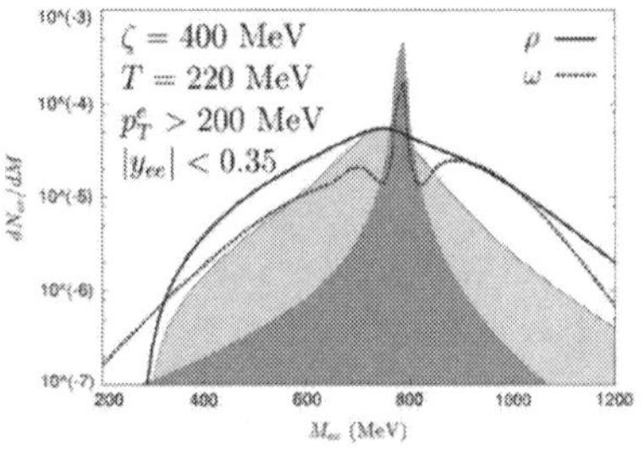

Figure 4. In-medium ρ and ω channels (solid and dashed line) and their vacuum contributions (light and dark shaded regions) for $\mu_5 = 290$ MeV.

We will define two angles as described in Fig. (4). In order to isolate the transverse polarizations, we will perform different cuts for each angle and study the variations of the ρ (and ω) spectral function. A quite visible secondary peak appears in a P-odd medium! Note

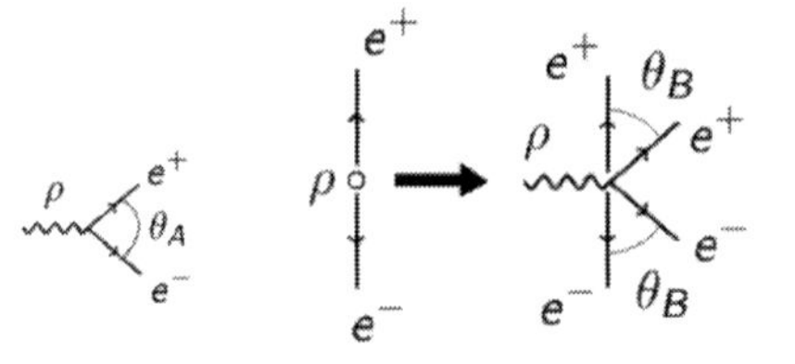

Figure 5. Case A: θ_A is the angle between the two outgoing leptons in the laboratory frame. Case B: θ_B is the angle between one of the two outgoing leptons in the laboratory frame and the same lepton in the dilepton rest frame.

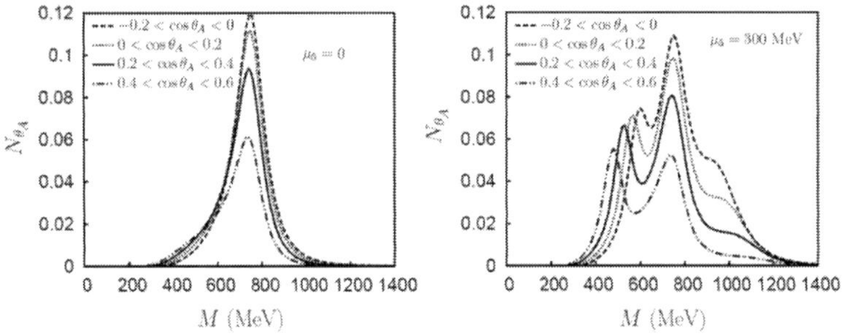

Figure 6. Angle θ_A between the two outgoing leptons in the laboratory frame. ρ spectral function depending on the dielectron invariant mass M in vacuum ($\mu_5 = 0$) and in a parity-breaking medium with $\mu_5 = 300$ MeV for different ranges of θ_A.

however that due to the cuts (needed to make the secondary peak visible) there is an important reduction of the number of events.

If the secondary peak is found for a particular angular coverage, its position would be an unambiguous measurement of the effective or mean value for μ_5. This value of course need not be the same for each HIC. It need not even be uniform. One only needs that domains are sufficiently large for the ρ meson to decay in the medium. The vacuum ρ peak hides the secondary one for $\mu_5 \simeq 100$ MeV due to its large width. For ω, all the peaks are visible. We

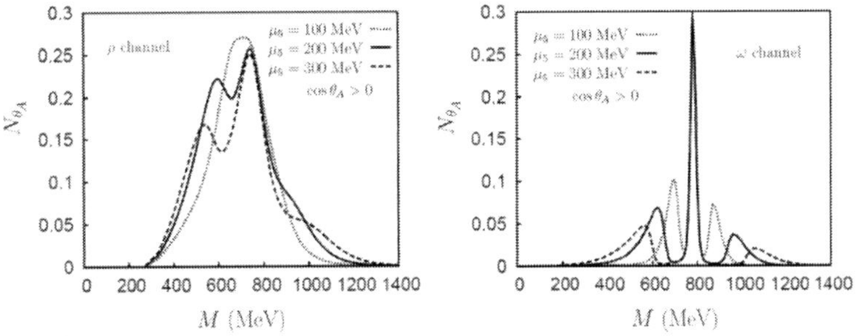

Figure 7. ρ and ω spectral functions depending on the invariant mass M and integrating $\cos\theta_A \geq 0$ for $\mu_5 = 100, 200$ and 300 MeV.

also present the combination of the ρ and ω channels. In this case, the total production is normalized to PHENIX data. It may seem questionable to assume that the ω decay inside the firewall (and this is one reason to carefully separate both contributions). However, local parity breaking should increase substantially the number of omega mesons decaying. For one thing there are the by now familiar parity considerations; parity is no longer a conserved number in such a medium. On the other hand, due to the different dispersion relations a number of ω mesons cannot leave the medium[22] and this obviously increases the probability of a decay inside the hadronic phase of the fireball. The discussion for angle B is rather similar. See [21]

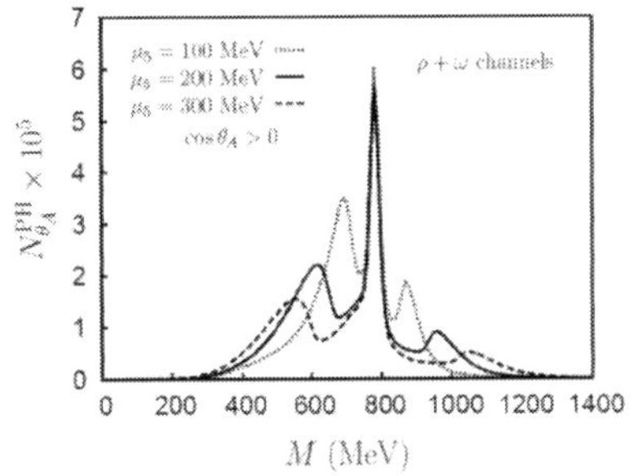

Figure 8. Combination of ρ and ω spectral functions depending on the invariant mass M and integrating $\cos\theta_A \geq 0$ for $\mu_5 = 100, 200$ and 300 MeV.

for more details. The main difference with θ_A is a slightly smaller number of events. The rest of the analysis is completely equivalent. Due to lack of space we do not show the corresponding plots here.

5. Conclusions and outlook

P-odd effects not forbidden by any physical principle in QCD at finite density or high temperature, particularly out of equilibrium. Topological fluctuations transmit their influence to hadronic physics via an axial chemical potential for light quarks only. Consequently in the light meson sector P- breaking leads to unexpected modifications of their in-medium properties. Local parity breaking has an influence on the observed dilepton spectrum in the low-mass region of PHENIX and STAR. We have been able to define some observables may allow us to establish P- and CP-breaking unambiguously.

Perhaps it is fair to end with some criticisms. To begin with, the possibility that domains with a non-vanishing chiral charge form and that they grow to a sufficiently large size is of course unproven. Without this assumption there could be no local parity breaking nor any of the effects discussed in this presentation. However, we think that it may be possible to understand this problem (and give an answer) within the glasma picture in the framework of relativistic hydrodynamics. On the other hand, the situation regarding Dalitz decays is clearly unsatisfactory. We have reasons to think that they may be substantially enhanced (and help explain e.g. the STAR data) but it seems very difficult to study this issue as the effective lagrangian approach as discussed here is not good at all.

All in all, the possibility that signals of parity breaking can be detected in a strong interaction context in HIC is truly fascinating. But detecting it is very challenging. We have proposed here a couple of observables that could possibly yield a signal if local parity breaking is present. Experimental collaborations should definitely check this possibility.

References

[1] Weingarten D *Phys. Rev. Lett.* **51**, 1830 (1983); Vafa C and Witten E *Phys. Rev. Lett.* **53** (1984) 535; Nussinov S *Phys. Rev. Lett.* **52**, 966 (1984); Espriu D, Gross M and Wheater J F *Phys. Lett.* B **146**, 67 (1984); for a review see Nussinov S and Lambert M *Phys. Rept.* **362** (2002) 193.

[2] Migdal A B *Zh. Eksp. Teor. Fiz.* **61** (1971); Lee T D and Wick G C *Phys. Rev.* D **9** , 2291 (1974); Vilenkin A *Phys. Rev.* D **22**, 3080 (1980.

[3] Andrianov A A and Espriu D *Phys. Lett.* B **663**, 450 (2008); Andrianov A A, Andrianov V A and Espriu D *Phys. Lett.* B **678**, 416 (2009).

[4] NICA White Paper, https://indico.cern.ch/event/275003/contribution/1/material/paper/0.pdf.

[5] Kharzeev D E, Pisarski R D and Tytgat M H G *Phys. Rev. Lett.* **81**, 512 (1998).

[6] Buckley K , Fugleberg T and Zhitnitsky A *Phys. Rev. Lett.* **84**, 4814 (2000); Kharzeev D E *Phys. Lett.* B**633**, 260 (2006); Ann. Phys. (NY) **325** (2010) 205; Kharzeev D E, McLerran L D and Warringa H J *Nucl. Phys.* A**803**, 227 (2008).

[7] Andrianov A A, Andrianov V A, Espriu D and Planells X *Phys. Lett.* B **710** (2012) 230.

[8] Buividovich P, Chernodub M, Luschevskaya E and Polikarpov M *Phys. Rev.* D **80**, 054503 (2009); Yamamoto A *Phys. Rev.* D **84** (2011) 114504; *Phys. Rev. Lett.* **107** (2011) 031601.

[9] For a review see e.g. Gelis F *Int. J. Mod. Phys.* A **28** (2013) 1330001.

[10] Fukushima K, Kharzeev D E and Warringa H J *Phys. Rev.* D **78**, 074033 (2008); *Phys. Rev. Lett.* **104** (2010) 212001; *Nucl. Phys.* A **836** (2010) 311; Toneev V D et al. *Phys. Atom. Nucl.* **75** (2012) 607.

[11] Andrianov A A, Andrianov V A, Espriu D and Planells X, arXiv:1010.4688 [hep-ph]; *AIP Conf.Proc.* 1343 (2011) 450-452.

[12] Sakurai J J *Annals Phys.* **11** (1960) 1.

[13] Carroll S M, Field G B and Jackiw R *Phys. Rev.* D **41**, 1231 (1990); Anddrianov A A and Soldati R *Phys. Rev.* D **51**, 5961 (1995); Andrianov A A, Espriu D, Giacconi P and Soldati R *JHEP* **0909**, 057 (2009).

[14] PHENIX Collaboration (Adare A et al.) *Phys. Rev.* C **81**, 034911 (2010).

[15] STAR Collaboration (Adamczyk L et al.) *Phys. Rev. Lett.* **113** (2014) 2.

[16] CERES Collaboration (Agakichiev G. et al.) *Phys. Rev. Lett.* **75** (1995) 1272; *Phys. Lett.* B **422** (1998) 405; *Eur. Phys. J.* C **41** (2005) 475.

[17] Arnaldi R et al.[NA60 Collaboration] *Phys. Rev. Lett.* **96**, 162302 (2006); *Eur. Phys. J.* C **61** (2009) 711.

[18] Andrianov A A, Espriu D and Planells X *Eur. Phys. J.* C **73** (2013) 1, 2294

[19] Planells X arXiv:1411.3283 [hep-ph].

[20] Andrianov A A, Espriu D and Planells X *Eur. Phys. J.* C **74** (2014) 2, 2776.

[21] Andrianov A A, Andrianov V A, Espriu D and Planells X *Phys. Rev.* D **90** (2014) 3, 034024.

[22] Andrianov A A, Kolevatov S S and Soldati R *JHEP* **1111**, 007 (2011).

Results of Beyond-the-Standard Model 2HDM Higgs Searches using the ATLAS detector in Run-I

Guillermo Nicolas Hamity
1 Jan Smuts Avenue, Braamfontein, Johannesburg, 2000, South Africa
E-mail: guillermo.nicolas.hamity@cern.ch

Abstract. One of the most alluring models used to probe Beyond the Standard Model physics in run-I at ATLAS stems from a simple extension of the Higgs sector. Models where the Higgs sector is assumed to have two Higgs doublets, known as two-Higgs-Doublet-Models (2HDMs), give rise to five Higgs bosons: two neutral CP-even (h,H) bosons, two charged ($H^{\pm}$) bosons, and one neutral CP-odd (A) boson. Direct searches for the additional Higgs bosons became an ever more promising prospect after the Higgs boson discovery, and various such searches have already been conducted in Run-I ATLAS data. The most recent 2HDM search is for the A boson in the sensitive Zh decay, where h is assumed to be the LHC discovered Higgs boson, within the mass range of $220 - 1000$ GeV. In such a search, upper limits on the cross section times branching ratio of the A boson decaying to Zh are set for various 2HDM scenarios. Where no excess is observed, exclusion limits are set on ranges of the 2HDM phase-space. In this note the search for $A \to Zh \to \ell\ell\tau\tau$ with fully hadronic tau decay is presented in detail, along with summaries from two of the most recent 2HDM searches in ATLAS.

1. Introduction

The Standard Model (SM) is the centrepiece describing the underlying principles of fundamental particle interactions at high energies. The model combines the fundamental electromagnetic, weak and strong forces, and is able to predict known particle interactions with great precision. The SM has withstood the test of multiple experimental observations, with it's *tour de force* being the recent discovery of the Higgs boson which is predicted in the SM by the ATLAS and CMS experiments at the LHC in Geneva [1, 2]. The discovered Higgs boson was found with a mass around 125 GeV and is thus far consistent with the SM zero-spin and positive-parity hypothesis [3, 4]. It therefore comes as no surprise that present ambiguities with the SM are not phenomenological, but stem rather from the it's 'collage-type' derivation [5]. The repetition of fermionic families and the un-quantification of quark charges suggest that the SM may be a low energy state of some underlying theory. More troublesome is that the SM gives no cause for the imbalance of matter over anti-matter, and excludes the gravitational force and dark matter. Models that describe Beyond the Standard Model (BSM) physics are generally aimed at answering some of these questions. In this light, searching for inconsistencies in the SM is crucial to searching for BSM phenomena.

One of the most basic extensions to the SM is by including an additional Higgs doublet in it's derivation [6]. The SM Higgs sector is only comprised of a simple scalar structure. When considering that the fermionic structure has multiple families and mixings, this assumption need no be the case. The extended models with two Higgs doublets are intuitively referred to as 2-Higgs-Doublet-Models (2HDMs). They include the Minimal Supersymmetric Standard Model (MSSM), an important candidate of Supersymmetry (SUSY), as well as axion [7] and baryogenesis [8] models. After including an additional Higgs doublet, as well as a few intuitive assumptions [6], 2HDMs give rise to five 'Higgs' bosons. There are two CP-even bosons, the light h and heavy H, distinguishable only by their mass, where $m_h < m_H$. Then there are two charged Higgs bosons, denoted as $H^{\pm}$, and a single CP-odd pseudoscalar, the A boson. One should note that the h and H bosons are candidates for the discovered boson, due to their zero-charge and even-parity nature. This allows for 2HDM searches where the mass of one of the CP-even bosons is that of the discovered Higgs. Generally in 2HDM searches one assumes that $m_h = 125$ GeV, although it is still feasible that the light Higgs lives somewhere in the $m_h < 125$ GeV range. Another common assumption made is that the remaining Higgs bosons have approximately equal masses $m_A \approx m_{H^{\pm}} \approx m_H$ [9]. With these assumptions the 2HDM phasespace is reduced considerably to three free parameters: the mass of one of the heavier Higgs bosons ($e.g\ m_A$), the mixing angle of the two CP-even Higgs bosons, α, and the ratio of the vacuum expectation values of the two Higgs doublets, β. The last two parameters are generally represented as $\tan\beta$ and $\cos(\beta-\alpha)$ for convenience. In direct searches for the 2HDM bosons, all thee parameters are allowed to vary. Furthermore the predicted couplings of the bosons to the vector bosons and fermions fluctuate in terms of $\tan\beta$ and $\cos(\beta-\alpha)$. An additional freedom in terms of the way that each Higgs doublet couples to vector bosons and fermions gives rise to four different 2HDM types. The ratio of the predicted coupling scale factors for the different 2HDM scenarios of the h boson to the SM Higgs boson couplings is given in table 1. Coupling ratios to the vector bosons (κ_V), up-type quarks (κ_u), down-type quark (κ_d), and leptons (κ_ℓ) are shown to vary as functions of β and α.

Table 1. Ratio of predicted coupling scale factors to SM particles of the four 2HDM h bosons to the predicted SM Higgs couplings [13].

Coupling scale factor	Type-I	Type-II	Type-III	Type-IV
κ_V	$\sin(\beta-\alpha)$	$\sin(\beta-\alpha)$	$\sin(\beta-\alpha)$	$\sin(\beta-\alpha)$
κ_u	$\cos(\beta)/\sin(\alpha)$	$\cos(\beta)/\sin(\alpha)$	$\cos(\beta)/\sin(\alpha)$	$\cos(\beta)/\sin(\alpha)$
κ_d	$\cos(\beta)/\sin(\alpha)$	$-\sin(\alpha)/\cos(\beta)$	$\cos(\beta)/\sin(\alpha)$	$-\sin(\alpha)/\cos(\beta)$
κ_ℓ	$\cos(\beta)/\sin(\alpha)$	$-\sin(\alpha)/\cos(\beta)$	$-\sin(\alpha)/\cos(\beta)$	$-\sin(\alpha)/\cos(\beta)$

Direct searches for 2HDM boson set upper limits on the cross section times decay branching ratio of a specific final state channel. In this way we are able to search for the existence of the 2HDM bosons and subsequent traces of BSM physics. As a bi-product of these studies one is able to exclude the existence of 2HDMs from certain areas of the m_A, $\tan\beta$, and $\cos(\beta-\alpha)$ phasespace if no sufficient excess above the upper limits is observed. Various searches have already been conducted, including the CMS searches for $H \to hh$ and $A \to Zh$ between $260 - 360$ GeV [10], and an independent search for $A \to Zh \to \ell\ell b\bar{b}$ [11]; and the $A \to \tau\tau$ search with ATLAS [12]. No excess was measured in the studies and large sections of the 2HDM phasespace were excluded. Indirect exclusion limits have also been placed on 2HDMs in the ($\tan\beta$, $\cos(\beta-\alpha)$) plane by interpreting the observed couplings of the discovered Higgs boson to other particles [13]. The ($\tan\beta$, $\cos(\beta-\alpha)$) phasespace is scanned for points where the measured Higgs couplings disagree with the 2HDM predictions, at 95% CL or greater, so as to exclude regions in ($\tan\beta$, $\cos(\beta-\alpha)$). There are however areas of phasepace for 2HDMs which have not been probed due to the low branching ratios of the various topologies, a task set for higher luminosity searches in Run-II.

2. The ATLAS detector

The LHC is currently the largest man-made particle accelerator in the world, capable of proton-proton (pp), heavy ion (lead-proton (Pb-p) and lead-lead (Pb-Pb)) collisions. It has a circumference of 27 km, a design center-of-mass energy of 14 TeV, and peak instantaneous luminosity of 10^{34} cm^{-2}s^{-1} for pp collisions. The ATLAS detector is one of the two general-purpose detectors at the LHC [14]. It was designed to accommodate a wide spectrum of different physics signatures and explore processes stemming from the TeV mass scale, where groundbreaking discoveries are expected. The ATLAS detector is a multi-purpose particle detector with approximately forward-backward symmetric cylindrical geometry. ATLAS uses a right handed coordinate system with the origin at the interaction point. The z-axis is along the beam pipe, x-axis points towards the center of the LHC ring and the y-axis points up. Polar coordinates (r, ϕ) are used in the transverse plane, ϕ being the azimuthal angle around the beam pipe. The pseudorapidity η is defined as $\eta = -\ln(\tan(\theta/2))$ where θ is the angle between the particle momentum 3-vector and the beam axis. The inner detector (ID) covers $|\eta| < 2.5$ and consists of silicon pixels, silicon micro-strips and a transition radiation tracker. The ID is surrounded by a superconducting solenoid providing a 2 T magnetic field. Surrounding the ID are two calorimeters, the inner electromagnetic (EM) and the outer hadron calorimeters. The EM calorimeter measures the energy and the position of EM showers within $|\eta| < 3.2$ as well as hadronic showers in the end-cap ($1.5 < |\eta| < 3.2$) and forward ($3.1 < |\eta| < 4.9$) regions. The hadronic calorimeter measures hadronic showers in the central region ($|\eta| < 1.7$). The muon spectrometer surrounds the calorimeters and consists of three subsystems: eight superconducting air-core toroid magnets, a system of tracking chambers and a fast tracking chamber for triggering. A three-level trigger system selects events to be recorded for offline analysis.

3. The $A \to Zh \to \ell\ell\tau_{had}\tau_{had}$ analysis

The $A \to Zh$ channel with the h decaying to $\tau\tau$ and Z to $\ell\ell$ is a new search that promises good sensitivity. In particular this channel probes areas of the Type-II and Type-IV 2HDMs which have not been accessible in previous searches. In this topology three orthogonal final states are available, depending on the either leptonic or hadronic decay of each tau lepton. They are $\tau_{lep}\tau_{lep}$, $\tau_{lep}\tau_{had}$ and $\tau_{had}\tau_{had}$. The production of the pseudoscalar A boson is predominantly driven by gluon-fusion. Contributions from b-associated production only become relevant for Type-II and Type-IV 2HDMs at high values of $\tan\beta$, dominating over gluon-fusion at around $\tan\beta = 10$ [9]. In most cases the ratio of b-associated A boson production to that of gluon-fusion is restricted to $< 3\%$. Two leading order Feynman diagrams of the gluon-fusion and b-associated A productions are shown in figure 4.

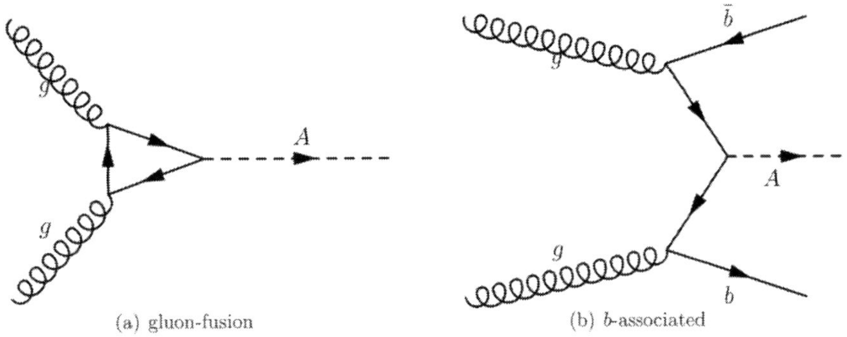

Figure 1. Feynman diagrams for (a) gluon-fusion and (b) b-associated A boson production.

Generally speaking, $A \to Zh$ is the dominant decay mode of A in any 2HDM when the mass of the A is less than two times the mass of the top-quark ($m_h + m_Z < m_A < 2m_t$). This makes the $A \to Zh$ channel a very promising search. At masses of $m_A > 2m_t$ the branching ratio BR($A \to Zh$) drops drastically due to the exposed $t\bar{t}$ kinematic region. In this paper the results from the search of the gluon-produced A boson decaying to $Zh \to \ell\ell\tau_{had}\tau_{had}$ are presented in the range of $220 < m_A < 1000$ GeV. The results are combined with the $\tau_{lep}\tau_{lep}$ and $\tau_{lep}\tau_{had}$ searches and finally interpreted in the various 2HDM scenarios in conjunction with a complementary $A \to Zh \to ffb\bar{b}$ analysis, where $f = \ell$, or ν.

A search for $A \to Zh(\tau\tau)$ is conducted with the 2012 ATLAS data at 8 TeV center of mass energy, where the light h boson is assumed to have a mass of $m_h = 125$ GeV. Signal Monte Carlo (MC) samples are produced with Pythia8 [15] for A masses at fixed working points: $m_A = 220, 240, 260, 300, 340, 350, 400, 500, 800, 1000$ GeV. As mentioned in the introduction, each tau in the analysis can decay leptonically (35%) or hadronically (65%). Since tau leptons are massive particles, they decay quickly in the detector and are never observed directly in ATLAS, but can only be reconstructed from their decay products. This makes differentiating light-leptons originating from leptonic-tau-decays to light-leptons originating from other processes virtually impossible. As such, tau leptons in the ATLAS detector are considered to be hadronically decaying taus. Tau leptons decay hadronically into either 1$-$ or 3$-$charged pions, with a ν_τ and possibly a neutral pion. Since the BR($A \to Zh$) is already relitavely small, and since the SM branching ratio BR($h \to \tau\tau$) is around 1000 times larger that BR($h \to \mu\mu$), searches in the $\tau_{lep}\tau_{lep}$ and $\tau_{lep}\tau_{had}$ channels can be conducted in $\tau\tau$ without worrying about contamination from $h \to \ell\ell$ ($\ell \neq \tau$). In all three analysis stradegies, the reconstructed A boson mass, m_A, is considered as the parameter of interest.

The fully hadronic final state is a particularly challenging analysis since it contains a large amount of missing transverse energy, E_T^{miss}, from the two neutrinos. In the case of $\tau_{had}\tau_{had}$, full reconstruction of the event topology requires solving a system of equations with 6 unknowns. This system is under-constrained and an exact solution is not possible. Therefore, the full event reconstruction, as well as the full reconstruction of the $m_{\tau\tau}$ mass, are impossible. The Missing Mass Calculator (MMC) method is therefore used to reconstruct $m_{\tau\tau}$ [16]. This method scans for the most likely $m_{\tau\tau}$ solution given the known conditions of the event, *i.e.* the visible taus and E_T^{miss} kinematics. The MMC is compared to alternate $m_{\tau\tau}$ reconstruction methods: the transverse mass and collinear approximation (CLA) methods [16], performing at a better efficiency and resolution than the CLA method.

The event selection in the $\tau_{had}\tau_{had}$ channel focuses on selecting a clean signal with exactly two leptons (e, or μ) and two visible tau jets. Events are triggered by single lepton triggers. The events are required to have exactly two same flavour opposite charge loose leptons with a combined invariant mass of $80 < m_{\ell\ell} < 100$ GeV, which corresponds to $m_Z = 90$ GeV. Track and calorimeter track isolation requirements are also imposed on the leptons. The event must also have exactly two opposite charge loose tau leptons, which need to pass electron and muon vetos, and have a MMC combined invariant mass of $75 < m_{tautau}^{MMC} < 175$ GeV, corresponding to $m_h = 125$ GeV. The large window here compensates for the resolution of the MMC algorithm. Two additional selection conditions are applied after an optimization study performed on different kinematic variables. The first of these is that the tau lepton with highest p_T should have a transverse energy of $E_T > 35$ GeV. Finally, events where the reconstructed di-lepton p_T is smaller than $0.64 \times m_A - 131$ GeV are discarded. This selection is applied to all events with a reconstructed A mass less than 400 GeV.

After the full event selection, the $\tau_{had}\tau_{had}$ analysis has a few surviving background processes, the most prominent of which is the Z+jets background. This background contains jets which are mis-identified as tau leptons. A small contribution from WZ di-boson processes also contributes to the fake background. Other important backgrounds include di-boson ZZ and

SM ZH production, which are comprised of true tau signatures, and are estimated from MC. All backgrounds caused by mis-identified jets are estimated via a data-driven template method prediction. This method is performed by constructing three orthogonal control regions. The first region is defined identically to the full signal selection, except that the leading tau is forced to fail the loose tau selection. The second region requires that the tau leptons have the same charge. The third region combines both the inverted loose tau and same tau charge requirements. When looking at data in these regions, the control regions are completely populated by fake tau jets. Data distributions in the control regions are extrapolated into the signal region and used to estimate the mis-identified tau background. The distribution is normalized using m_A sidebands.

Systematic uncertainties due to the fake background normalization are considered by taking into account the fluctuations in normalization values when choosing alternate control regions. This uncertainty constitutes the largest uncertainty, approximately 25%. Uncertainties due to detector calibration effects, and theoretical uncertainties are also included. The m_A distribution of the $\tau_{had}\tau_{had}$ channel after full event selection is given in figure 2, showing the predicted backgrounds, the 340 GeV mass signal signal with a cross section times branching ratio of 50 fb, and the observed data.

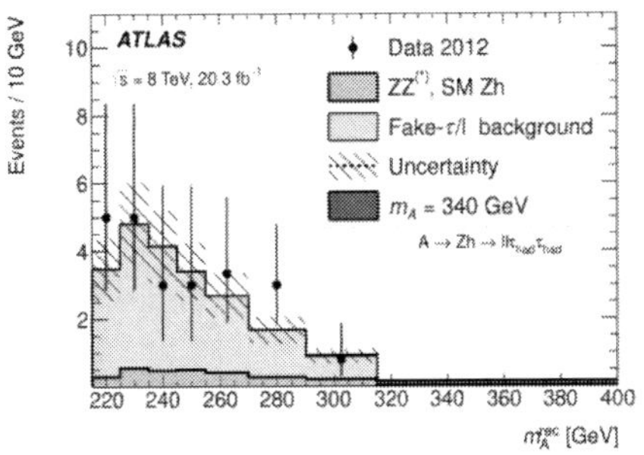

Figure 2. The reconstructed A boson mass, m_A of the $\tau_{had}\tau_{had}$ channel after full event selection [17].

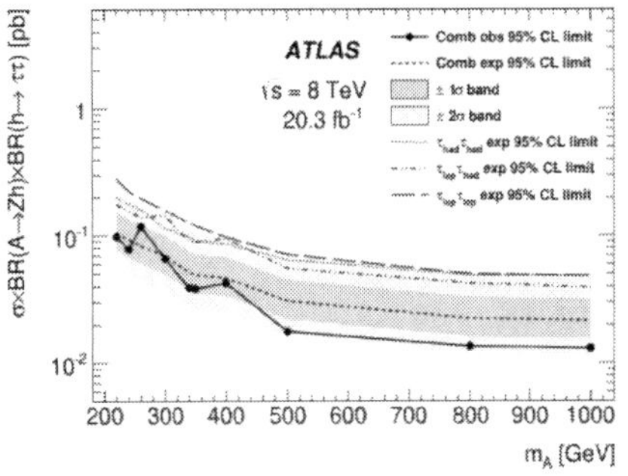

Figure 3. Expected and observed 95% CL upper limits of the gluon-fusion cross section times branching ratio of the combined $A \to Zh \to \ell\ell\tau\tau$ [17].

A binned likelihood ratio test, along with an asymptotic approximation, is used in order to plot the 95% CL upper limits on the cross section times branching ratio of $\ell\ell\tau\tau$. Figure 3 shows the expected and observed upper limits of gluon-fusion $\sigma \times \mathrm{BR}(A \to Zh) \times \mathrm{BR}(h \to \tau\tau)$ for the combined $A \to Zh \to \ell\ell\tau\tau$ analysis. No assumptions are made on the branching ratio of $h \to \tau\tau$ in order to maintain the result model independent. The observed combined upper limit is within 2σ of the expected limit. The expected 95% CL for the $\tau_{lep}\tau_{lep}$, $\tau_{lep}\tau_{had}$ and $\tau_{had}\tau_{had}$ are shown in dashed lines. The drop in the limits at higher values are from the lack of data observed in the highest m_A bins. The 95% CL upper limits on the gluon-fusion $\sigma \times \mathrm{BR}(A \to Zh) \times \mathrm{BR}(h \to \ell\ell\tau\tau)$ range from 0.098–0.013 pb in a range of $220 \leq m_A \leq 1000$ GeV. No excess is observed outside the 2σ. The low observed limits at the higher mass points is caused by a lack of events in the final 315-2000 GeV m_A^{rec} bin.

The final limits are combined with limits from a complementary $A \to Zh \to ff b\bar{b}$ analysis, where $f = \ell, \nu$ [17]. The results are presented as model independent limits on the $A \to Zh$ cross section times branching ration and no excess is observed outside of the uncertainties. They are interpreted in the context of the four 2HDM scenarios us in order to exclude sections of the phasespace. The final paper includes the first published search for an A boson in the $\ell\ell\tau\tau$ channel.

4. Other recent 2HDM searches with ATLAS in Run-I

4.1. $H^{\pm}$ to τ+jets

A search for a charged Higgs boson decaying via $H^{\pm} \to \tau^{\pm}\nu$ is conducted in the hadronic final state at 8TeV center of mass energy pp collisions with an integrated luminosity of 19.5fb^{-1} [18]. The search is conducted between the mass intervals of $m_{H^+} \in (80\,\text{GeV}, 160\,\text{GeV})$ and $m_{H^+} \in (180\,\text{GeV}, 1\,\text{TeV})$, split by the top quark mass ($m_t \approx 173.3$ GeV) around which the physics processes are not well understood by MC. In the low mass region ($m_{H^+} < m_t$) the H^+ production is dominated by di-top production via gluon-fusion, where one t-quark decays to H^+ and a b-quark jet, while the other decays to W^-b producing additional jets in the final state. In the high mass region ($m_{H^+} > m_t$) H^+ is produced via t-quark associated production, with at least one b-quark jet in the final state.

Event selection in the $H^{\pm}$ to τ+jets search optimizes on events with the following criteria in the final state: one hadronically decaying tau; a large amount of E_T^{miss}; two quark jets from hadronic W^- decay; and at least one(two) b-jets for the high(low) mass case. The backgrounds originate from processes with either a true or a mis-identified tau lepton. The true tau background is predicted by using an embedding method in μ+jets data, and replacing the μ leptons with true MC tau leptons which are allowed to decay hadronically though the detector. The amount of background originating from mis-identified jets is predicted by calculating the identification efficiency of taus in control regions. A small amount $(1-2\%)$ of background from mis-identified e and μ leptons is taken from MC. The final discriminating variable is chosen to be the H^+ transverse mass.

The 95% CL expected limits on the cross section and/or branching ratios of $H^+ \to \tau^+\nu$ are derived using an asymptotic approximation. Limits in the low mass region on $B(t \to bH^+) \times B(H^+ \to \tau^+\nu)$ via di-top production are found to be between 0.23% and 1.3%. For the high mass range the limits on the cross section times branching ratio, $\sigma(pp \to tH^+ + X) \times B(H^+ \to \tau^+\nu)$, are between 0.76 pb and 4.5 fb. No Excess is observed and the observed limits agree with the SM predictions within systematic uncertainties.

4.2. Di-Higgs resonant searches

In the SM, direct di-Higgs production and the Higgs self coupling to di-Higgs have very low cross sections. In the best case, gluon-fusion produced di-Higgs has a predicted cross section of around 10 fb at 8TeV [19]. These processes where therefore not accessible during the Run-I data taking period at ATLAS. The measurement of the di-Higgs production cross section will be one of the main tasks in Run-II at higher center of mass energy. However, in the light of 2HDMs the cross section of di-Higgs production may be significantly boosted [20]. Depending on the chosen 2HDM parameters, two light Higgs may originate from a heavy Higgs boson with a cross section times branching ratio of up to a few pb. This gave rise to various searches in ATLAS during Run-I for hh production. One such search is a model independent non-resonant/resonant search for $hh \to bb\gamma\gamma$ [21]. The search observed 5 events from an expected 1.5 events, 1.3 ± 0.5 of which came from fitted background predictions, while 0.17 ± 0.04 came from SM Higgs production. Figure 4 shows the expected and observed 95% CL limits on the resonant cross section times branching ratio, $\sigma_X \times \text{BR}(X \to hh)$. The expected cross section times branching ratio of a Type-I 2HDM with $\tan\beta = 1$ and $\cos(\beta - \alpha) = -0.05$ is superimposed for comparison. The fluctuation at $m_X = 300$ GeV corresponds to a fluctuation of 3σ. When considering the look elsewhere effect this is brought down to 2.1σ. Currently the di-Higgs production can not be excluded around the $m_X = 300$ GeV mass, and a $\sigma_X \times \text{BR}(X \to hh)$ analysis will be conducted at a higher center of mass energy when the LHC reopens in Run-II.

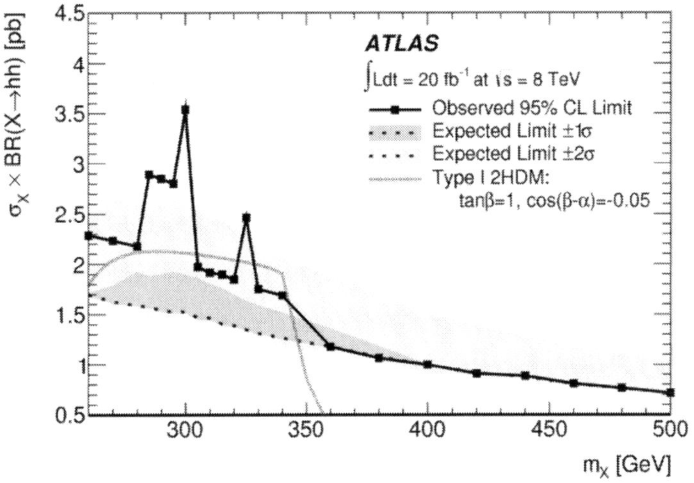

Figure 4. Expected and observer 95% CL upper limits on the cross section times branching ratio of a resonant search for di-Higgs production in the $bb\gamma\gamma$ final state [21]. An arbitrary Type-I 2HDM scenario is supper-imposed for comparison.

5. Conclusion

A few 2HDM searches which where conducted during Run-I data taking at ATLAS where summarized. The focus is place on the most recent search of the pseudoscalar Higgs in the $A \to Zh$ channel. The analysis techniques in the fully hadronic channel are presented in detail. Further 2HDM motivated searches for the charged ($H^+ \to \tau^+\nu$) and heavy Higgs ($X \to hh$) bosons are also presented, none of which show any conclusive evidence for the existence of the bosons. The analysis strategies include placing 95% CL upper limits on their cross sections and/or branching ratios. Where no excess is observed, regions of the 2HDM phase-space are excluded. The list of 2HDM searches in ATLAS is not exhausted here and readers are encouraged to visit results from $h/A/H \to \tau\tau$ [22] and $hh \to bbbb$ [23] searches, to name but a few. These ATLAS searches will continue at higher center of mass energies in Run-II in order to place greater restrictions on the 2HDM phase-space, or alternitavely reveal some BSM physics phenomena.

References

[1] ATLAS Collaboration, 2013 Measurements of Higgs boson production and couplings in diboson final states with the ATLAS detector at the LHC *Phys.Lett. B* **726** 88
[2] CMS Collaboration, 2013 Combination of standard model Higgs boson searches and measurements *CMS-PAS-HIG-13-005*
[3] ATLAS Collaboration, 2013 Evidence for the spin-0 nature of the Higgs boson using ATLAS data *Phys.Lett. B* **726** 120
[4] CMS Collaboration, 2013 Study of the Mass and Spin-Parity of the Higgs Boson Candidate Via Its Decays to Z Boson Pairs *Phys.Rev.Lett.* **110** 081803
[5] Langacker P, 1981 Grand unified theories and proton decays *Phys.Rev.Lett.* **72** 185
[6] Branco G C, *et al.*, 2012 Theory and phenomenology of two-Higgs-doublet models *Phys.Rept.* **516** 1
[7] Kim J E, 1987 Light Pseudoscalars, Particle Physics and Cosmology *Phys.Rept.* **150** 1
[8] Joyce M, Prokopec T and Turok N, 1994 Nonlocal electroweak baryogenesis. Part 2: The Classical regime *Phys.Rev.* **D53** 2958
[9] ATLAS Collaboration, 2013 Beyond-the-Standard-Model Higgs boson searches at a High-Luminosity LHC with ATLAS *ATLAS-PUB-2013-016*

[10] CMS Collaboration, 2013 Search for extended Higgs sectors in the H to hh and A to Zh channels in $\sqrt{(s)} = 8$ TeV pp collisions with multileptons and photons final states *CMS-PAS-HIG-13-025*

[11] CMS Collaboration, 2014 Search for a pseudoscalar boson A decaying into a Z and an h boson in the llbb final state *CMS-PAS-HIG-14-011*

[12] ATLAS Collaboration, 2014 Search for neutral Higgs bosons of the minimal supersymmetric standard model in pp collisions at $\sqrt{s} = 8$ TeV with the ATLAS detector *J. High Energy Phys.* JHEP11(2014)056

[13] ATLAS Collaboration, 2014 Constraints on New Phenomena via Higgs Coupling Measurements with the ATLAS Detector *ATLAS-CONF-2014-010*

[14] CERN, 1999 ATLAS: Detector and physics performance technical design report. Volume 1 *CERN-LHCC-99-14*

[15] Sjostrand T, Mrenna S and Skands P, 2006 PYTHIA 6.4 Physics and Manual *JHEP* **0605** 026

[16] Elagin A, Murat P, Pranko A and Safonov A, 2011 A New Mass Reconstruction Technique for Resonances Decaying to di-tau *Nucl.Instrum.Meth.* **A654** 481

[17] ATLAS Collaboration, 2015 Search for a CP-odd Higgs boson decaying to Zh in pp collisions at $\sqrt{s} = 8$ TeV with the ATLAS detector *CERN-PH-EP-2015-013*

[18] ATLAS Collaboration, 2014 Search for charged Higgs bosons decaying via $H^{\pm} \to \tau^{\pm}\nu$ in hadronic final states using pp collision data at $sqrts = 8$ TeV with the ATLAS *ATLAS-CONF-2014-050*

[19] Baglio J, and Djouadi A, Grober R, *et al.*, 2013 The measurement of the Higgs self-coupling at the LHC *JHEP 1304 (2013) 151*

[20] Nathaniel C, and Jamison G, and Scott T, 2013 Searching for Signs of the Second Higgs Doublet *RU-NHETC-2013-07*

[21] ATLAS Collaboration, 2014 Search For Higgs Boson Pair Production in the $\gamma\gamma b\bar{b}$ Final State using pp Collision Data at $\sqrt{s} = 8$ TeV from the ATLAS Detector *HIGG-2013-29*

[22] ATLAS Collaboration, 2014 Search for neutral Higgs bosons of the minimal supersymmetric standard model in pp collisions at $\sqrt{s} = 8$ TeV with the ATLAS detector *JHEP11(2014)056*

[23] ATLAS Collaboration, 2014 A search for resonant Higgs-pair production in the $b\bar{b}b\bar{b}$ final state in pp collisions at $\sqrt{s} = 8$ TeV *ATLAS-CONF-2014-005*

Journal of Physics: Conference Series **623** (2015) 012014 doi:10.1088/1742-6596/623/1/012014

Performance of the ATLAS Tile Calorimeter

Louise Heelan, on behalf of the ATLAS Collaboration
University of Texas at Arlington
E-mail: louise.heelan@cern.ch

Abstract. The ATLAS Tile hadronic calorimeter (TileCal) provides highly-segmented energy measurements of incoming particles. It is a key detector for the measurement of hadrons, jets, tau leptons and missing transverse energy. It is also useful for identification and reconstruction of muons due to good signal to noise ratio. The calorimeter consists of thin steel plates and 460,000 scintillating tiles configured into 5000 cells, each viewed by two photomultipliers. The calorimeter response and its readout electronics is monitored to better than 1% using radioactive source, laser and charge injection systems. The calibration and performance of the calorimeter have been established through test beam measurements, cosmic ray muons and the large sample of proton-proton collisions acquired in 2011 and 2012. Results on the calorimeter performance are presented, including the absolute energy scale, timing, noise and associated stabilities. The results demonstrate that the Tile Calorimeter has performed well within the design requirements and it has given essential contribution to reconstructed objects and physics results. In addition, the data quality procedures used during the LHC data-taking are described and the outcome of the detector consolidation in the maintenance period is also presented.

1. The Tile calorimeter in ATLAS

ATLAS [1] is a general purpose detector at one of four collision sites at the Large Hadron Collider (LHC) at CERN, Geneva, Switzerland. The experiment was designed to search for new physics with a wide range of final states, including signatures with missing transverse momentum, jets and leptons. The Tile calorimeter (TileCal) is the hadronic calorimeter in the central region of the detector, a sampling calorimeter composed of alternating pieces of plastic scintillator and steel absorber [2, 3]. To achieve the physics goals desired by ATLAS the energy resolution of the TileCal is expected to be: $\sigma/E = 50\%/\sqrt{E} \oplus 3\%$. In addition the response is expected to be linear within 2% for jets up to 4 TeV.

As shown in Figure 1, the TileCal is composed of three physical volumes; a long barrel (LB) that spans $|\eta| < 1.0$, and two extended barrels each spanning $0.8 < |\eta| < 1.7$. Between the LB and EB there are gap and crack scintillators to help reconstruct energy otherwise lost in these transition regions. The TileCal provides full azimuthal coverage by further segmenting each partition into 64 wedge-shaped modules each spanning $\Delta\phi = 0.1$. There are three radial layers, with layer A closest to the beam line, followed by layers BC and D moving radially outward, as can be seen in Figure 2.

Light generated in the scintillating tiles is collected on two sides of the tile and transmitted via wavelength shifting fibers to photomultiplier tubes (PMTs) located together with the front-end electronics in long mechanical drawers on the outer radius of each module. The signal from one side of the tile is read by one PMT, connected to one electronic channel. Two channels,

Journal of Physics: Conference Series **623** (2015) 012014 doi:10.1088/1742-6596/623/1/012014

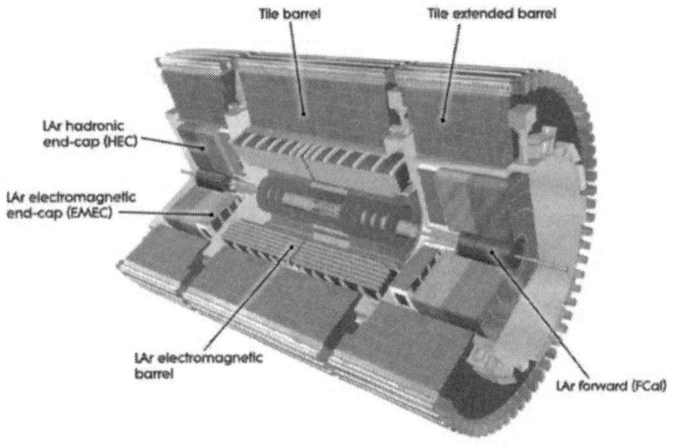

Figure 1. The calorimeter subsystems of the ATLAS detector. Shown is the Liquid Argon (LAr) calorimeter system, and the Tile calorimeter barrel and extended barrel partitions. The total coverage provided by the ATLAS calorimeters extends to $|\eta| < 4.9$ [1].

which collect light from either side of the scintillator tile, are read out as one unit called a cell. There are a total of approximately 10k channels (or equivalently approximately 5k cells).

The analog signal of the PMT is processed by a so-called 3-in-1 card which provides signal shaping and amplification in bi-gain mode (with gain ratio of 1:64). The dynamic range covers energy deposits from 10 MeV to 800 GeV, where the upper limit corresponds to the low gain saturation, and the lower limit to the high gain precision (1 ADC counts). After amplification the signal from each gain is sent to a digitizer where the pulse is sampled every 25 ns, in sync with the LHC clock cycle. If the event is accepted by the trigger these samples are sent to the Read-out Driver (ROD) located with the off-detector back-end electronics where data integrity checks are made, along with real-time reconstruction of the signal.

2. Signal reconstruction and calibration

The TileCal recorded almost 30 fb^{-1} of data during Run 1 at the LHC. The digital data from each channel, sampled every 25 ns, were reconstructed using the Optimal Filtering (OF) algorithm [5].

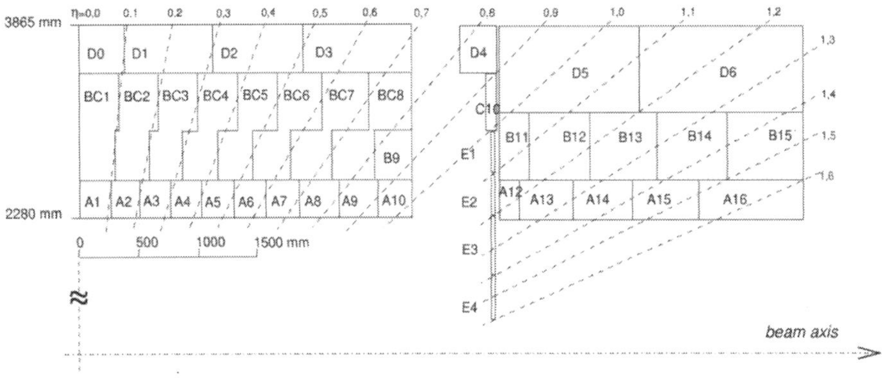

Figure 2. The TileCal cell layout for $\eta > 0$; the geometry is mirrored for $\eta < 0$.

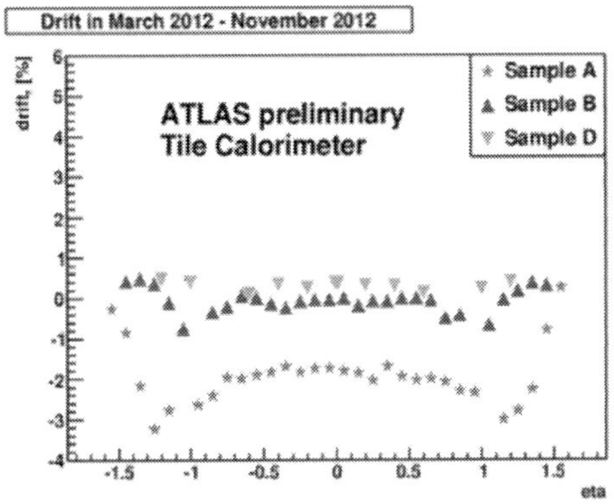

Figure 3. Change in response of cells to the cesium calibration source (in %) as a function of cell η from March 2012 to November 2012, corresponding to the 2012 LHC proton collision run. Sample refers to the radial layer, with A being the inner-most layer.

The channel amplitude, A, and time, τ, were reconstructed according to:

$$A = \sum_{i=1}^{n=7} a_i S_i \quad , \qquad A\tau = \sum_{i=1}^{n=7} b_i S_i \tag{1}$$

where the weights a_i and b_i were derived using the known pulse shape and sample autocorrelation matrix. The weights were applied based on the expected time of the pulse peak (where $\tau = 0$). The phase τ was measured with respect to time-of-flight of the particle. For events where $\tau \neq 0$ the reconstructed amplitude underestimated the true amplitude by a known function, and hence a correction was applied to recover the true pulse amplitude. Within a phase $\tau \pm 10$ ns the average difference in amplitude after correction was seen to be $< 1\%$.

A series of calibration constants, C_i, were used to the convert the reconstructed channel peak amplitude in ADC-counts (A) to energy (E):

$$E = A \cdot C_{\mathrm{ADC}\to\mathrm{pC,CIS}} \cdot C_{\mathrm{pC}\to\mathrm{GeV,TB}} \cdot C_{\mathrm{Cs}} \cdot C_{\mathrm{laser}} \tag{2}$$

The constant $C_{\mathrm{pC}\to\mathrm{GeV,TB}} = 1.05$ pC/GeV, was the global electromagnetic scale of the TileCal, set using 11% of the modules in a test beam of electrons [4]. The other three calibration constants, $C_{\{\mathrm{CIS,Cs,laser}\}}$ were derived using dedicated calibration systems, and maintained this electromagnetic scale across the calorimeter.

The conversion between ADC counts and pC, represented by $C_{\mathrm{ADC}\to\mathrm{pC,CIS}}$, was made for each channel using a charge injection system, where a known charge was injected into the 3-in-1 cards and the response was measured. The injected charge spanned a range of values scanning both the high gain and low gain range. This system was also used to monitor the front-end electronics and correct for non-linearities. The charge injection calibration was performed twice per week, and over the LHC Run 1 showed the front-end electronics were extremely stable. In 2012 the variation of the charge injection response was 0.04% across all channels.

The constant C_{Cs} was extracted using a ^{137}Cs source that was moved through every tile, emitting 0.662 MeV photons to illuminate the scintillator. This system and associated constant were used to calibrate the scintillators, PMTs, and correct for residual cell differences. Cesium calibrations were taken monthly and were shown to have a precision of 0.3%. Figure 3 shows the change in response of the cells to the cesium source as a function of η for each radial layer as a function of time in 2012. Cell A13 ($|\eta| = 1.3$) was the cell most exposed to radiation from proton collisions (from the cells shown in the plot), and displayed a maximum variation of 3.5% across the 2012 LHC run period.

The final calibration constant in Equation 2, C_{laser}, was calculated using a laser calibration that system sent pulses of light directly to the PMTs. It was used to monitor and measure the individual PMT gain variations between the monthly cesium scans. Laser calibrations were typically taken twice per week and during collision data-taking laser pulses were sent to the PMTs in LHC bunch crossings without protons to monitor channel timing. The stability of the calibration system was better than 0.5% over the period of one month.

Special care was taken to calibrate the gap and crack scintillators, cells E1-E4 in Figure 2, as they were exposed to the most amount of radiation of any TileCal cells. The cesium source could not enter the crack scintillators, cells E3 and E4. However using the laser calibration system the cell E4 exhibited a variation of 8% in response over the 2012 period.

All of these variations in response were monitored with the respective calibration systems and corrected such that the global electromagnetic scale of the calorimeter was maintained.

The initial channel time was set using laser events, where all channels were illuminated at the same time, and with single LHC beam events. In beam "splash" events a single LHC proton beam was incident on a collimator 150 m upstream of the ATLAS detector. The result of this impact was a spray of many high energy particles depositing significant amounts of energy in all channels. Using the beam splash events taken in 2011 and after correcting for particle time of flight the reconstructed cell times across the entire calorimeter were found to be within 0.5 ns of each other. This timing precision was verified using jets from proton collisions.

3. Run 1 operations

Table 1 shows a summary of the colliding proton beam conditions delivered to ATLAS by the LHC during Run 1. In 2011 and 2012 the bunch spacing between beam bunches filled with protons was 50 ns, so out-of-time pileup effects were visible in the calorimeter due to the concomitant presence of signals from nearby proton bunch collisions in the TileCal readout window. The instantaneous luminosity was increased with time, and with it an increase in the average number of interactions per bunch crossing. Multiple interactions in one bunch crossing gave rise to in-time pileup effects.

In ATLAS the data were grouped according run numbers, where one run number was typically assigned to a proton fill, with intermediate changes to a new run in the case of significant detector configuration changes. Within one run the data were split into units called luminosity blocks, typically one minute long. Each ATLAS subsystem had a team of experts and automatic scripts reviewing the data taken by their respective system in real time, and promptly thereafter. For situations of poor data quality, where the data reconstructed was not judged of sufficient quality to be used for physics analysis, luminosity blocks were flagged as unusable.

Table 1. LHC beam conditions during the Run 1 proton collision program. In the table B.C. stands for bunch crossing. Not shown is proton collision data from 2009 where 14 nb^{-1} of data were recorded with a center of mass energy at 900 GeV.

	2010	2011	2012
Max. center of mass energy [TeV]	7	7	8
Delivered int. luminosity [pb^{-1}]	0.0481	5.5	22.8
Min. bunch spacing [ns]	150	50	50
Max. avg. interactions per B.C.	4	17	36

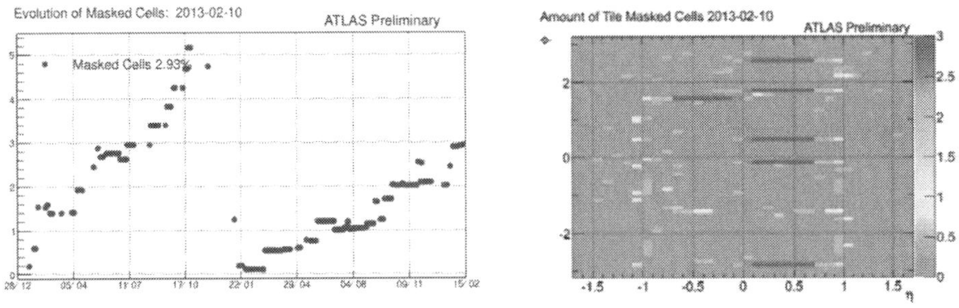

Figure 4. Percentage of faulty cells (called masked channels) that were not reconstructed from TileCal as a function of time across Run 1 (left); and the $\eta - \phi$ location of these masked cells at the end of Run 1 (right), where the z-axis represents the actual number of cells.

3.1. Tile calorimeter operation during Run 1

The fraction of good data that was recorded by TileCal was 100% in 2010, 99.2% in 2011, and 99.6% in 2012. The largest causes of the inefficiency were due to problems causing four or more consecutive modules to not be properly read out due to busy processes inside the back-end electronics or trips of PMT high voltage power-supply sources that power PMTs in four consecutive modules. Another large source of data integrity problems during Run 1 was a channel timing problem affecting an entire partition, where channels were reconstructed with a delay of 25 ns. The loss of data due to this channel timing problem was recovered when the data were re-reconstructed with updated detector conditions correcting for the known time shift.

There were additional operational problems that did not cause a total loss of data. These were situations where some fraction of the data from the detector was compromised but the information regarding the event could be recovered using information from neighboring modules. One such problem was failures of the low voltage power supplies (LVPS), that were located at the end of the electronics drawers on the detector, and provided power to the front-end electronics in that one module. Two types of failures were possible, one case where the LVPS failed to turn back on and required a hardware intervention for replacement and/or repair, and a second case where the LVPS simply tripped and was restarted successfully.

For the first case, in 2011 the TileCal lost approximately one LVPS per month, and in 2012 one LVPS was lost every two months. In these cases the channels associated to the problematic module were removed from the reconstruction, and energy from neighboring modules was interpolated to assign energies to the missing channels. The fraction of dead cells as a function of time in Run 1 is shown in Figure 4 (left). Large jumps in the fraction of dead cells represented moments when a LVPS turned off and could not be restarted. Smaller steps correspond to single cell problems arising from other less dramatic sources of channel failures. Maintenance periods allowed replacement of the LVPS and/or repair of the faulty readout cells, corresponding to steep reductions in the figure. At the end of Run 1 there were six LVPS that were off, dominating the 2.9% of cells that were faulty. The physical location of these faulty channels can be seen in Figure 4 (right).

During Run 1 there were frequent trips of the LVPS that were correlated with the integrated luminosity. Studies showed the rate of LVPS trips were approximately one every picobarn in 2011, and one every two picobarns in 2012. As in the case of permanent faulty channels, energy was interpolated from neighboring cells to account for the energy deposited in cells belonging to a module with a tripped LVPS. An automatic recovery mechanism was in place to recover tripped LVPS during physics runs. During the 2011-2012 maintenance period 40 new versions of the LVPS were installed, and of the 14k LVPS trips in 2012 only one was from one of these new LVPS.

4. Run 1 performance with single particles

The performance of the TileCal, and hence the methods used to reconstruct, calibrate, and correct for problematic regions of the detector, was evaluated with single isolated particles.

Single isolated muons were used to study the electromagnetic scale across the detector, using the ratio of the energy deposited to the path length traveled by the muon in a cell, dE/dl. For muon energies below 100 GeV this ratio is approximately constant. This quantity was studied across the physical detector volume, between cells in the same layer, between layers, by its stability over time, and compared with Monte Carlo (MC) predictions.

Figure 5 shows the dE/dl distribution for cosmic muon data taken in 2008 and MC simulations for one typical cell. The cosmic muons were selected with momentum in the range $10 < p < 30$ GeV to ensure the dominant energy loss was from ionization. For all cells in a given layer the variation in the response was found to be $\pm 2\%$, validating the calibration corrections implemented to account for cell-to-cell variations.

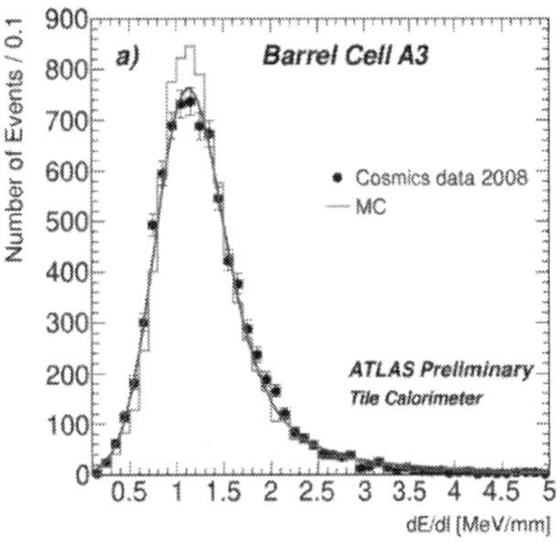

Figure 5. The amount of energy deposited per unit length, dE/dl, in the cell A3 in the long barrel. Shown in filled black dots is data from cosmic ray muons taken in 2008, and shown with the red line is results from MC simulations.

For all cells in a given layer the truncated mean of dE/dl was computed for data and MC. The truncated mean was defined such that 1% of the entries in the upper side of the dE/dl were neglected from the calculation, in order to reduce contributions from rare (non-ionizing) energy loss mechanisms. This ratio of data/MC is shown in Table 2 for each layer in each partition for the three years of cosmic muon data studied. The results show the layers were well inter-calibrated, except for LB layer D, which was approximately 4% over calibrated with respect to the other layers. The response to cosmic ray muons was constant in time, validating the calibration constants that corrected for the scintillator and PMT response variations.

The performance of the TileCal was also studied using single isolated charged hadrons, using the ratio of the energy measured by the TileCal, E, to the momentum as measured by the inner detector, p, for a single track. A signal compatible with a minimum ionizing particle in the LAr calorimeter was required, to avoid any bias of the TileCal response due to interactions in upstream material. The E/p distributions are shown in Figure 6 for data taken in 2011 and MC simulations as a function of η (left) and momentum (right). The data and MC agree within 3 %, except in the transition region between the LB and EB ($0.8 < |\eta| < 1.1$) where

Table 2. The data/MC ratio for the truncated mean of dE/dl averaged over all cells of a given layer, shown as a function of time from cosmic ray muons collected in 2008, 2009, and 2010. The error bars represent the statistical and systematic uncertainties.

Radial layer	R_{2008}	R_{2009}	R_{2010}
LB-A	0.966 ± 0.012	0.972 ± 0.015	0.971 ± 0.011
LB-BC	0.976 ± 0.015	0.981 ± 0.019	0.981 ± 0.015
LB-D	1.005 ± 0.14	1.013 ± 0.014	1.010 ± 0.013
EB-A	0.964 ± 0.043	0.965 ± 0.032	0.988 ± 0.014
EB-B	0.977 ± 0.018	0.966 ± 0.016	0.988 ± 0.014
EB-D	0.986 ± 0.012	0.975 ± 0.012	0.982 ± 0.014

deviations up to 10% are seen. As a function of particle momentum the agreement deviates above about 15 GeV, which is the transition point for the gain of the electronics readout. The transition region between the low energy parametrization model and quark-gluon string model in the Geant4 physics list (QGSP_BERT) used in the simulation happens for hadrons with energies in the range 12-25 GeV [6, 7].

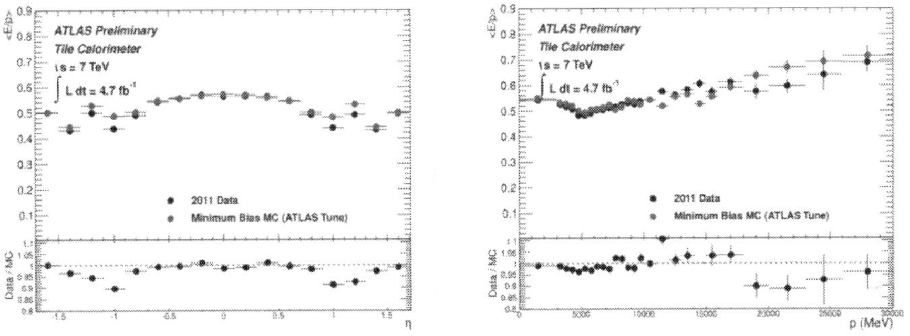

Figure 6. Ratio of the energy deposited in TileCal (E) to the track momentum (p) for isolated charged tracks as a function of η (left) and particle momentum (right).

5. Activities during the LHC long shutdown 1

In March 2013 the LHC entered a Long Shutdown (LS) period lasting for nearly two years so the accelerator could be commissioned to higher center of mass energy and luminosity. During this time the TileCal community took the opportunity to repair and improve many aspects of the subsystem. On the front-end, each of the 256 modules was opened, consolidated, inspected, and/or repaired to address problems detected during Run 1, such as corrupted data, channels with large noise, etc. Once the module was closed an extensive sign-off procedure was employed to ensure all problems identified were addressed and that no new problems were introduced in the process. There was a mobile test bench used to test the front-end electronics for a single module on the detector, there was a detector verification system used to test the full readout of a single module, and finally full calibration sets using the charge injection and laser systems were taken twice a week to extensively check and monitor every channel in TileCal.

All of the LVPS were replaced by the newer version which has increased tolerance to radiation and as a consequence fewer trips, as demonstrated by the 40 LVPS installed in 2012. For Run 2 these new LVPS will significantly reduce the number of trips, and therefore also reduce instances of corrupt data that sometimes followed the recovery of a LVPS. These new LVPS also exhibit lower electronic noise that follows more closely a single Gaussian distribution. The electronic noise for cells in layer A as a function of η is shown in Figure 7 for the old (nominal Run 1) and new LVPS.

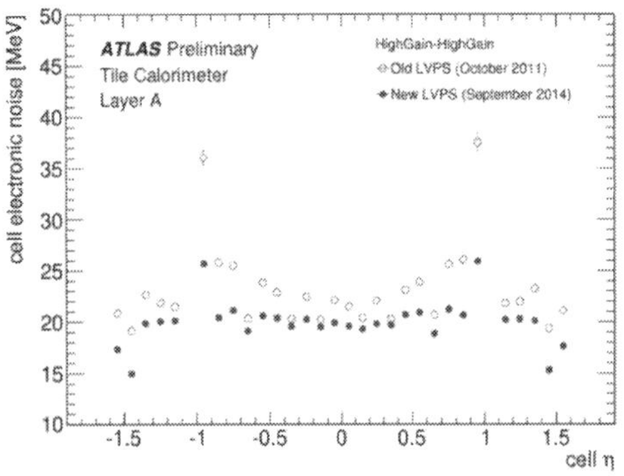

Figure 7. The electronic noise as a function of cell η with the original LVPS (2011) and with LVPS installed during the LS period from 2013-2015. The cell noise is shown for the case where both channels contributing to the cell were read out in high gain.

Other activities undertaken by the TileCal community during the LS were an update of the mechanical aspects of the cesium system, and the laser calibration system to improve the light mixing to avoid non-unformities in the light distribution thereby yielding more precise calibration constants. In Run 2 eight additional E3 and E4 cells (per EB) will be read out, where in Run 1 those readouts were used for the Minimum Bias Scintillator Trigger (MBTS) system. In addition trigger systems were updated to allow the TileCal layer D to be used in coincidence with the muon trigger system to reduce the muon trigger fake rate.

6. Summary and outlook

The ATLAS TileCal performed very well in the LHC Run 1, both in operation, calibration, performance and stability. Analysis of single particles demonstrate the success of the reconstruction and calibration methods to correct for changes over time from irradiated scintillators, aging PMTs, and cell-to-cell variations. During the LHC LS (2013-2015) the TileCal community has been quite active in upgrading and improving many components of the system. The precision of the calibration constants directly contributes to the energy resolution of reconstructed objects, such as jets, and missing transverse momentum, and correct modeling of the calorimeter in MC is essential as many searches use MC for background estimation techniques. The LHC Run 2 is just around the corner, and the TileCal will play an essential role for quests for new physics at higher energy regimes.

References

[1] ATLAS Collaboration 2008 *JINST* **3** S08003
[2] ATLAS Collaboration 1996 *CERN-LHCC-96-42*
[3] ATLAS Collaboration 2010 *Eur. Phys. J. C* **70** 1193
[4] ATLAS/TileCal Collaboration 2009 *NIM* **A606** 362-94
[5] Cleland W E and Stern E G 1994 *NIM* **A338** 467-97
[6] Agostinelli S *et al.* 2003 *NIM* **A506** 250-303
[7] Allison J *et al.* 2006 *IEEE Transactions on Nuclear Science* **53** 270-8

NLO Heavy Quark Energy Loss in Strongly-Coupled Quark-Gluon Plasmas

W. A. Horowitz

Department of Physics, University of Cape Town, Private Bag X3, Rondebosch 7701, South Africa

E-mail: wa.horowitz@uct.ac.za

Abstract. Improvements to heavy and light flavor energy loss models based on strong-coupling physics in quark-gluon plasma lead to qualitative agreement between the models and data from RHIC and LHC within the regimes of applicability of the calculations. Thus it is possible to describe self-consistently the dynamics of quark-gluon plasma from the lowest observed momentum modes to the highest from the AdS/CFT correspondence.

1. Introduction

In high-energy nuclear physics we seek a self-consistent theoretical picture of the dynamics of the quark-gluon plasma (QGP) that accurately describes the data observed from heavy ion collisions at facilities such as the Relativistic Heavy Ion Collider (RHIC) at Brookhaven National Laboratory and the Large Hadron Collider (LHC) at CERN. With such a description in hand we hope to claim an understanding of the properties of the QGP and thus map out simultaneously theoretically and experimentally a major portion of the phase diagram of quantum chromodynamics (QCD) [1].

Two of the major sets of observables associated with heavy ion collisions are those associated with low-p_T physics described by relativistic hydrodynamics with a very small viscosity-to-entropy ratio η/s [2, 3, 4, 5] and those associated with high-p_T probes such as the suppression of jets and open heavy flavor [6, 7]. The very small η/s is most naturally explained by assuming the dynamics of low momentum modes in the QGP are best described by strong-coupling physics as derived using the methods of the anti-de-Sitter/conformal field theory (AdS/CFT) correspondence [8, 9, 10], but not by the methods of weakly-coupled plasma using the methods of perturbative QCD (pQCD) [11]. On the other hand, a huge variety of observables related to high-p_T probes are accurately described using leading order methods of pQCD [6, 7, 12, 13], but are not described using the leading order methods of AdS/CFT [14, 15].

Since we seek a self-consistent description of QGP physics, we hope to either extend the weak-coupling description of the QGP down to low-momentum observables or, alternatively, extend the strong-coupling description up to high-momentum observables. We will demonstrate in this proceedings that it is possible to do the latter, to extend the AdS/CFT description of QGP physics up to the highest momentum probes of the medium [16, 13].

2. AdS/CFT and Hard Probes

It is natural to ask why one might hope that AdS/CFT could be used to describe the physics of hard probes of heavy ion collisions. Since QCD is an asymptotically free theory, one might naturally expect that the physics of hard probes is dominated by that of pQCD. However, it is not clear that the pQCD perspective is a self-consistent one. Despite the fact that the QGP exists at a temperature of a trillion degrees [2, 3, 4, 5], the scale set by the temperature of the plasma T_{QGP} is not asymptotically large compared to the natural scale of high-energy nuclear physics, $\Lambda_Q CD$; in fact the two scales are the same order of magnitude, $T_{QGP} \sim \Lambda_{QCD}$. In pQCD, the dominant energy loss mode at high momentum is that of radiation, gluon bremsstrahlung [17]. This radiation process is necessarily third order in the strong coupling α_s, and it is not yet clear at what scale(s) these couplings run: while it is likely that large scales such as the momentum of the probe P is an important factor in the strength of the coupling of the parton to the medium, it is also likely that T_{QGP} is a critically important scale, too. In fact, as the coupling grows non-perturbatively large at low momenta, it is not unreasonable to think that energy loss physics is actually dominated by scales on the order of T_{QGP} rather than P.

In the work that follows we will assume that the strong coupling physics dominates the energy loss processes associated with high-momentum probes of the QGP. One may then hope that the AdS/CFT correspondence will provide qualitative, perhaps even quantitative, insight into the properties of the QGP created in heavy ion collisions. We will show that under this assumption of the dominance of strong-coupling physics we can describe a number of observables from RHIC and LHC associated with hard probes of QGP [16, 13].

2.1. Light Flavor

At strong coupling the relevant degrees of freedom of the sQGP are not quasiparticles. Therefore it is difficult to connect AdS/CFT calculations to light flavor single particle observables. Rather, it is much more natural to compare to composite objects such as jets [18]. Even so, it is still highly non-trivial to find the correct description in the dual string theory from AdS/CFT of the relevant object from the field theory. For jets, the original proposal was that the part of the string within some physical distance Δx to the string endpoint corresponded to a jet in the field theory [18]. With this description, however, one finds that a fully stopped string corresponds to a jet that still has a sizable fraction $\gtrsim 30\%$ of its original energy; i.e. the jet is not fully thermalized in the medium despite its complete lack of motion. Inspired by the scale separation used in thermal field theory calculations, an alternative jet prescription was proposed in [16]: the part of a string above some Δu threshold in the fifth dimension corresponds to the jet while the part of the string below the threshold corresponds to thermalized, low-momentum medium modes. With this prescription and a reasonable separation scale of ~ 500 MeV a fully stopped jet is completely thermalized: the jet has no energy in the high-momentum modes; all its energy is in low-momentum modes associated with the medium.

We found that the new jet prescription led to the recovery of the Bragg peak in the energy loss of light flavor in QGP [16]. Incorporating the full numerical jet energy loss calculation from AdS/CFT in a simple geometrical model for the medium produced in heavy ion collisions led to a theoretical prediction [16], shown in Fig. 1 (a), that was significantly oversuppressed compared to data[1]. However, despite a better description of thermalized jets, one can see from Fig. 1 (a) that the proposed jet description still does not correspond perfectly to the physics in QCD: with the energy scale separation prescription, jets propagating in a $T = 0$ "vacuum" QGP still lose

[1] The observable shown in the figure is known as the nuclear modification factor, or R_{AA}, which is defined as the differential yield of some observable in $A+A$ collisions divided by the number observed in $p+p$ collisions scaled by the expected number of binary (or $p+p$-like) collisions in the $A+A$ collisions. An $R_{AA}=1$ implies that the QGP medium has no effect on the investigated probe; null control measurements of known weakly-coupled probes from the electroweak sector have $R_{AA}\sim 1$.

a significant fraction of their energy; see Fig. 1 (a). Since R_{AA} measures the difference between vacuum and medium physics, it is natural (and critical) to properly account for this "vacuum" energy loss by subtracting it in some way from the "in-medium" energy loss. Fig. 1 (b) shows the results when this renormalization was performed by taking the ratio of R_{AA} to R_{pp} [16]. (A similar result emerges when one defines a renormalized energy loss $\Delta E_{renorm} = \Delta E_{AA} - \Delta E_{pp}$ and then computes R_{AA}.)

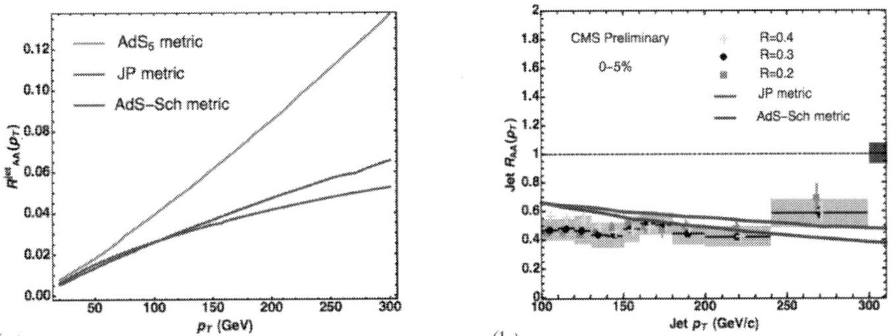

(a) (b)

Figure 1: (a) $R_{AA}(p_T)$ for light flavor jets in static ("AdS-Sch"), Bjorken expanding ("JP"), and $T_{QGP} = 0$ ("AdS$_5$") plasmas from AdS/CFT [16]. (b) The renormalized jet $R_{AA}(p_T)$ from AdS/CFT [16] compared to preliminary data from the CMS collaboration [19].

2.2. Heavy Flavor

It is much easier to directly connect the strong-coupling picture to heavy quark observables: the heavy quark can be treated theoretically as an external probe [20, 21, 22]. Early leading order derivations of the mean momentum loss rate of a massive quark in a strongly-coupled medium [20, 21] implemented in an energy loss model showed a quantitative agreement with data measured at RHIC [23, 24]. However, the application of this same model to LHC led to a prediction in contradiction with data: the model predicted significant oversuppression compared to observation [15].

These early leading order energy loss derivations were improved by the calculation of the momentum fluctuations associated with energy loss. It was shown in [25] that the three-momentum p^i of an on-shell heavy quark moving at constant velocity in a thermal bath evolves as:

$$\frac{dp_i}{dt} = -\mu\, p_i + F_i^L + F_i^T, \tag{1}$$

where the drag coefficient [20, 21, 22] of a heavy quark of mass M_Q in a plasma of temperature T with 't Hooft coupling λ is

$$\mu = \frac{\pi\sqrt{\lambda}T^2}{2M_Q} \tag{2}$$

and the longitudinal (L) and transverse (T) fluctuating momentum kicks are correlated as

$$\langle F_i^L(t_1)F_j^L(t_1)\rangle = \kappa_L \hat{p}_i\hat{p}_j g(t_2 - t_1) \tag{3}$$

$$\langle F_i^T(t_1)F_j^T(t_1)\rangle = \kappa_T(\delta_{ij} - \hat{p}_i\hat{p}_j) g(t_2 - t_1), \tag{4}$$

where $\hat{p}_i = p_i/|\vec{p}|$,

$$\kappa_T = \pi\sqrt{\lambda}T^3\gamma^{1/2} \tag{5}$$

$$\kappa_L = \gamma^2\kappa_T, \tag{6}$$

γ is the usual Lorentz boost factor, and g is a function known only numerically.

It is important to estimate the regime where it is appropriate to use the above formulae to compute the energy loss of a heavy quark in a strongly-coupled plasma. In the setup for the derivation of the above formulae, the heavy quark is assumed to move at an (approximately) constant velocity. There are two obvious possible speed limits for the heavy quark that then result. The first speed limit is well-known, and is associated with the calculation of the drag coefficient, which is to say the mean energy loss rate [25, 26]. For the string dual of a finite mass heavy quark to have a time-like string endpoint motion, the heavy quark velocity is restricted to

$$\gamma < \gamma_{crit}^{sl} = \left(1 + \frac{2M_Q}{\sqrt{\lambda}T}\right)^2 \sim \frac{4M_Q^2}{\lambda T^2}. \tag{7}$$

There is another speed limit, which comes from requiring that the accumulated momentum fluctuations for a particular heavy quark are not large compared to its natural momentum relaxation; i.e. by requiring that the momentum picked up via fluctuations over the time scale set by the drag coefficient is small in comparison to the total momentum of the heavy quark one finds an upper limit on the speed of the heavy quark given by

$$\gamma \lesssim \gamma_{crit}^{fluc} = \frac{M_Q^2}{4T^2}. \tag{8}$$

One can see that the speed limit set by the mean loss rate, Eq. (7) is parametrically smaller than that set by the fluctuations, Eq. (8). However, for finite $\lambda \lesssim 12$, it turns out that the speed limit from fluctuations is numerically smaller than that set by the mean loss rate. At the temperatures reached by the medium at RHIC and LHC, the speed limit from fluctuations leads to a maximum momentum for which the calculations apply to charm quarks (of mass ~ 1.5 GeV) of $p_T \sim 10$ GeV/c. We will see explicitly this breakdown in the calculation in the model calculations that follow.

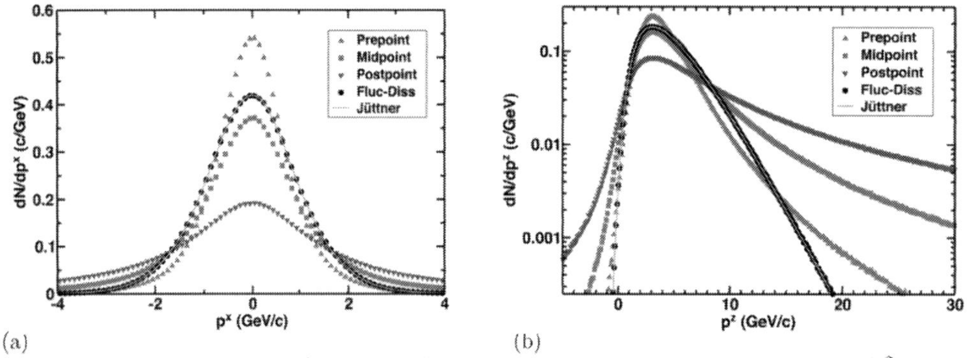

(a) (b)

Figure 2: The p^x-differential (a) and p^z-differential (b) distribution of $m_c = 1.5$ GeV/c^2 charm quarks strongly-coupled to a strongly-coupled plasma in three spatial dimensions at $T = 400$ MeV that is moving at $0.9c$ in the z direction. The effect of the thermal drag and fluctuations on the heavy quarks from strong coupling is found from implementing Eq. (1) using the Itô (pre-point), Stratonovich (mid-point), and Hänggi-Klimontovich (post-point) stochastic integrals. An implementation for which the fluctuation-dissipation theorem holds is also shown and compared to the result expected, the Jüttner distribution.

A major complication of the momentum fluctuation calculation of Eqs. (2)–(6) is that the momentum kicks are multiplicative; i.e. the strength of the fluctuations is a function of momentum itself. Stochastic integration in which the fluctuations are multiplicative have an irreducible ambiguity: the result depends explicitly on the time within each time step that the integrand is evaluated [27]. (Compare this ambiguity to usual Riemann integration for which the result converges to a unique result as timestep size goes to zero *irrespective* of the exact time within a timestep that the integrand is evaluated.) Fig. 2 illustrates the significant difference of result from stochastic integration for different (common) choices of time within a timestep to evaluate the integrand. The ambiguity in the stochastic integration procedure can be resolved with the use of the Wong-Zakai theorem [28]: the solution of any Langevin system with an autocorrelation time that goes to zero is given by the Stratonovich stochastic integral. For the problem at hand, the autocorrelation time is small compared to the natural time scale of the problem set by the drag coefficient so long as the heavy quark obeys the speed limit set by the mean energy loss calculation, Eq. (7).

In order to compare with data from $A+A$ collisions, we must first demonstrate that we have the production mechanism in $p+p$ collisions under control. Fig. 3 (a) compares the spectrum of electrons from the decay of heavy flavor mesons from RHIC and from FONLL [29, 30, 31]. The spectra of open heavy quarks is readily available from an online generator [32], and the FONLL D and B meson fragmentation functions are well documented [31]. The decay of the mesons to electrons in FONLL is not as well documented, and the results of this work [13] as shown in Fig. 3 (a) differ significantly from those shown in [33], with the results presented here diverging much more from the experimental results at low, $p_T \lesssim 3$ GeV/c. Thus the comparison of electrons from $A+A$ is restricted from above by the speed limit Eq. (7) and from below by the currently inaccurate production/fragmentation function mechanism.

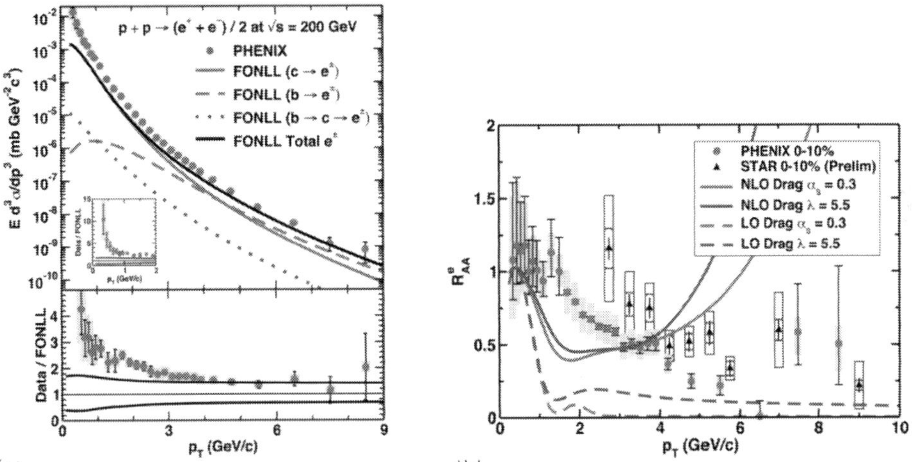

(a) (b)

Figure 3: (a) FONLL predictions [29, 30, 31] for the spectrum of electrons from heavy flavor decay at RHIC compared to data from PHENIX [33]. (b) Strong-coupling energy loss predictions at leading (dashed) and next-to-leading order (solid) for two schemes (red, blue) of translating QCD parameters to $\mathcal{N}=4$ SYM parameters [13] compared to data from PHENIX [33] and preliminary data from STAR [34].

Nevertheless, we show in Fig. 3 (b) the comparison of the strong-coupling energy loss model predictions for electrons from heavy flavor decay to RHIC data [33, 34], and then D and B meson suppression to LHC data [35, 36] in Fig. 4. The plots include both the leading and next-to-leading order predictions from AdS/CFT [13]. One of the subtleties involved in comparing the AdS/CFT calculations to data is the map between the parameters (α_s, N_c, and T) in

QCD and the parameters in $\mathcal{N} = 4$ SYM. Two schemes for translating between the theories were proposed in [25]; we report the results of both schemes here as an attempt to provide a reasonable exploration of the parameter space. (One would hope that a theory dual to QCD would produce results somewhere between the predictions presented here.) In the range of applicability, 3 GeV/c $\lesssim p_T^e \lesssim$ 4 GeV/c, the theory and data agree for electron suppression. Similarly, in the range of $p_T^D \lesssim$ 10 GeV/c and for all current B meson measurements, the theory and data agree. In fact, the fast rise in R_{AA} for the observables for momenta above the scale set by the speed limit is precisely the sign of the calculation breaking down. A more correct treatment of the above equations would include autocorrelations, which would mitigate some of that spurious rapid rise. Additionally, a derivation of the energy loss that allowed the heavy quark to dynamically slow down (a highly non-trivial calculation) would include the Larmor-like energy loss experienced by accelerating probes in AdS/CFT [37, 38]. It is not yet known how much of a quantitative difference including this additional energy loss channel would make as a function of momentum, but an implementation of this improved energy loss theory would almost certainly compare more favorably with data.

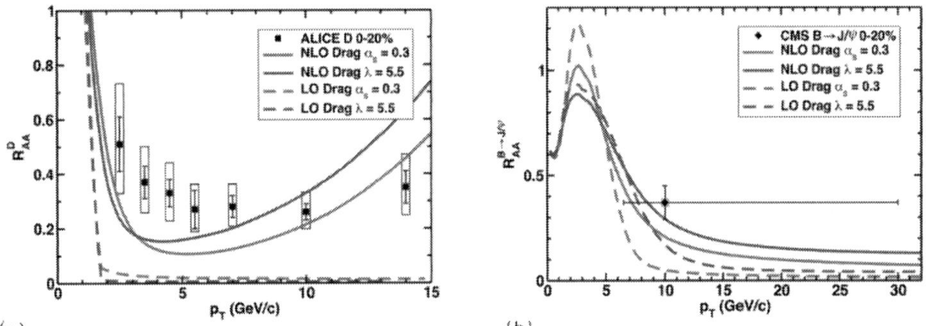

(a) (b)

Figure 4: $R_{AA}(p_T)$ predictions from AdS/CFT at leading (dashed) and next-to-leading (solid) order for two schemes (red, blue) of translating QCD parameters to $\mathcal{N} = 4$ SYM parameters [13] compared to (a) D meson data from ALICE [35] and (b) non-prompt J/ψ mesons compared to data from CMS [36].

3. Conclusions

We showed in this proceedings that strong-coupling energy loss models based derivations from the AdS/CFT correspondence—in which a high-momentum probe is strongly-coupled to a strongly-coupled medium—can qualitatively describe both the light and heavy flavor suppression data measured at RHIC and LHC. Thus one can view the physics of a wide range of observables related to the quark-gluon plasma produced in these colliders self-consistently within a single theoretical framework. Future theoretical work includes refining the treatment of energy loss of light flavors and increasing the regime of applicability of heavy quark energy loss. The latter will likely be accomplished by considering the fluctuations in energy loss of dynamical string solutions (as opposed to fluctuations on the static, approximately infinite mass solutions). Further comparisons with data are of course desired to more stringently test the strong-coupling energy loss theory. Of especial utility are observables related to b quarks for which the theory is better controlled. Additionally, as the fluctuations are predicted to increase significantly with the speed of the heavy quark, heavy flavor correlations measurements will likely provide a valuable discriminator between strong- and weak-coupling energy loss physics.

4. Acknowledgments
The author wishes to thank the South African National Research Foundation and SA-CERN for support.

5. References
[1] The Frontiers of Nuclear Science, A Long Range Plan 2008 (*Preprint* 0809.3137)
[2] Hirano T, Huovinen P and Nara Y 2011 *Phys.Rev.* **C84** 011901 (*Preprint* 1012.3955)
[3] Shen C, Heinz U, Huovinen P and Song H 2011 *Phys.Rev.* **C84** 044903 (*Preprint* 1105.3226)
[4] Qiu Z, Shen C and Heinz U 2012 *Phys.Lett.* **B707** 151–155 (*Preprint* 1110.3033)
[5] Gale C, Jeon S, Schenke B, Tribedy P and Venugopalan R 2013 *Phys.Rev.Lett.* **110** 012302 (*Preprint* 1209.6330)
[6] Majumder A and Van Leeuwen M 2011 *Prog.Part.Nucl.Phys.* **A66** 41–92 (*Preprint* 1002.2206)
[7] Horowitz W 2013 *Nucl.Phys.* **A904-905** 186c–193c (*Preprint* 1210.8330)
[8] Gubser S, Klebanov I R and Peet A 1996 *Phys.Rev.* **D54** 3915–3919 (*Preprint* hep-th/9602135)
[9] Kovtun P, Son D T and Starinets A O 2005 *Phys.Rev.Lett.* **94** 111601 (*Preprint* hep-th/0405231)
[10] Casalderrey-Solana J, Liu H, Mateos D, Rajagopal K and Wiedemann U A 2011 (*Preprint* 1101.0618)
[11] Danielewicz P and Gyulassy M 1985 *Phys.Rev.* **D31** 53–62
[12] Djordjevic M, Djordjevic M and Blagojevic B 2014 *Phys.Lett.* **B737** 298–302 (*Preprint* 1405.4250)
[13] Horowitz W 2015 (*Preprint* 1501.04693)
[14] Horowitz W and Gyulassy M 2011 *J.Phys.* **G38** 124114 (*Preprint* 1107.2136)
[15] Horowitz W 2012 *AIP Conf.Proc.* **1441** 889–891 (*Preprint* 1108.5876)
[16] Morad R and Horowitz W 2014 *JHEP* **1411** 017 (*Preprint* 1409.7545)
[17] Wicks S, Horowitz W, Djordjevic M and Gyulassy M 2007 *Nucl.Phys.* **A784** 426–442 (*Preprint* nucl-th/0512076)
[18] Chesler P M, Jensen K, Karch A and Yaffe L G 2009 *Phys.Rev.* **D79** 125015 (*Preprint* 0810.1985)
[19] Collaboration C (CMS) 2012
[20] Gubser S S 2006 *Phys.Rev.* **D74** 126005 (*Preprint* hep-th/0605182)
[21] Herzog C, Karch A, Kovtun P, Kozcaz C and Yaffe L 2006 *JHEP* **0607** 013 (*Preprint* hep-th/0605158)
[22] Casalderrey-Solana J and Teaney D 2006 *Phys.Rev.* **D74** 085012 (*Preprint* hep-ph/0605199)
[23] Akamatsu Y, Hatsuda T and Hirano T 2009 *Phys.Rev.* **C79** 054907 (*Preprint* 0809.1499)
[24] Horowitz W A 2010 (*Preprint* 1011.4316)
[25] Gubser S S 2008 *Nucl.Phys.* **B790** 175–199 (*Preprint* hep-th/0612143)
[26] Casalderrey-Solana J and Teaney D 2007 *JHEP* **0704** 039 (*Preprint* hep-th/0701123)
[27] He M, van Hees H, Gossiaux P B, Fries R J and Rapp R 2013 *Phys.Rev.* **E88** 032138 (*Preprint* 1305.1425)
[28] Wong E and Zakai M 1965 *Ann.Math.Stat.* **36** 1560–1564
[29] Cacciari M, Greco M and Nason P 1998 *JHEP* **9805** 007 (*Preprint* hep-ph/9803400)
[30] Cacciari M, Frixione S and Nason P 2001 *JHEP* **0103** 006 (*Preprint* hep-ph/0102134)
[31] Cacciari M, Frixione S, Houdeau N, Mangano M L, Nason P *et al.* 2012 *JHEP* **1210** 137 (*Preprint* 1205.6344)
[32] http://www.lpthe.jussieu.fr/~cacciari/fonll/fonllform.html
[33] Adare A *et al.* (PHENIX) 2011 *Phys.Rev.* **C84** 044905 (*Preprint* 1005.1627)
[34] Mustafa M (STAR) 2013 *Nucl.Phys.* **A904-905** 665c–668c (*Preprint* 1210.5199)
[35] Abelev B *et al.* (ALICE) 2012 *JHEP* **1209** 112 (*Preprint* 1203.2160)
[36] Chatrchyan S *et al.* (CMS) 2012 *JHEP* **1205** 063 (*Preprint* 1201.5069)
[37] Mikhailov A 2003 (*Preprint* hep-th/0305196)
[38] Fadafan K B, Liu H, Rajagopal K and Wiedemann U A 2009 *Eur.Phys.J.* **C61** 553–567 (*Preprint* 0809.2869)

Radiation hardness of plastic scintillators for the Tile Calorimeter of the ATLAS detector

H Jivan[1,2], B Mellado[1], E Sideras-Haddad[1,2], R Erasmus[1], S Liao[1], M Madhuku[3], G Peters[1], O Solvyanov[4]

1: University of the Witwatersrand, 1 Jan Smuts Avenue, Braamfontein 2000, Johannesburg.
2: DST-NRF Centre for Excellence in Strong Materials
3: iThemba LABS, North, Empire Road, Braamfontein 2000, Johannesburg.
4: European Organization for Nuclear Research (CERN), CH-1211 Genève 23, Suisse

Email: harshna.jivan@gmail.com

Abstract The radiation damage in polyvinyl toluene based plastic scintillator EJ200 obtained from ELJEN technology was investigated. This forms part of a comparative study conducted to aid in the upgrade of the Tile Calorimeter of the ATLAS detector during which the Gap scintillators will be replaced. Samples subjected to 6 MeV proton irradiation using the tandem accelerator of iThemba LABS, were irradiated with doses of approximately 0.8 MGy, 8 MGy, 25 MGy and 80 MGy. The optical properties were investigated using transmission spectroscopy whilst structural damage was assessed using Raman spectroscopy. Findings indicate that for the dose of 0.8 MGy, no structural damage occurs but a breakdown in the light transfer between base and fluor dopants is observed. For doses of 8 MGy to 80 MGy, structural damage leads to hydrogen loss in the benzene ring of the PVT base which forms free radicals. This results in an additional absorptive component causing increased transmission loss as dose is increased.

1. Introduction

The Tile Calorimeter of the ATLAS detector, is a hadronic calorimeter responsible for recording the energy and trajectory of hadrons, taus as well as jets of quarks and gluons that result from the proton-proton collisions within the Large Hadron Collider of CERN. The Tile Calorimeter consists of a central barrel and two extended barrels which each contain 64 azimuthal modules. Each module contains a matrix of steel plates with tiles of plastic scintillators sandwiched in between. The steel plates act as an absorber medium, converting the incoming jets to a 'shower' of particles whilst the scintillator tiles absorb energy from the incoming particles and fluoresce to emit light. This light is passed through wavelength shifting optical fibres and detected by photomultiplier tubes. The signal is further processed with readout electronics in order to digitize the data for analysis thereafter. In addition, the GAP region between the central and extended barrels, where many of the read-out electronics components are housed, contain additional scintillator plates. These assist with reconstruction of missing transverse energy within this region. [1]

Plastic scintillators in specific are employed within the detector due to their properties of high optical transmission and fast rise and decay times. This enables efficient data capture since fast signal pulses can be generated [2]. In addition, plastic scintillators are easier to manufacture as compared to inorganic crystals which require special growing methods, and are more cost effective [3] for covering the typically large detector areas. The main problem encountered by plastic scintillators however, is radiation damage incurred due to their interactions with the ionizing particles to be detected. This damage leads to a significant decrease in the light yield of the scintillator and introduces an error into the time-of flight data acquired.

With the LHC gearing up to run proton collisions at increased center of mass energy and higher luminosities, the radiation environment within the ATLAS detector is expected to become much harsher. The Tile Calorimeter has therefore implemented a series of upgrades in order to ensure that the detector performance can be sustained for several years to come. Part of phase two of this upgrade will be implemented in 2018 where the current crack scintillators from the GAP region of the Tile Calorimeter will be replaced with more radiation hard plastics. To aid in the choice of a replacement candidate, this comparative study is underway which examines the radiation hardness of several commercially available plastic scintillators. In addition, the study aims to further understand the type of structural damage that these plastics undergo which may assist in future developments of radiation tolerant plastic scintillators. In these proceedings, we present preliminary results for EJ200, a polyvinyl toluene based plastic obtained from ELJEN Technologies.

2. The Scintillation mechanism

Plastic scintillators primarily consist of organic fluors suspended in a polymer base. The polymer base that is employed generally contains some type of aromatic ring structure which gives rise to a delocalized π-electron structure within the molecule. When ionizing radiation impinges the scintillator, some of its kinetic energy may be absorbed by these delocalized π-electrons, resulting in molecular excitations. The absorption is typically exhibited in the visible and ultra-violet regions corresponding to excitation of the singlet π-electron state. The energy level diagram of a π-electron is shown in Figure 1 below. For transitions from the lowest vibrational first excited state to the ground state, prompt fluorescence occurs whereby energy is emitted in the form of light of a characteristic wavelength. This fluorescence process forms the basic mechanism for scintillation in organic plastic scintillators. [2]

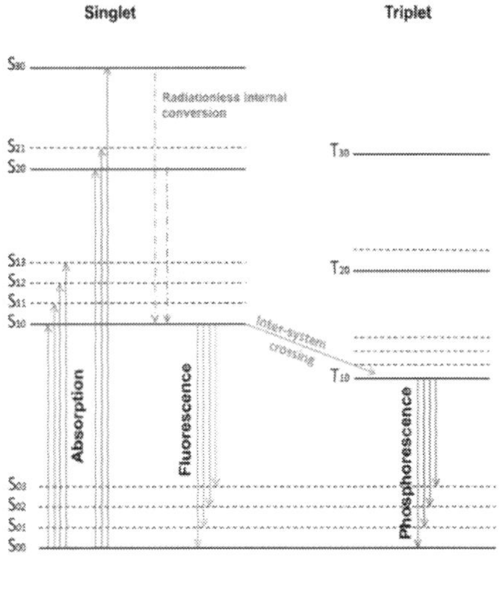

Figure 1: Energy level diagram of an organic molecule with π-electron structure, adapted from[2].The singlet states with spin 0 are labelled Sab and the triplet states of spin 1 are labelled Tab where a indicates the excitation level and b indicates the vibrational state of that level.

The energy of the fluorescent light is generally less than the energy required for absorption, however a small overlap between the absorption and emission spectra may occur. In order to prevent self-absorption of scintillation light, small concentrations of primary fluors are added to the plastic polymer. These ensure that the plastic scintillator is transparent to its own scintillation light. Occasionally, a secondary fluor may be added which acts specifically as a wavelength shifter. This is used in order to ensure that the wavelength of light leaving the scintillator is suitable for detection by optical fibres which lead to photomultiplier tubes. [4]

Plastic scintillator EJ200 is composed of a polyvinyl toluene base and 3% of added organic fluors. Polyvinyl toluene consists of long chains of vinyl toluene molecules that encompass a benzene ring (C_6H_4) bonded to a methyl group (CH_3) and a vinyl group (CH_2-CH-). Information on the exact composition of the doped fluors are not made public by the manufacturers. Some important properties of EJ200 are listed in table 1 below.

Table 1: Properties of plastic scintillator EJ200 as provided by the manufacturers [5]

	EJ200
Light Output (% Anthracene)	64
Wavelength of maximum emission (nm)	425
Rise time (ns)	0.9
Decay time (ns)	2.1
Refractive index	1.58

3. Experimental Procedure

The tandem accelerator of iThemba LABS in Gauteng was used to irradiate samples with 6 MeV protons. To ensure that the study simulated a similar type of particle-scintillator interaction as observed in the Tile Calorimeter, protons would be required to pass through the samples whilst imparting energy primarily through ionization losses. In order to ensure this, SRIM simulations were performed to determine the stopping range of 6 MeV protons within the plastic material. Based on these simulations, several samples were cut and polished to dimensions of 0.5 cm by 0.5 cm, with thickness of ~350 μm. A polishing procedure based on standard metallographic techniques was employed. Thereafter, samples were irradiated to doses of 0.8 MGy, 8 MGy, 25 MGy and 80 MGy. Quadrupole magnets were used to focus the proton beam to a spot size of ~20-30 μm and the beam was then scanned in the x and y plane to achieve an irradiated area of approximately 1.8 mm by 1.8 mm.

The effects of proton damage on the optical properties of the samples were characterized by conducting light transmission experiments. Transmission spectroscopy was conducted using the Varian Carry 500 spectrophotometer. Light transmission was measured relative to transmission in air over a range of 200-800 nm. Transmission spectra were collected prior to irradiation as well as over several days after.

The structural damage undergone by the scintillators were then investigated using Raman spectroscopy. Raman spectra were obtained for the un-irradiated control samples as well as on the 0.8 MGy and 8 MGy irradiated samples using the Horiba Jobin-Yvon Raman spectrograph. An Argon laser was used to provide a 515 nm excitation wavelength. Raman studies were conducted 3 days, 10 days and 4 weeks after irradiation.

In addition, light yield studies on the proton damaged samples are to be carried out whereby the scintillator's response to 0.5 MeV gamma rays emitted by a ^{90}Sr source will be measured using the light box set-up at CERN. In these experiments, the sample will be coupled to two Kuraray Y-11(200) optical fibers which will be connected to a standard Tile Calorimeter photomultiplier tube. As light yield experiments are to be conducted several weeks after irradiation, this will provide time for partial recovery in the scintillators.

4. Light Transmission Results and Analysis

The ratio of transmission in each irradiated EJ200 sample relative to its transmission before irradiation for several doses as a function of wavelength is shown in Figure 2 below. Absorption in the PVT base occurs over 240 nm to 300 nm whilst absorption in the fluors is expected to occur over 270 nm to 400 nm. For the dose of 0.8 MGy, a transmission dip at lower wavelengths is observed, followed by a peak over 320-400 nm. This feature indicates that a competitive absorptive process has developed between the primary and secondary fluors.

As the dose increases, we observe the formation of an absorptive tint whereby the absorption drop-off shifts to higher wavelengths. This implies that additional absorbing species such as free radicals are forming within the scintillator and leads to a further decrease in light emitted by the scintillator. This additional absorption component may also reduce the attenuation length of the scintillation light as described by C. Zorn (1993) [6].

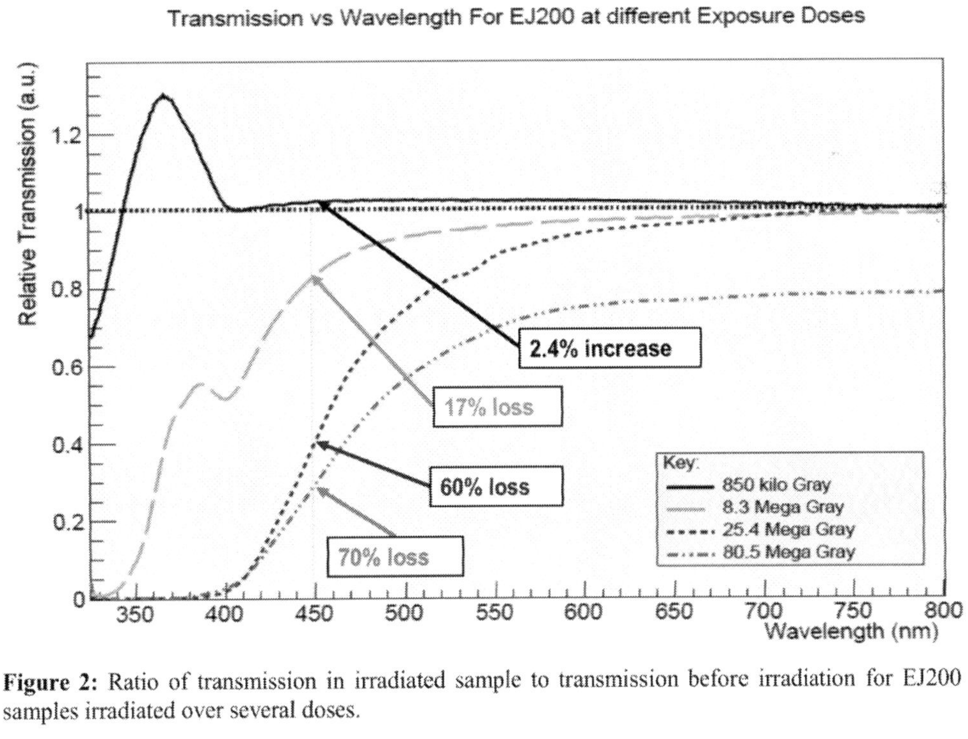

Figure 2: Ratio of transmission in irradiated sample to transmission before irradiation for EJ200 samples irradiated over several doses.

5. Raman Spectroscopy Results and Analysis

Raman spectra were obtained for the 0.8 MGy samples, 8 MGy samples as well as for un-irradiated control samples. The samples irradiated to 0.8 MGy maintained their structure and very little damage was observed. In the 8 MGy sample, an increased amount of background fluorescence could be observed which may be due to free radicals interacting with the excitation light. Raman spectroscopy could not be performed for the high dose of 80 MGy because this background fluorescence becomes too large to obtain an experimentally viable spectrum. Figure 3 shows the background subtracted Raman spectrum for the 8 MGy irradiated sample taken over several days.

The peaks observed are allocated to the characteristic functional groups or vibrational groups using the Raman Peak assignment datasheet from Horiba Jobin-Yvon. The peak assignment is summarized in Table 2. An additional peak (X) was found to form after irradiation but disappeared over time as the scintillator healed. In order to gauge changes in the amount of specific species present, the intensities of the various peaks were plotted relative to peak 13 which represents the C-C bond typically found in benzene rings. This plot is shown in Figure 4.

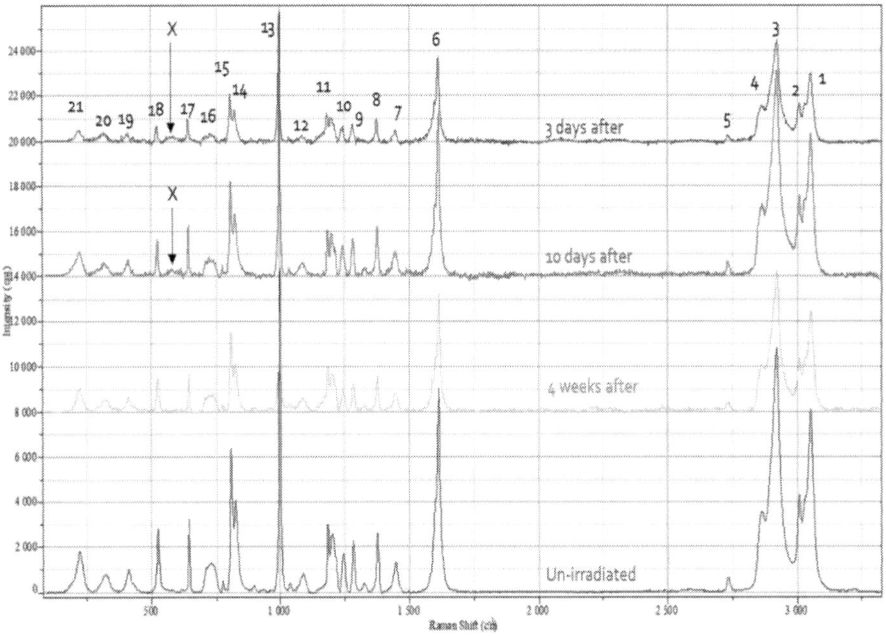

Figure 3: Background subtracted Raman spectra for 8 MGy irradiated EJ200.
N.B Spectra overlap but intensities have been shifted in order to have better visible representation.

Table 2: Allocation of peaks to characteristic functional groups or vibrational groups.

Functional Group/ Vibration	Peak Assignment
δ(C-C) aliphatic	20-21
ν(C-C) alicyclic or aliphatic chain vibrations	9-12, 14-19, X
ν(C-C) aromatic ring chain vibrations	13
δ(CH3)	8
δ(CH2) or δ(CH3)asymmetric	7
ν(C=C)	6
ν(C-H)	3-4
ν(=(C-H))	1-2

The ratio of species typically found in the vinyl backbone of the sample versus that in the benzene ring shows an increase after irradiation. A small amount of additional alicyclic rings and aliphatic chains are found with the additional peak X being typical of these bonds. These bonds represent single carbon rings and chains stripped of hydrogen. This indicates that structural damage occurs within the scintillator whereby bonds in the benzene ring are broken. Hydrogen may be lost by the benzene rings to form free radicals. After 4 weeks of healing, the scintillator appears to substantially recover its structure.

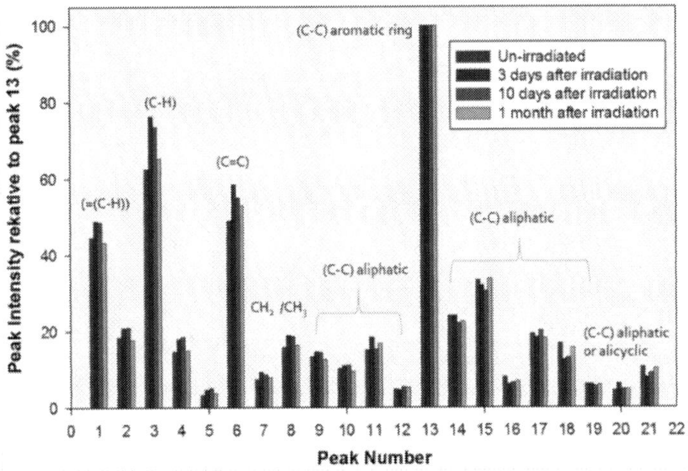

Figure 4: Plot of the intensities of peaks relative to peak 13 representing the C-C aromatic ring bond.

6. Conclusion

According to the results obtained in this study, irradiation to high doses in plastic scintillator EJ200 using 6 MeV protons causes radiation damage and leads to a reduced performance in its scintillation capabilities. At a dose of approximately 0.8 MGy, the structure of the scintillator remains robust, however scintillation light loss may still occur since light transmission results indicate that a competitive absorptive process develops between the primary and secondary fluors.

At a dose of 8 MGy, the scintillator becomes less transparent to its own light. The formation of an absorptive tint indicates that free radicals may be absorbing light that would otherwise be collected from the scintillator. These free radicals may arise from hydrogen that is lost through bonds breaking within the benzene rings of the polyvinyl toluene base and hence leading to additional aliphatic chains and alicyclic rings observed in the Raman spectra. The Raman spectra further indicate that after 4 weeks, the scintillator significantly recovers its structure. As the irradiation dose is increased, a greater loss to light transparency is observed. This can also be visibly noticed in the increased yellowing of the irradiation spot among samples with progressing higher doses. Increasing dose leads to greater structural damage and which in turn will lead to a greater reduction in the light output of the scintillator.

Overall, EJ200 can be considered as substantially radiation hard since for a dose of 0.8 MGy, it still maintains its structural composition and remains fairly transparent. It is only at doses higher than 8 MGy that permanent structural damage occurs. The highest dose currently expected within the Tile Calorimeter over the next several years is 10 kGy and therefore EJ200 may be a potential candidate for consideration during the 2018 replacement of the GAP scintillators. Light yield studies will need to be taken into account however, before this conclusion can be confirmed.

7. Upcoming work

This study is being carried out on several other plastic scintillators including PVT based EJ208, EJ260 and BC-408 obtained from ELJEN technologies and Saint Gobain Crystals respectively. These commercial scintillators will be compared to the polystyrene based scintillators presently used within the Tile Calorimeter. In addition to the techniques looked at in this paper, healing in the scintillators as well as the effect of dose rate on the damage induced are being looked at. Electron Paramagnetic resonance (EPR) and nuclear magnetic resonance (NMR) studies are also being conducted in order identify the types of free radicals formed due to radiation damage.

References

[1] ATLAS Collaboration, "The ATLAS Experiment at the CERN Large Hadron Collider," IOP Publishing and SISSA, 2008.

[2] G. F. Knoll, "Chapter 8: Organic Scintillators," in *Radiation Detection and Measurement, Third Edition*, Michigan, John Wiley & Sons, Inc., 1999, pp. 220-222.

[3] M. Chen, "Queen's University, PHYS 352: Measurement, Instrumentation and Experiment Design," 4 May 2011. [Online]. Available: http://www.physics.queensu.ca/~phys352/lect19.pdf

[4] Universität Heidelberg, "Scintillation Detectors: Particle Detection via Luminescence," 8 May 2011. [Online]. Available: http://www.kip.uni-heidelberg.de/~coulon/Lectures/Detectors/Free_PDFs/Lecture4.pdf

[5] ELJEN Technology, "Products: Plastic Scintilators," [Online]. Available: http://www.eljentechnology.com/index.php/products/plastic-scintillators. [Accessed 12 October 2013].

[6] C. Zorn, "A PEDESTRIAN'S GUIDE TO RADIATION DAMAGE IN PLASTIC SCINTILLATORS," *Radiat. Phys. Chem.*, vol. 41, no. 1/2, pp. 37-43, 1993.

Journal of Physics: Conference Series **623** (2015) 012017 doi:10.1088/1742-6596/623/1/012017

Double Higgs production at FCC-he and prospects for measurements of self-coupling

Mukesh Kumar[⋆,1,†], **Xifeng Ruan**[2], **Alan S. Cornell**[1], **Rashidul Islam**[3] **and Bruce Mellado**[2]

[1]National Institute for Theoretical Physics, School of Physics, University of the Witwatersrand, Wits 2050, South Africa.

[2]University of the Witwatersrand, School of Physics, Private Bag 3, Wits 2050, South Africa.

[3]Department of Physics, University of Calcutta, 92, Acharya Prafulla Chandra Road, Kolkata 700009, India.

E-mail: ⋆`mukesh.kumar@cern.ch`

†Talk Presented by Mukesh Kumar at the *Workshop on Discovery Physics at the LHC (KRUGER 2014)*.

Abstract. The measurement of the triple Higgs boson coupling is one of the most important goals of the Higgs physics program in present and future collider experiments. This would provide the first direct information on the Higgs potential, which is responsible for electroweak symmetry breaking mechanism. We present a double Higgs production scenario at the Large Hadron-Electron Collider (LHeC) and Future Circular Hadron-Electron Collider (FCC-he) through e^-p collisions, which will provide information about trilinear coupling and the possibility of probing new physics, if there is any. The LHeC will provide e^- beams to collide head-on with proton beams of 7 TeV from the Large Hadron Collider (LHC). The prospect of replacing the LHC with the high energy FCC, with proton beams of 50 TeV, is used for FCC-he studies. Energy of the e^- is taken to be 60 GeV for both LHeC and FCC-he. Effects of non-standard CP-even and CP-odd couplings for hhh, hWW and $hhWW$ have been studied and constrained at a 95% C.L .

1. Introduction

The long awaited scalar particle in the Standard Model (SM) of particle physics, known as the Higgs boson, is here to stay as is evident from the results of Run I of the ATLAS and CMS collaborations of the Large Hadron Collider (LHC). It is certain that the new particle is a Higgs boson. But still there are some doubts whether it is *the* Higgs boson of the SM. To confirm, that one needs to understand and probe its properties and interactions with other existing fermions and bosons. Especially for the Higgs self coupling (λ), which rests on a very high pedestal as far as its importance in understanding the scalar sector goes. It will help us understand the electroweak symmetry breaking (EWSB) mechanism. And if we consider it as the SM Higgs boson then one could only *see* this interaction through multi-Higgs productions. A major constraint in the extraction of such signals with respect to the backgrounds is that it is highly dependent on the type of collider. Each collider, be it the FCC-hh (Future Circular Hadron-Hadron Collider), LHC, ILC (International Linear Collider) or something else, has its own challenges in detection of double Higgs signals. Therefore, one should look for all the possibilities, even those beyond these colliders. Here we consider double Higgs production in high energy e^-p collisions for the first time. Numerous studies show that, at hadron colliders like the LHC, it would be a very difficult task to extract the precise information about λ. And we hope that the high energy e^-p collider, on top of the LHC, will be used to probe these properties, including others like the CP behaviour of effective hWW and $hhWW$ couplings.

2. LHeC and FCC-he

One of the proposed colliders, in the new energy frontier at the LHC, is to inject an electron beam which will collide head-on with the available proton beam. This e^-p collision option with center of mass energy $\sqrt{s} \approx 1.3$ TeV (where e^- energy $E_e = 60$ GeV, and the proton beam energy $E_p = 7$ TeV) is known as the *Large Hadron Electron Collider* (LHeC). Its many physics goals, like QCD precision, electroweak physics, probing high parton densities and physics beyond the SM (BSM), can be found in Ref. [1], along with the technical details of collider configuration. The Ring-Ring (RR) option assumes $E_e = 50 - 100$ GeV, whereas the Linac-Ring (LR) option can be realised with $E_e = 60 - 140$ GeV for the LHeC. The design for the new collider is still going through intense scrutinization, the details of which can be found in Ref. [2].

However, further upgrades to the high-energy LHC (HE-LHC) would provide proton beam energies up to 50 TeV. This is another available option, popularly known as the *Future Circular Hadron-Electron Collider* (FCC-he). This upgrade, with $E_p = 50$ TeV, assumes a maximum of $E_e = 250$ GeV in the RR option and the working energy in the LR option for electrons assumes potential reuse of the LHeC with $E_e = 60$ GeV. In this proceeding we present the double Higgs production at the FCC-he with $\sqrt{s} \approx 3.5$ TeV and the prospects of measurements of the involved couplings with preferred LR configuration.

3. The Higgs boson at the LHeC

As is evident from the introduction, the Higgs boson searches are of utmost importance for quite a few past and present colliders, and its properties are needed to be measured accurately to get more precisely insight into the laws of nature. In this respect the future colliders will play a very important role in sectors where the LHC would have limited access even after its upgrades. In one such study, the authors of Ref. [3] investigated the bottom Yukawa coupling at the LHeC. They used the forward jet tagging to improve the Higgs boson signal significantly in the $h \to b\bar{b}$ decay mode from deep inelastic electron-proton scattering at the LHC. In the process we see that the requirement of forward jet tagging in the charged current events strongly enhances the signal-to-background ratio in their study. It is also to be noted that the charged current process at the LHeC proceeds through W-vector boson fusion. Here, in the absence of the contamination from the hZZ coupling, one could directly measure hWW coupling strength as well. CP properties of the Higgs boson can be determined considering an effective five-dimensional vertex given by [4]

$$\Gamma_{\mu\nu}(p,q) = \frac{g}{m_W}\left[\lambda_C(p \cdot q g_{\mu\nu} - p_\nu q_\mu) + i\lambda'_V \epsilon_{\mu\nu\rho\sigma} p^\rho q^\sigma\right], \tag{1}$$

where λ_C and λ'_V are the effective coupling strengths for the anomalous CP-conserving and the CP-violating operators respectively. They have shown that the azimuthal angle between missing energy and non-b jet, $\Delta\phi_{MET-J}$, is a powerful and unambiguous probe of the anomalous hWW couplings, both for CP-conserving and violating type.

4. Double Higgs boson production at the FCC-he

In our present work [5], we would like to see the double Higgs production through charged currents in e^-p collisions and the representative leading order (LO) Feynman diagrams are shown in Fig. 1. At the LHeC, the center of mass energy is such that the double Higgs production is suppressed by phase space. As a result, the signal efficiency would be too low to detect the

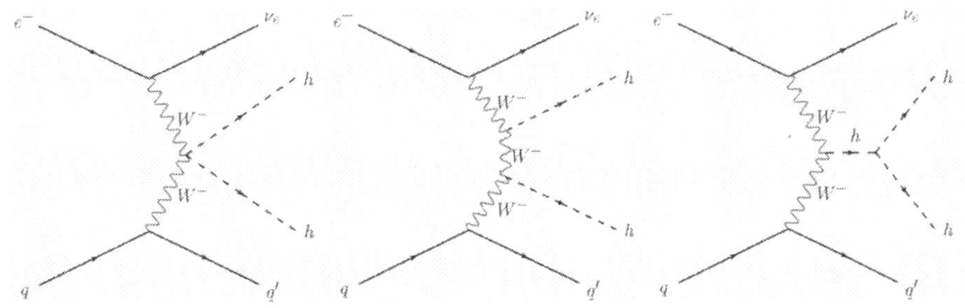

Figure 1: *Representative Feynman diagrams contributing to the process $e^- p \to \nu_e hhj$ in leading order. In the above diagrams $q \equiv u, c, \bar{d}, \bar{s}$ and $q' \equiv d, s, \bar{u}, \bar{c}$ respectively.*

events. But there is a further plan for the HE-LHC that would provide 50 TeV protons. This would increase the LHeC center of mass energy up to ~ 3.5 TeV with 60 GeV electrons. This energy would present an opportunity to probe Higgs self coupling strength (λ of the introduction) through double Higgs boson production.

4.1. Theoretical Framework

For our study we have parametrized all the bosonic couplings (since the possible fermionic couplings involved are highly constrained) that appear in the processes of Fig. 1 by the following effective Lagrangians:

$$\mathcal{L}^{(3)}_{hhh} = \frac{m_h^2}{2v}(1 - g^{(1)}_{hhh})h^3 + \frac{1}{2} g^{(2)}_{hhh} h\partial_\mu h \partial^\mu h, \tag{2}$$

$$\mathcal{L}^{(3)}_{hWW} = -\frac{g}{2m_W} g^{(1)}_{hWW} W^{\mu\nu} W^\dagger_{\mu\nu} h - \frac{g}{m_W}\left[g^{(2)}_{hWW} W^\nu \partial^\mu W^\dagger_{\mu\nu} h + \text{h.c.}\right] - \frac{g}{2m_W}\tilde{g}_{hWW} W^{\mu\nu}\tilde{W}^\dagger_{\mu\nu} h, \tag{3}$$

$$\mathcal{L}^{(4)}_{hhWW} = -\frac{g^2}{4m_W^2} g^{(1)}_{hhWW} W^{\mu\nu} W^\dagger_{\mu\nu} h^2 - \frac{g^2}{2m_W^2}\left[g^{(2)}_{hhWW} W^\nu \partial^\mu W^\dagger_{\mu\nu} h^2 + \text{h.c.}\right] - \frac{g^2}{4m_W^2}\tilde{g}_{hhWW} W^{\mu\nu}\tilde{W}^\dagger_{\mu\nu} h^2. \tag{4}$$

Here $g^{(1)}_{hhh}$ is defined such that it appears as a multiplicative constant to λ_{SM}, i.e., $\lambda \to g^{(1)}_{hhh}\lambda_{SM}$ in the expression of EWSB potential in the SM:

$$V(\Phi) = \mu^2 \Phi^\dagger \Phi + \lambda(\Phi^\dagger\Phi)^2 \to \frac{1}{2} m_h^2 h^2 + \lambda v h^3 + \frac{\lambda}{4} h^4, \tag{5}$$

with $\lambda = \lambda_{\text{SM}} = m_h^2/(2v^2) \approx 0.13$. The effective couplings appearing in the above mentioned Lagrangian $g^{(1)}_{hhh}$, $g^{(2)}_{hhh}$, $g^{(1)}_{hWW}$, $g^{(2)}_{hWW}$, $g^{(1)}_{hhWW}$, and $g^{(2)}_{hhWW}$ are CP-conserving, whereas $\tilde{g}_{hWW}$ and $\tilde{g}_{hhWW}$ are CP-violating couplings. m_h and m_W are respectively masses of the Higgs and W-bosons, $W^{\mu\nu} = \partial^\mu W^\nu - \partial^\nu W^\mu$. The vaccum expectation value (*vev*) for the SM Higgs-doublet is $v \approx 246$ GeV and g is the electroweak coupling in the SM.

Here we consider the speculated detector parameters, and cut-based analysis, to get charged current signals with respect to all possible charged/neutral current and photo-production backgrounds. A statistical analysis is also performed to find the sensitivity of $g^{(1)}_{hhh}$ with other effective couplings.

Process	CC (fb)	NC (fb)	PHOTO (fb)
Signal:	2.40×10^{-1}		
$bbbbj$:	8.20×10^{-1}	$3.60 \times 10^{+3}$	$2.85 \times 10^{+3}$
$b\bar{b}jjj$:	$6.50 \times 10^{+3}$	$2.50 \times 10^{+4}$	$1.94 \times 10^{+6}$
$zzj(z \to b\bar{b})$:	7.40×10^{-1}	1.65×10^{-2}	1.73×10^{-2}
$t\bar{t}j$(hadronic):	3.30×10^{-1}	$1.40 \times 10^{+2}$	$3.27 \times 10^{+2}$
$t\bar{t}j$(semi-leptonic):	1.22×10^{-1}	$4.90 \times 10^{+1}$	$1.05 \times 10^{+2}$

Table 1: *Cross sections (in fb): $E_e = 60$ GeV, $E_p = 50$ TeV, $j = gu\bar{u}d\bar{d}s\bar{s}c\bar{c}$. Initial cuts: $|\eta| \leq 10$ for jets, leptons and b, $P_T \geq 10$ GeV, $\Delta R_{\min} = 0.4$ for all particles.*

4.2. Cross section, Detector setup and cut-based analysis

We present the fiducial cross sections for the signal as well as the backgrounds, before cut-based analysis, in Table 1. For the signal we consider the charged current process $pe^- \to \nu_e hhj, h \to b\bar{b}$. Photo-production backgrounds are very important, with other charged/neutral-current backgrounds, and those backgrounds are estimated through "Equivalent photon approximation structure functions"; which is calculated with the "Improved Weizsaecker-Williams formula" [6].

In the detector setup the maximum rapidity $|\eta|$ range is up to 7. For the b-tagging, the jets with $|\eta| < 5$ and transverse momentum $p_T > 15$ GeV is taken. The fake rate for a c-initiated jet and a light jet to the b-jet is 10% and 1% respectively. The weight corresponding to the b-tagging efficiency, or fake rate, is assigned to each event. Furthermore the following cut flows are taken for analysis:

- Select 4 b + 1-jet: $p_T^{jet} > 20$ GeV, $|\eta| < 7$ for *non-b*-jets, $|\eta| < 5$ for b-jets. The four b jets must be well separated within $\Delta R > 0.7$ [1] in case of overlapped truth matching in the b-tagging.
- Rejecting leptons with $p_T^{e^-} > 10$ GeV (to suppress the neutral-current process).
- Rapidity of the forward jet $\eta_{forward-jet} > 4.0$, the forward jet as defined as the *non-b*-jet which has the largest p_T after selecting at least 4 b-jets.
- The missing transverse energy $E_T^{miss} > 40$ GeV and $\Delta\Phi_{E_T^{miss},leadingjet} > 0.4$, $\Delta\Phi_{E_T^{miss},subleadingjet} > 0.4$ [2].
- Pair the four b-jets into two pairs and calculate the invariant masses of each pair. The composition of the pair which has the smallest variance of mass to $(m_H - 40)$ GeV is chosen. The first pair is defined as $90 < M_1 < 125$ GeV, which must have the leading b-jet. The other pair is defined as $75 < M_2 < 125$ GeV.
- Choosing the invariant mass of all four b-jets greater than 280 GeV.

And the signal significance is calculated with a Poisson distribution [3], considering the expected signal (S) and background (B) yields at 10 ab^{-1} luminosity. After performing these cut based analysis the signal events are ~ 63 with respect to total background events ~ 35 and $s = 8.7$. A 5% systematic error is introduced into signal and backgrounds.

[1] The ΔR is defined as the distance between two objects i and j in the rapidity-azimuthal plane: $\Delta R = \sqrt{(\phi_i - \phi_j)^2 + (\eta_i - \eta_j)^2}$, where ϕ_i and η_i are the azimuthal angle and the rapidity of the object i.

[2] $\Delta\Phi_{ij}$ is the azimuthal angle difference of two objects i and j. The (sub)leading jet is the *non-b* jet defined with (second-)highest p_T.

[3] $s = \sqrt{2\left((S + B)\log(1 + S/B) - S\right)}$

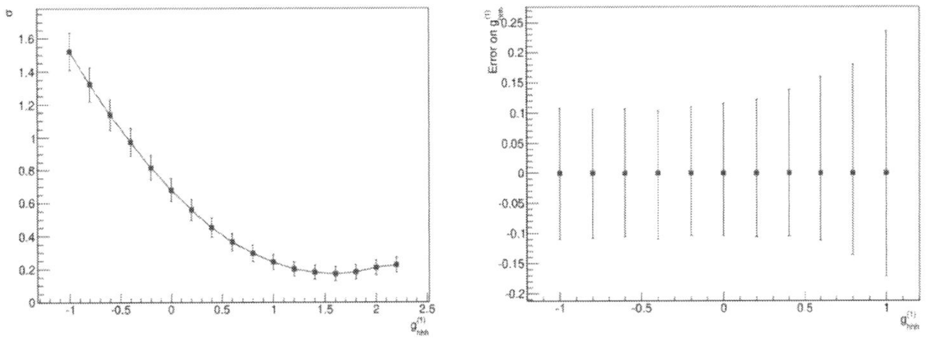

(a) (b)

Figure 2: *(a) Variation of cross section of the signal process $\sigma(pe^- \to \nu_e hhj, h \to b\bar{b})$ with respect to $g^{(1)}_{hhh}$ with error bar at each value of $g^{(1)}_{hhh}$, (b) local error through linear interpolation at each value of $g^{(1)}_{hhh}$.*

4.3. Statistical analysis

Following the method given in Ref. [7], exclusion limits for $g^{(1)}_{hhh}$ are calculated. Fig. 2 shows significant behaviour of cross section variation with respect to $g^{(1)}_{hhh}$ which is expected due to interference between resonant and non-resonant Higgs mediation in the charge-current signal process. The 95% upper limits of all effective couplings appear in the Lagrangian Eqs. (2), (3) and (4), due to cross section influence are also calculated and shown in Fig 3. The sensitivity of CP-even and odd HWW effective couplings, $g^{(1)}_{hWW}$ and $\tilde{g}_{hWW}$, are of the same order as $g^{(1)}_{hhh}$ $\sim 10^0$. However for $g^{(2)}_{hhh} \sim 10^{-3} - 10^{-4}$ and $g^{(2)}_{hWW}$, $g^{(1,2)}_{hhWW}$, $\tilde{g}_{hhWW}$, these are of the order $\sim 10^{-2}$.

5. Conclusion

We performed a detailed analysis of the double-Higgs production at the FCC-he, an e^-p machine with $\sqrt{s} \approx 3.5$ TeV. Considering the speculated detector parameters, we found the feasibility of such machine for Higgs physics scenarios. It is shown that this machine will provide high precision for the measurement of not only Higgs-self -coupling strengths, but also other effective couplings, which appear in the production mechanism of double-Higgs through e^-p collisions. Furthermore, we approached two aspects of this project:

- Taking different channels of Higgs decays $h \to c\bar{c}, \mu^+\mu^-, \tau^+\tau^-$ and
- Performing a global fit-analysis with the involved Wilson coefficients of the effective six dimension operators appearing in combinations with the effective couplings in this study.

As such, we hope all the above works not only provide the physics potential of e^-p machine to study the Higgs-physics, but also provide information on the quantitative value of the parameters needed to build detectors at this facility.

References

[1] J. L. Abelleira Fernandez *et al.* [LHeC Study Group Collaboration], J. Phys. G **39**, 075001 (2012) [arXiv:1206.2913 [physics.acc-ph]].
[2] LHeC web page with recent papers, talks and workshop documentation: `http://cern.ch/lhec`

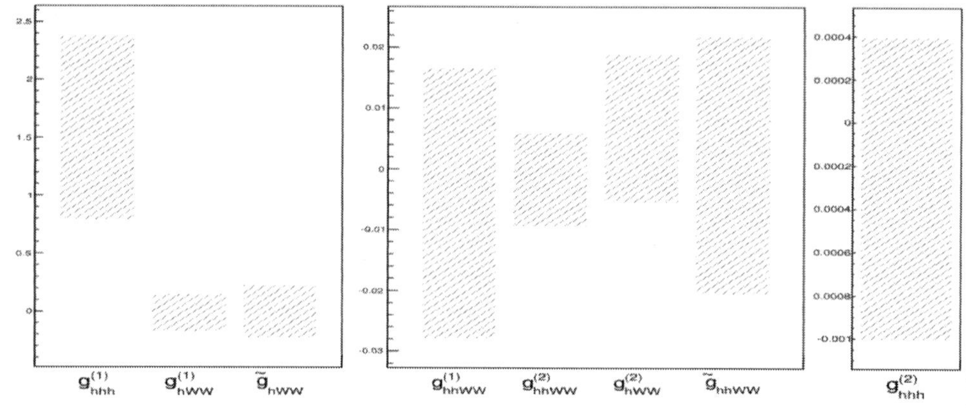

(a)

Figure 3: *The limits on the coupling strength, derived at 0.4* ab^{-1}. *The* $g_{hhh}^{(2)}$ *has only the upper limit because the cross section dependence is monotonic in this region.*

[3] T. Han and B. Mellado, Phys. Rev. D **82**, 016009 (2010) [arXiv:0909.2460 [hep-ph]].
[4] S. S. Biswal, R. M. Godbole, B. Mellado and S. Raychaudhuri, Phys. Rev. Lett. **109**, 261801 (2012) [arXiv:1203.6285 [hep-ph]].
[5] In preparation: M. Kumar, X. Ruan, A. Cornell, R. Islam, M. Klein, U. Klein, B. Mellado
[6] V. M. Budnev, I. F. Ginzburg, G. V. Meledin and V. G. Serbo, Phys. Rept. **15** (1975) 181.
[7] G. Cowan, K. Cranmer, E. Gross and O. Vitells, Eur. Phys. J. C **71** (2011) 1554 [Erratum-ibid. C **73** (2013) 2501] [arXiv:1007.1727 [physics.data-an]].

International Workshop on Discovery Physics at the LHC (Kruger2014) IOP Publishing
Journal of Physics: Conference Series **623** (2015) 012018 doi:10.1088/1742-6596/623/1/012018

SUSY Searches in the ATLAS Experiment

Lawrence Lee
CERN B104-1-B18, CH-1211 Geneva 23, Switzerland
E-mail: Lawrence.Lee@CERN.ch

Abstract. Recent results from the ATLAS experiment at the LHC in the search for supersymmetry are presented. A focus is placed on searches requiring nonstandard techniques in particularly challenging final states. All searches presented here analyze pp collisions with $\sqrt{s} = 8$ TeV collected in 2012. Within these results, in the absence of significant disagreement from the Standard Model expectations, limits are placed on various models in both strong and weak sparticle production as well as on models with long-lived sparticles.

1. Introduction

The Standard Model (SM) leaves one begging for a more complete description of the universe. Despite its successes, with the aesthetic and empirical shortcomings of the theory, new physics entering at or below the TeV scale is highly desirable. The most theoretically elegant way to introduce such physics is via a weak scale supersymmetry (SUSY). Such a theory has the potential to deflate the so-called *naturalness* problem, explain the dark matter content of the universe, and even tackle issues such as providing a mechanism for neutrino masses. However, in the absence of a SUSY at or below the TeV scale, within the reach of the LHC, the naturalness motivations for SUSY are weakened. It is therefore reasonable for one to hope that if SUSY exists, it enters at these scales.

However, there have been nothing but very strong exclusions placed on the SUSY parameter space since its inception roughly 50 years ago. Most recently, Run I of the LHC has produced the most stringent limits on SUSY to date. Various limits on supersymmetric models can be seen in Figure 1 showing large regions of SUSY parameter space excluded by ATLAS [1].

Despite these strong limits, portions of unexcluded SUSY parameter space exist where the theory is still well motivated. Furthermore, models with signatures that are more difficult to find become more compelling. The results presented here will mainly focus on those signatures that often require the use of interesting, nonstandard techniques.

2. Strongly Produced Sparticles

Supersymmetry can present itself in very different phenomenologies depending on the allowed couplings and the sparticle mass spectra. At a hadron collider like the LHC, the strong production of gluinos and squarks will dominate initial sensitivity given that they are of a mass attainable at the given collision energy. The most likely way to find SUSY is via very inclusive searches for strongly produced sparticles.

A combined analysis is performed searching for excesses in final states containing jets, missing transverse energy ($E_{\mathrm{T}}^{\mathrm{miss}}$), and at least one isolated lepton [2]. Diagrams for example processes can be found in Figure 2. Three classes of signal regions are constructed with separate techniques used for each class.

The first of these channels requires a single hard lepton ($p_T > 25$ GeV) and is sensitive to mid- to high-mass-splittings between the pair-produced sparticles and those through which the cascade decay occurs. A veto is placed on events with multiple leptons where no second lepton with transverse

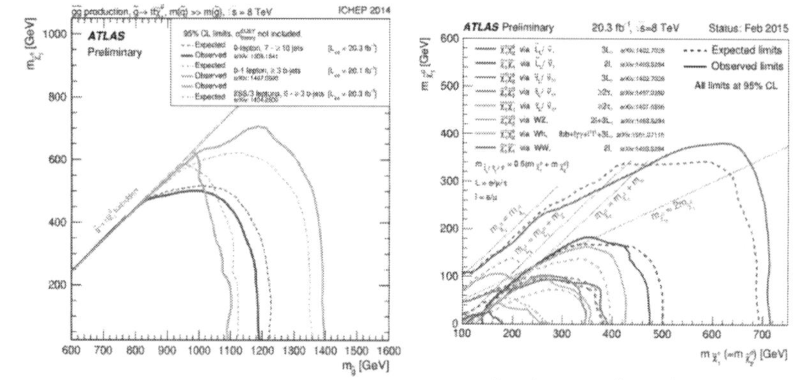

(a) Gluino-mediated stop production

(b) Electroweak Production

Figure 1: Example summaries of ATLAS exclusions in simplified SUSY model spaces are shown.

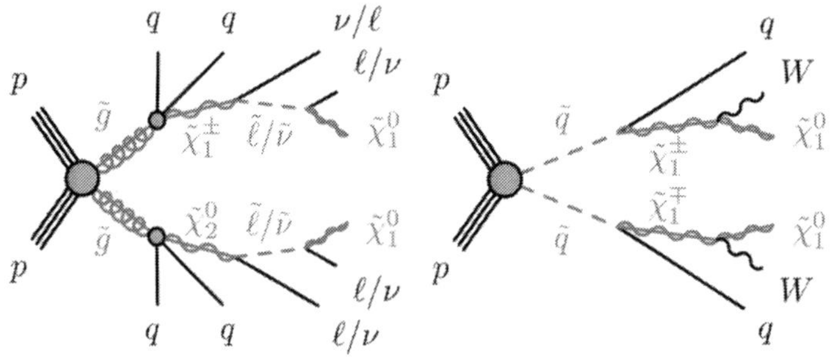

(a) Example gluino pair production

(b) Example squark pair production

Figure 2: Example diagrams are shown for squark and gluino pair-production processes with cascade decays that can include isolated leptons in the final state.

momentum above 10 GeV may exist. The primary discriminating variable of choice is the magnitude of the $E_{\mathrm{T}}^{\mathrm{miss}}$ or the mass variable

$$m_{\mathrm{eff}}^{\mathrm{inc}} = \sum_{i=1}^{N_l} p_{\mathrm{T},i}^{l} + \sum_{j=1}^{N_{jet}} p_{\mathrm{T},j} + E_{\mathrm{T}}^{\mathrm{miss}}. \tag{1}$$

An example signal region within this channel additionally requires at least six jets above various p_{T} requirements, $E_{\mathrm{T}}^{\mathrm{miss}} > 350$ GeV, the ordinary transverse mass $m_{\mathrm{T}} > 150$ GeV, and $m_{\mathrm{eff}}^{\mathrm{inc}} > 150$ GeV. An example discriminating distribution for this region can be seen in Figure 3a showing no significant excess above expected backgrounds.

Alternatively, this analysis also looks in a channel with soft leptons to allow for sensitivity to compressed SUSY spectra. This analysis uses a $E_{\mathrm{T}}^{\mathrm{miss}}$-significance measure defined by $E_{\mathrm{T}}^{\mathrm{miss}}/m_{\mathrm{eff}}^{\mathrm{inc}}$ where the denominator normalizes the quantity to reflect the change in the $E_{\mathrm{T}}^{\mathrm{miss}}$ resolution of the detector. This variable is therefore used to remove events with large $E_{\mathrm{T}}^{\mathrm{miss}}$ that is the result of a poorly reconstructed jet.

Signal regions with a single soft lepton require one electron (muon) with transverse momentum between 7 (6) GeV and 25 (25) GeV. If a second lepton exists in the event with transverse momentum above 7 (6) GeV, the event is vetoed. An example signal region used in this analysis additionally

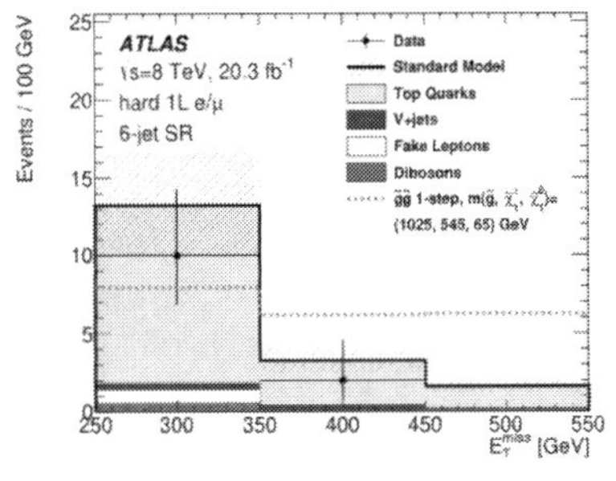

(a) Hard Lepton Channel

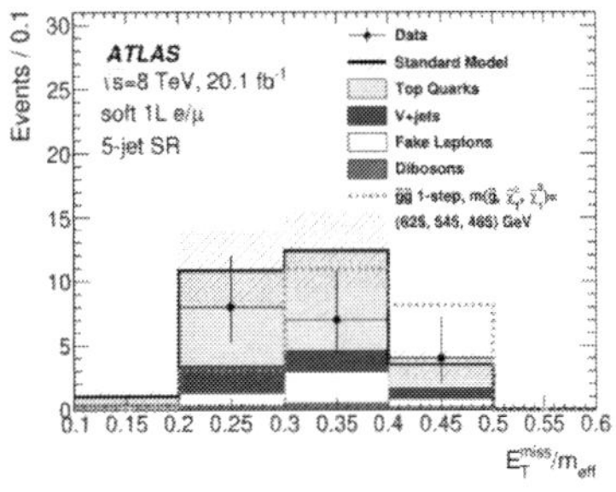

(b) Soft Lepton Channel

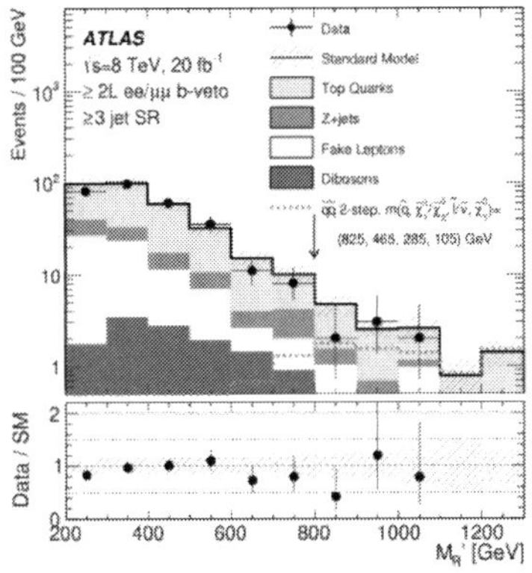

(c) Razor Dilepton Channel

Figure 3: Example signal region distributions for each of the channels used in the squark and gluino search requiring leptons.

requires at least five jets above various momentum requirements, $E_{\mathrm{T}}^{\mathrm{miss}} > 300$ GeV, $m_{\mathrm{T}} > 100$ GeV, and $E_{\mathrm{T}}^{\mathrm{miss}}/m_{\mathrm{eff}}^{\mathrm{inc}} > 0.3$. The results from this region can be seen in Figure 3b. Additionally, a region with two soft leptons is studied but not discussed here.

Finally, a class of events with two hard leptons is used to probe these models employing the so-called *Razor* variables [illegible]. These variables that attempt to characterize the mass of a new sparticle by performing boosts into a proxy "R" frame can provide significant sensitivity to new physics. All visible particles are organized into two "megajets" in a configuration that minimizes the sum of the squared masses of the two "megajet" four-momenta. Two masses are constructed in this "R" frame, one representing the longitudinal momentum M_R' of the system and the other representing the transverse information M_T^R with assumptions on the momenta of the escaped particles. M_R' is defined as

$$M_R' = \sqrt{(j_{1,E} + j_{2,E})^2 - (j_{1,L} + j_{2,L})^2} \quad (2)$$

Finally, the Razor variable "R" is constructed from the ratio of the two that tends to peak at low values for SM-like events and tends to be uniformly distributed in the interval [0,1] for SUSY-like events.

In this channel, the ratio R is used to reduce SM backgrounds, while the final binned variable is the mass M_R'. Two opposite-sign leptons are required with the highest p_{T} above 14 GeV and the second leading $p_{\mathrm{T}} > 10$ GeV. In an example signal region, at least three jets with $p_{\mathrm{T}} > 50$ GeV are required. In this region, additional requirements for final signal sensitivity are $R > 0.35$ and $M_R' > 800$ GeV. The binned distribution of M_R' for this region can be seen in Figure 3c.

In the absence of a significant excess, limits are placed in the gluino/squark vs. lightest neutralino mass plane. Results from the three channels above are statistically combined into a final exclusion

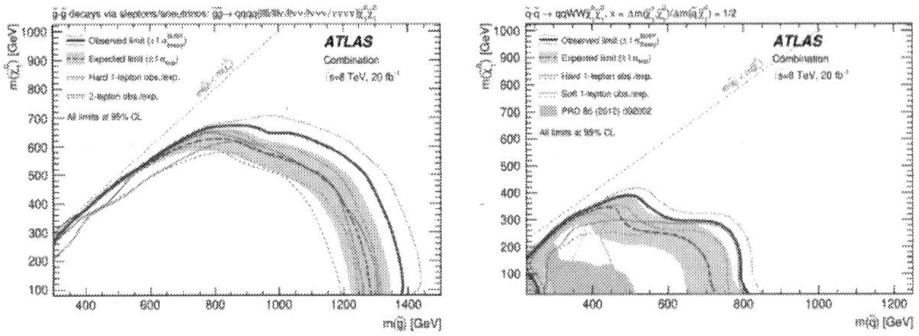

(a) Gluino Production (b) Squark Production

Figure 4: Example mass exclusion contours are shown for the channels targeting strong production of squarks and gluinos in leptonic final states.

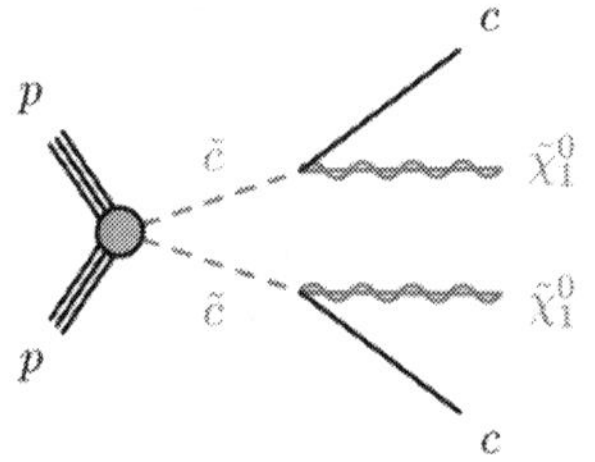

Figure 5: An example diagram is shown for scharm pair production with decays to a final state of two c-jets and two lightest neutralinos.

result in each of these plots and models are excluded for a very wide range of sparticle masses. These results can be seen in Figure 4.

Supersymmetry deflates the naturalness problem by canceling off those quadratically divergent corrections to the Higgs mass. For this reason, this cancellation must be most effective for heavy SM fermions. It is theoretically desirable to have lower sparticle masses for the super partners of the heavier quarks on naturalness grounds. More dedicated searches are also performed focusing on the strong production of such squarks.

While there are ever increasing limits on the stop and sbottom squarks, there are few dedicated constraints on low mass scharm pair production. The analysis presented here looks for pair produced scharm quarks which each decay to a charm quark and a neutralino lightest supersymmetric particle (LSP) in the process depicted in Figure 5. This analysis employs the novel technology of lifetime-based charm tagging to reduce various backgrounds. The algorithms used here are analogous to lifetime-based btaggers that have become commonplace in identifying those jets that contain B hadrons. The algorithm used here is an instance of an existing ATLAS btagger that has been reoptimized for charm efficiency, and calibrated for 2-dimensional use to simultaneously reject b-jets, and jets originating from light quarks and gluons. Multidimensional rejection and efficiency performance of this algorithm can be seen in Figure 6. Efficiencies for selecting b- and c-jets are calibrated in data using $t\bar{t}$ events and jets containing D^* hadrons, respectively. The working point used for this analysis represents a tagging efficiency of 19% (12.5%, 0.5%) for tagging c-jets (b-jets, light flavor and gluon jets) in $t\bar{t}$ events.

Control regions are used to constrain background estimations in signal regions. This procedure is validated in orthogonal validation regions.

Signal regions are constructed with $E_{\mathrm{T}}^{\mathrm{miss}} > 150$ GeV – as events are recorded as the result of $E_{\mathrm{T}}^{\mathrm{miss}}$ triggers – and two c-tagged jets with $p_{\mathrm{T}} > 130, 100$ GeV. The azimuthal angle between the $E_{\mathrm{T}}^{\mathrm{miss}}$ and any jet is required to be > 0.4 to reduce the effects of fake $E_{\mathrm{T}}^{\mathrm{miss}}$ from jet mismeasurement. A requirement on the $E_{\mathrm{T}}^{\mathrm{miss}}$ significance ($E_{\mathrm{T}}^{\mathrm{miss}}/m_{\mathrm{eff}} > 0.25$) is placed in a similar way to the previous analysis. A requirement is placed on the invariant mass of the di-c-jet system to reduce backgrounds

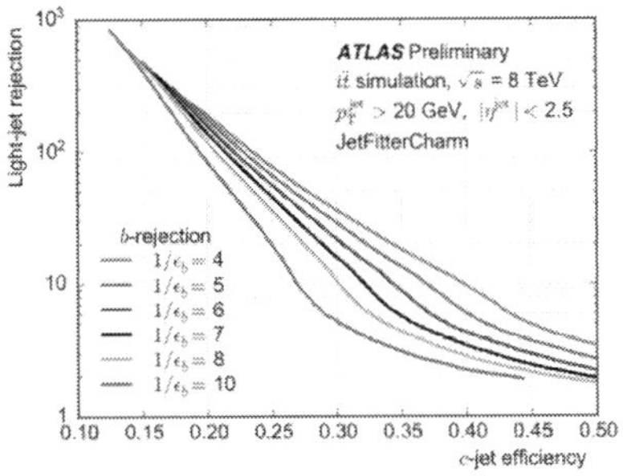

(a) Tagger performance for various b-jet efficiencies

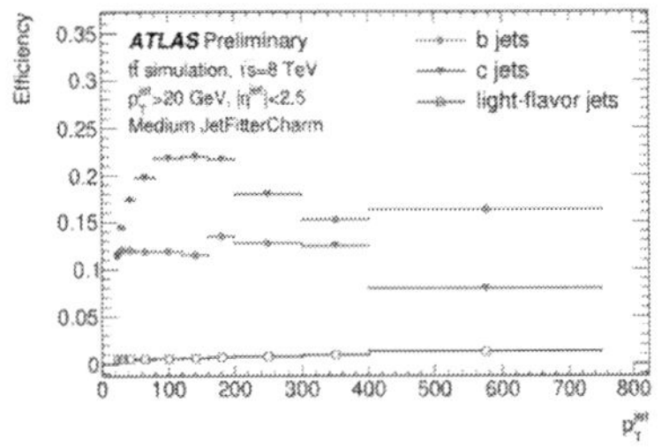

(b) Tagger performance as a function of jet p_T

Figure 6: Rejection and efficiency curves are shown for the c-tagger JetFitterCharm, used in the analysis described in the text.

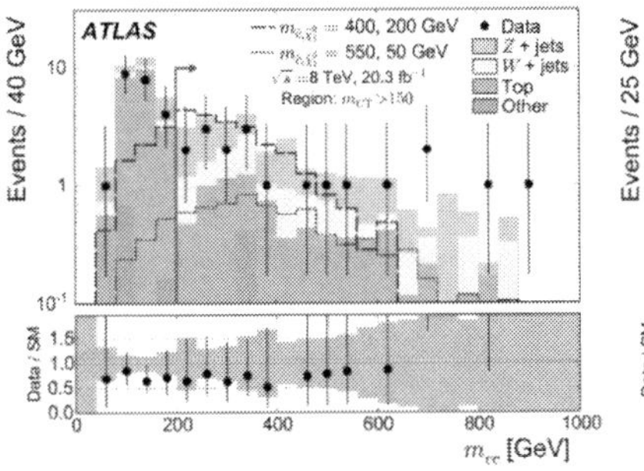

(a) m_{cc}

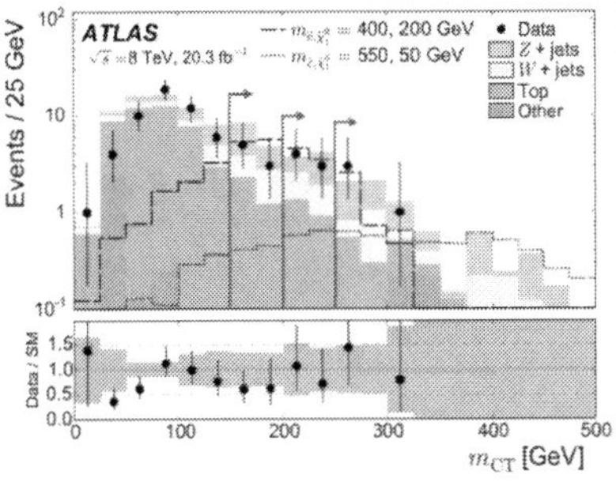

(b) m_{CT}

Figure 7: Signal region distributions for the charm-tagged scharm search are shown.

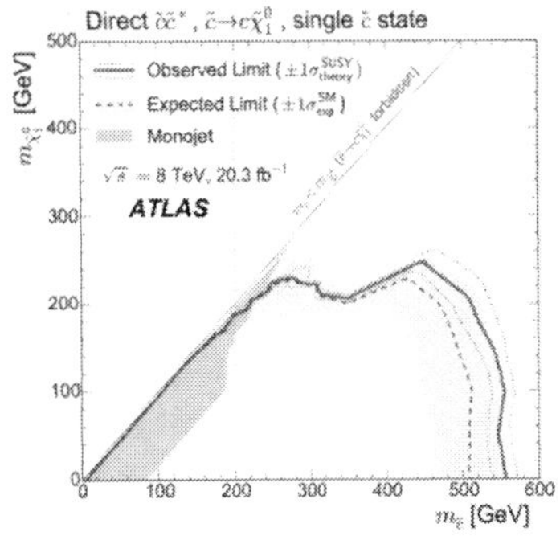

Figure 8: Mass exclusion contours in the scharm vs. lightest neutralino mass plane.

from $g \to c\bar{c}(b\bar{b})$ splittings.

The definition of signal and control regions for this analysis uses the cotransverse mass m_{CT} [4] to reduce backgrounds. Processes with characteristic scales exhibit a kinematic endpoint in this variable above which contributions are suppressed. In $t\bar{t}$ events, if the identified c-jets are the real b-jets from top decays, this variable is expected to have a kinematic endpoint at around 135 GeV. As a result, regions are constructed with requirements of $m_{\mathrm{CT}} > 150, 200, 250$ GeV to reduce most backgrounds significantly. An additional signal region is constructed from the results of a search for stop squarks [5]

Journal of Physics: Conference Series **623** (2015) 012018 doi:10.1088/1742-6596/623/1/012018

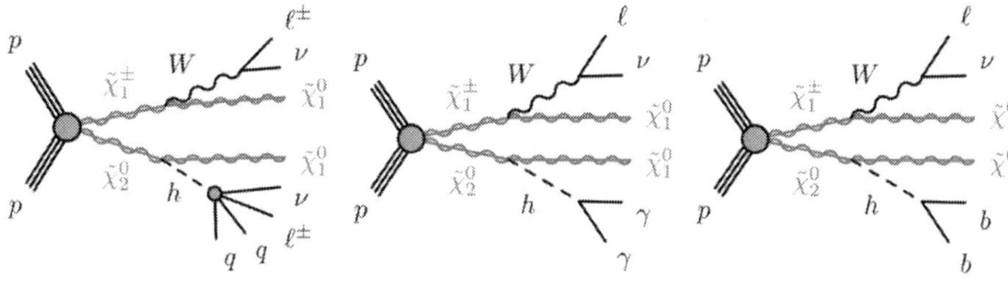

(a) $h \to qql\nu$ (b) $h \to \gamma\gamma$ (c) $h \to bb$

Figure 9: Diagrams representing each higgs decay targeted by a dedicated channel in this analysis are shown.

targeting compressed spectra with a selection for monojet-like events.

The results of this analysis can be found in Figure 7. In the absence of a significant excess, limits are placed on scharm pair production as a function of the masses of the scharm and lightest neutralino and can be found in Figure 8.

3. Electroweak Sparticle Production

In the limit that the strongly produced sparticles are most often produced offshell, production of the EWKinos may dominate and will most likely be found via multilepton final states. A relatively new handle in such searches is the now known mass of the SM Higgs. Given this knowledge, targeted searches can be performed for higgsino-like neutralinos where additional kinematic constraints can be used to suppress backgrounds. The search presented here [6] uses several decay modes of the Higgs to search for associated production of neutralinos and charginos giving final states via a 125 GeV Higgs and a W boson in each event. Diagrams for example processes can be found in Figure 9.

This analysis also employs three search channels, each focusing on a particular Higgs decay. The W is always assumed to decay leptonically, and studied decays of the h are to bb, $\gamma\gamma$, and $WW \to qql\nu$ final states.

In a way similar to the other analyses, individual significant backgrounds are estimated by obtaining normalizations from dedicated control regions. A transverse mass variable is used to suppress background in the lepton plus two b-jet channel. The results of this channel can be seen in Figure 10a and no excess over the SM expectation is observed. The lepton plus diphoton channel performs an unbinned fit in the diphoton invariant mass spectrum. The expected contribution from the Higgs within the window around 125 GeV is taken from simulation while other background contributions are taken from the sidebands of this distribution. The results of this channel can be seen in Figure 10b. Finally, for when the Higgs decays to $WW \to qql\nu$, the two isolated leptons in the event are required to be of the same sign to reduce SM backgrounds. The sensitive variable used here is the partially reconstructed Higgs mass m_{ljj}. When the two jets and lepton from the Higgs decay are properly identified, this mass should be strictly less than the mass of the Higgs. Other backgrounds, such as combinatoric effects or background processes, will result in masses generally larger than the 125 GeV mass of the Higgs. In this way, a signal region can be constructed requiring a low value of this m_{ljj} invariant mass. An example signal region can be seen in Figure 10c. Each of these channels comprises several signal regions and a small sample of them are shown here for illustration.

In the absence of a significant excess, limits are set on chargino and neutralino production. These limits can be seen in Figure 11. For a massless lightest neutralino, EWKino masses up to roughly 250 GeV are excluded. Increasing this lightest neutralino mass to around 40 GeV reduces the exclusion to roughly 170 GeV in EWKino mass.

4. Long Lived Particles

New particles that are long-lived or stable can also exist in many models and provide a very different phenomenology for collider experiments. Searches for these effects require unique analysis techniques

Journal of Physics: Conference Series **623** (2015) 012018 doi:10.1088/1742-6596/623/1/012018

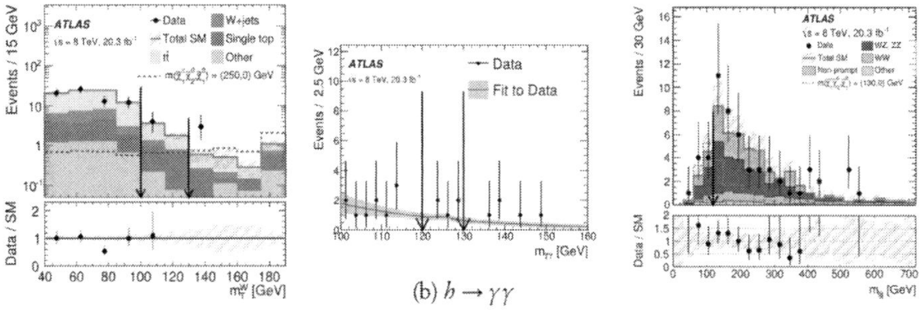

(a) $h \to bb$ (b) $h \to \gamma\gamma$ (c) $h \to qql\nu$

Figure 10: Example signal region distributions for the various channels of the electroweak-production analysis assuming the existence of a 125 GeV Higgs boson.

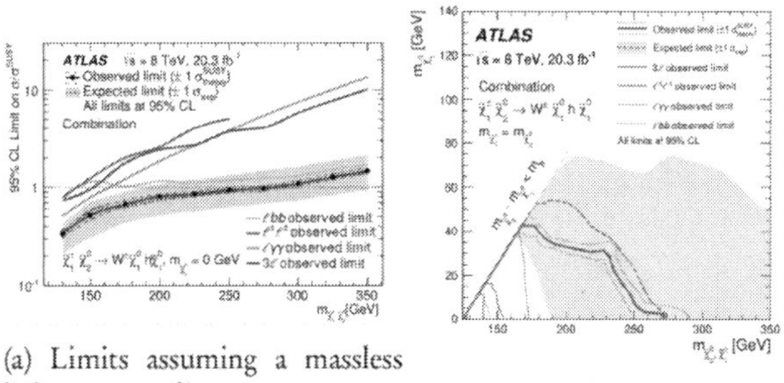

(a) Limits assuming a massless lightest neutralino

(b) Limits as a function of lightest neutralino mass

Figure 11: Mass exclusion limits on the EWKinos are shown.

and use modern detectors in ways they were not designed for. The ATLAS experiment has performed several such searches looking for SUSY signatures that would otherwise skirt existing limits.

The first of these discussed is a search for photons that appear to not have been produced in the hard interaction [7]. Such a photon can present itself either as a non-pointing photon, using the high granularity EM calorimeter of ATLAS to determine the pointing angle of EM showers, or as a late photon as seen by a late-arriving EM shower resolved by the very sensitive timing of the calorimeter.

The non-pointing nature of a photon is quantified in the variable z_{origin} which is the distance along the beam axis between the detector origin and the intersection of the axis with the projected photon momentum. The sensitive variable is

$$\Delta z_{\gamma} = |z_{\text{origin}} - z_{\text{PV}}| \tag{3}$$

where z_{PV} is the z location of the identified primary vertex. The resolution on this pointing is measured in $Z \to ee$ events for low values. Simulated signal processes show agreement in this region and resolutions for higher distances are shown in Figure 12a.

The arrival time of the EM shower also provides discrimination for certain signals. This is measured as t_{γ} directly from the EM calorimeter. The resolution on this time is roughly around 300 ns, though much of this resolution is from a collision timing spread of roughly 220 ns. This resolution can be seen in Figure 12b as a function of cell energy.

These handles are used to probe processes in which a long-lived particle decays to photons and escaped energy. The signature discussed here contains two non-pointing photons and large $E_{\text{T}}^{\text{miss}}$. The results are specifically interpreted in a GMSB model (SPS8) where the gravitino is the LSP.

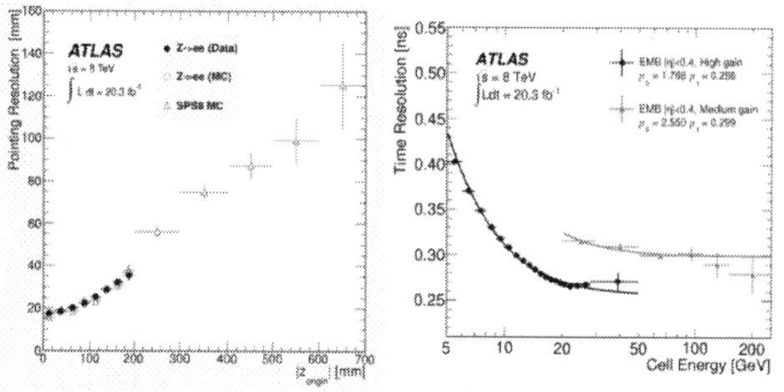

(a) Pointing Resolution (b) Timing Resolution

Figure 12: Expected and validated resolutions on the measured pointing and timing of photons is shown.

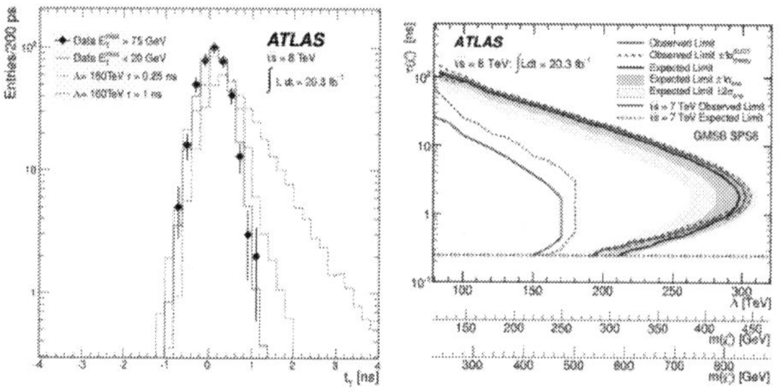

(a) Signal Region Distributions (b) Mass Exclusions

Figure 13: Signal region results and associated mass exclusions are shown for the nonpointing and delayed photon analysis.

Low $E_{\mathrm{T}}^{\mathrm{miss}}$ control regions are used and $Z \to ee$ calibration regions show good agreement in t_γ. Signal regions are constructed by requiring $E_{\mathrm{T}}^{\mathrm{miss}} > 75$ GeV as shown in Figure 13a. These distributions agree well with the control region predictions and no significant deviation is observed. As a result, limits can be set on GMSB model-space excluding lightest neutralino masses up to 430 GeV for lifetimes of order 1 ns. These exclusions can be seen in figure Figure 13b.

Another analysis focuses on particles that are long-lived enough to pass through some portion of the detector [8]. Such particles are motived by many beyond the SM theories - splitSUSY, GMSB, UED, etc. The goal is to identify particles that are moving measurably slower than the speed of light.

If such a particle is electrically charged and passes through layers of the ATLAS pixel detector, these can be identified by exotic Bethe-Bloch contours in dE/dx vs qp space. This gives a measurement of the particle's relativistic $\beta\gamma$. The detector limits this range to within roughly the interval [0.2,1.5] with about a 20% resolution. Alternatively, if the particle arrives at the calorimeter system, precise calorimeter timing can provide a measurement of the particle's relativistic β. If the particle lives long enough to also pass through the muon system, located furthest away from the interaction, additional timing resolution is created from this additional timing information. This resolution can be found in Figure 14a.

This analysis looks at mass spectra defined by

$$m_{\beta\gamma} = \frac{p}{\beta\gamma} \tag{4}$$

for measurements from the pixel detector, and

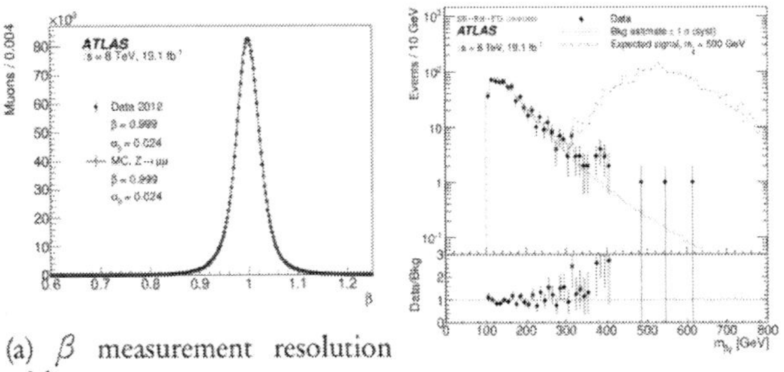

(a) β measurement resolution with muon system

(b) Example mass distribution

Figure 14: Distributions are shown showing the resolution on the β measurement when including the muon system and an example mass distribution are shown.

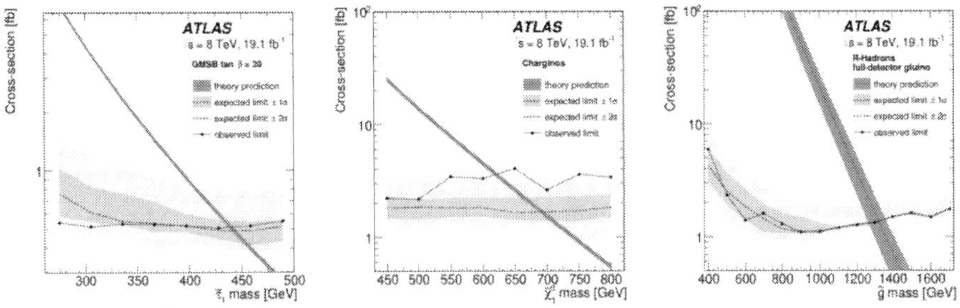

(a) GMSB $\tilde{\tau}$

(b) Stable Chargino

(c) Gluino R-Hadron

Figure 15: Cross-section exclusions for various kinds of long-lived particles are shown.

$$m_\beta = \frac{p}{\beta} \tag{5}$$

for measurements solely based on the detector timing resolutions. Backgrounds are primarily composed of mismeasured muons. In order to get a handle on this contribution, these mass variables are decomposed. The momentum measurement is due to the position of tracking detector hits. The methods for measuring β and $\beta\gamma$ are completely independent from the momentum measurement. Because of this, random pairings of muon momenta and timing measurements are used to build up the expected fake background contributions. An example distribution can be seen in Figure 14b.

In the absence of a significant excess, limits are placed on various models that give rise to this signature. Long-lived staus in GMSB contexts are excluded for masses up to around 380-440 GeV for various values of $\tan\beta$. Stop R-hadrons, bound states including a stop, are excluded up to masses of roughly 880 GeV and gluino R-hadrons are excluded up to roughly 1250 GeV. Long-lived chargino masses are excluded up to 620 GeV. These exclusions, which represent just a portion of the exclusions performed by the analysis, can be seen in Figure 15.

5. Indirect Searches

Finally, in the absence of direct evidence for supersymmetry, it may still be that new particles are hidden among the SM spectrum. In such cases, indirect searches can be useful. It is expected that for a stop with mass at or near the top mass, one of the most sensitive probes is the measurement of the $t\bar{t}$ spin correlation which is sensitive to contributions from the scalar nature of the stop.

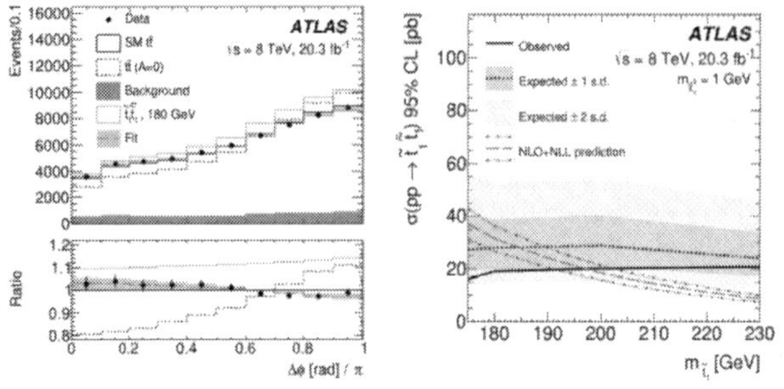

(a) Azimuthal angle between leptons (b) Stop pair-production cross section limits

Figure 16: Results and exclusions for the $t\bar{t}$ spin correlation stop search are shown.

In this analysis, a standard dilepton $t\bar{t}$ selection is used where the azimuthal angle between the two leptons is used as the measure of spin correlation [9]. The angular asymmetry given by this spin correlation is found to be in good agreement with the Standard Model. Near-degenerate stop contributions to this spin correlation can deviate significantly from the pure-top distribution. A standard dileptonic $t\bar{t}$ selection is applied and the measured asymmetry $A_{\text{helicity}} = 0.38 \pm 0.04$ agrees well with the SM prediction of 0.318 ± 0.005. The sensitive angular distribution can be seen in Figure 16a. Given this agreement, assuming a massless neutralino and a 100% branching ratio of $\tilde{t} \to t + \chi_1^0$, stop masses are excluded between the top mass and 191 GeV as seen in Figure 16b.

6. Conclusions

Many results have been presented across the SUSY search program at the ATLAS experiment. A wide range of final states have been explored utilizing novel techniques and extending the ATLAS detector to provide measurements it was not designed to make.

While the results presented here contain nothing but exclusions of SUSY, LHC Run II holds much promise in finding a weak scale supersymmetry providing sensitivity to the majority of the most motivated model spaces with an increased collision energy. As collisions go above high mass production thresholds, production cross sections greatly increase allowing for new discovery potential. So despite the negative landscape at hand, we are still in a hopeful era for supersymmetry.

7. References

[1] ATLAS collaboration. The ATLAS Experiment at the CERN Large Hadron Collider. *JINST*, 3:S08003, 2008.

[2] ATLAS collaboration. Search for squarks and gluinos in events with isolated leptons, jets and missing transverse momentum at $\sqrt{s} = 8$ TeV with the ATLAS detector. 2015.

[3] Christopher Rogan. Kinematical variables towards new dynamics at the LHC. 2010.

[4] Daniel R. Tovey. On measuring the masses of pair-produced semi-invisibly decaying particles at hadron colliders. *JHEP*, 0804:034, 2008.

[5] ATLAS collaboration. Search for pair-produced third-generation squarks decaying via charm quarks or in compressed supersymmetric scenarios in pp collisions at $\sqrt{s} = 8$ TeV with the ATLAS detector. *Phys.Rev.*, D90(5):052008, 2014.

[6] ATLAS collaboration. Search for direct pair production of a chargino and a neutralino decaying to the 125 GeV Higgs boson in $\sqrt{s} = 8$ TeV pp collisions with the ATLAS detector. 2015.

[7] ATLAS collaboration. Search for nonpointing and delayed photons in the diphoton and missing transverse momentum final state in 8 TeV pp collisions at the LHC using the ATLAS detector. *Phys.Rev.*, D90(11):112005, 2014.

[8] ATLAS collaboration. Searches for heavy long-lived charged particles with the ATLAS detector in proton-proton collisions at $\sqrt{s} = 8$ TeV. *JHEP*, 1501:068, 2015.

[9] ATLAS collaboration. Measurement of Spin Correlation in Top-Antitop Quark Events and Search for Top Squark Pair Production in pp Collisions at $\sqrt{s} = 8$ TeV Using the ATLAS Detector. 2014.

Journal of Physics: Conference Series **623** (2015) 012019 doi:10.1088/1742-6596/623/1/012019

The Energy Momentum Tensor associated with Hard Parton production in Finite Time

Ben Meiring and W A Horowitz

Department of Physics, University of Cape Town, Lovers Walk St, Cape Town, South Africa

E-mail: mrnben002@myuct.ac.za, wa.horowitz@uct.ac.za

Abstract. We use the Schwinger-Keldysh finite-time formalism applied to an interacting scalar field theory to derive perturbative expressions detailing the system which exists during the initial stages of a high energy collision. Further, we define the QFT conditional expectation value in an effort to describe the initial conditions associated with jet production in heavy ion physics.

In this paper we calculate $\langle\phi\rangle(\mathrm{x})$ for a scalar Yukawa model, demonstrate the finiteness of the energy momentum tensor for $\lambda\phi^4$ to leading order, and derive an expression for the conditional expectation value of operators to aid in the description jet-like behaviour in interacting theories.

1. Introduction

AdS/CFT computations have been used to describe the energy loss of QCD-like particles moving through a strongly coupled plasma[1, 2], but little is understood regarding the initial conditions of these jets. The intention of this research is to understand the behaviour of particles in an interacting theory as a function of spacetime, during the initial moments after a high energy collision. Asymptotic freedom in QCD suggests that a perturbative approach with weak coupling will be well justified in this regime. A natural object to study is the expectation value of the energy momentum tensor. This can be done in the Schwinger-Keldysh formalism, matching the full interacting theory states to asymptotically free states at $t = \infty$ in the interaction picture. We will refer to these asymptotic states as $|\mathrm{in}\rangle$ states. For some operator $\hat{O}(t, \vec{x})$ the expectation value is given by

$$\langle \hat{O}_{\mathrm{Heisenberg}}\rangle(t) = \langle \underrightarrow{T} \exp\left(i\int_{-\infty}^{t} dz_1\, \hat{H}_{\mathrm{I}}^{-}(z_1)\right)\hat{O}_{\mathrm{I}}(t)\underleftarrow{T}\exp\left(-i\int_{-\infty}^{t} dz_1\, \hat{H}_{\mathrm{I}}^{+}(z_1)\right)\rangle \tag{1}$$

The operator here is spacetime dependent, and so information regarding the evolution of the system can be retrieved. The "+" and "−" superscripts are used to disintguish the Hamiltonians of the time ordered and anti-time ordered exponentials, but the same Hamiltonians are used. It can be shown that this object is equivalent to the contour ordered exponential

$$\langle T_C \left(e^{-i\int_{-\infty}^{\infty} dz_1\left(\hat{H}_I^{+}(z_1)-\hat{H}_I^{-}(z_1)\right)}\hat{O}_I(t)\right)\rangle, \tag{2}$$

where T_C is the contour ordering operator which orders the fields by their position along the Schwinger-Keldysh contour given below.[1]

[1] The superscript + indicates a field on the top path of the contour, which will be evaluated before any field with index − which is located on the bottom path.

Figure 1: Time evolution indicated by the Schwinger-Keldysh contour.

This system can be solved in a similar way to usual diagrammatic calculations, now with four possible propagators due to contractions between " + " and " − " field operators. For some field $\phi(x)$ the contour ordered contractions $\langle 0|T_C\{\phi^i(x_1)\phi^j(x_2)\}|0\rangle$ will give propagator contributions of the form

$$D^{ij}(x_1,x_2) = \begin{cases} \langle 0|\underleftarrow{T}\{\phi(x_1)\phi(x_2)\}|0\rangle & \text{if } i,j=+,+ \\ \langle 0|\{\phi(x_2)\phi(x_1)\}|0\rangle & \text{if } i,j=+,- \\ \langle 0|\{\phi(x_1)\phi(x_2)\}|0\rangle & \text{if } i,j=-,+ \\ \langle 0|\underrightarrow{T}\{\phi(x_1)\phi(x_2)\}|0\rangle & \text{if } i,j=-,- \end{cases}$$

Important relations $D_R(x_1-x_2) = \sum\limits_{m=\{+,-\}} mD^{+m}(x_1,x_2)$ and $\sum\limits_{n,m=\{+,-\}} nmD^{nm}(x_1,x_2) = 0$ can be derived through this definition.

2. Expectation of $\phi(x)$

To gain an intuition for spacetime dependent Schwinger-Keldysh calculations, we find $\langle\phi(x)\rangle$. For now we will drop the T_C path ordering operator and assume that every expectation value is contour ordered. First we note that $\langle \text{in}_1|\text{in}_2\rangle \sim \delta_{\text{in}_1,\text{in}_2}$ in Schwinger-Keldysh, which can be seen by finding the expectation of the identity operator. In general we can choose the operator $\hat{O}$ to be evaluated on the + or − contour without changing the result. Choosing $\phi(x)$ on the + contour, we will have

$$\langle\phi(x)\rangle_{\text{Heisenberg}} = \sum_{n=0}^{\infty}\frac{(-i)^n}{n!}\langle\phi^+(x)\prod_{i=1}^{n}\sum_{m_i=\{+,-\}} m_i H^{m_i}\rangle_{\text{interaction}} \tag{3}$$

If we choose the incoming states to be the same $|\text{in}_1\rangle$ then contracting $\phi(x)$ with an external state will allow us to factorize the operator out. Thus the rest of diagram will depend on the expectation $\langle \text{in}_1|\mathbb{I}|\text{in}_2\rangle$ with $|\text{in}_1\rangle \neq |\text{in}_2\rangle$. Thus these diagrams will contribute 0, and $\phi(x)$ will only contract with $\phi(z_i)$ operators from the interaction Hamiltonians.

We will choose $H^{m_i} = \int d^4z\,\left(g\psi\phi\psi^{m_i} - \delta_g\psi\phi\psi^{m_i} + \frac{1}{2}\psi\left(\delta_\psi\partial^2 - \delta_{m_\psi}\right)\psi^{m_i} + \frac{1}{2}\phi\left(\delta_\phi\partial^2 - \delta_{m_\phi}\right)\phi^{m_i}\right)$. The counter terms here simply cancel the divergences that come from the loop integrals, so we will not consider their effect in too much detail. Focusing on the $g\psi\phi\psi^{m_i}$ term, we explicitly write out the possible contractions.

$$\begin{aligned}\langle\phi(x)\rangle_{\text{Heisenberg}} &= \sum_{n=0}^{\infty}\frac{(-ig)^n}{n!}\langle\phi^+(x)\int d^4z_1...d^4z_n\sum_{m_i=\{+,-\}} m_1\psi\phi\psi^{m_1}...\sum_{m_n=\{+,-\}} m_n\psi\phi\psi^{m_n}\rangle \\ &= \sum_{n=0}^{\infty}\frac{(-ig)^n n}{n!}\int d^4z_n\langle\sum_{m_n=\{+,-\}} m_n\overline{\phi^+(x)\phi^{m_n}}(z_n)\,(\psi\psi^{m_n}(z_n))\prod_{i}^{n-1}\int d^4z_i\sum_{m_i=\{+,-\}} m_i\psi\phi\psi^{m_i}\rangle \\ &= (-ig)\int d^4z\left(\sum_{m=\{+,-\}} mD^{+m}(x,z)\right)\sum_{n=0}^{\infty}\frac{(-ig)^{n-1}}{(n-1)!}\langle(\psi\psi^m(z))\prod_{i}^{n-1}\int d^4z_i\sum_{m_i=\{+,-\}} m_i\psi\phi\psi^{m_i}\rangle \\ &= (-ig)\int d^4z\left(\sum_{m=\{+,-\}} mD^{+m}(x,z)\right)\langle\psi\psi^m(z)\rangle_{\text{Heisenberg}}\end{aligned}$$

Including the counter terms in this analysis gives us the same result, now with all loop contributions being renormalized. $\langle\psi\psi^m(z)\rangle_{\text{Heisenberg}}$ is just the expectation value of some operator. It does not matter if this operator is evaluated on the + or − contour, the answer will be the same through the m sum.

$$\langle\phi(x)\rangle_{\text{Heisenberg}} = (-ig)\int d^4z \left(\sum_{m=\{+,-\}} mD^{+m}(x,z)\right)\langle\psi\psi(z)\rangle_{\text{Heisenberg}}$$
$$= (-ig)\int d^4z\, D_R(x-z)\langle \text{in}|\psi\psi(z)|\text{in}\rangle_{\text{Heisenberg}}$$

2.1. Leading order emission from a single particle

The above result will hold in arbitrary dimensions.

$$\langle\phi(x)\rangle = -ig\int d^{n+1}z\, D_R(x-z)\langle\psi\psi(z)\rangle \tag{4}$$

Notice that this is the solution to the equations of motion given by classically coupled Lagrangian $\mathcal{L} = \frac{1}{2}\partial_\mu\phi\partial^\mu\phi - \frac{1}{2}m_\phi^2\phi^2 - g\phi\langle\psi\psi\rangle$. In general if we are able to determine the expectation $\langle\psi\psi\rangle$, we will have this expectation exactly. This gives this quantity the interpretation of being a propagation of the ϕ field from the current given by the expectation of the square of the ψ. In weak coupling this case be considered as a statistical ensamble of the possible virtual/real particles off which the ϕ field can be emitted.

Suppose we have $|\text{in}\rangle = |\psi\rangle$. To leading order we will find $\langle\psi|\psi\psi|\psi\rangle(z) = 2\langle\psi|\psi\psi|\psi\rangle(z)$ Using wavepackets $\tilde{\psi}(k)$ we can represent the asymptotic states as

$$|\psi\rangle = \int \frac{d^nk}{(2\pi)^{n/2}\sqrt{2E_k}}\tilde{\psi}(k)|\vec{k}\rangle. \tag{5}$$

with normalisation condition $\int \frac{d^nk}{(2\pi)^n}|\tilde{\psi}(k)|^2 = 1$. We find

$$\langle\psi|\psi\psi|\psi\rangle(z) = \int \frac{d^nk_1d^nk_2}{(2\pi)^n\sqrt{E_{k_1}E_{k_2}}}\tilde{\psi}(k_1)\tilde{\psi}^*(k_2)e^{-iz(k_1-k_2)}. \tag{6}$$

Then by substituting the above expression into equation 4, representing $D_R(x-z)$ in Fourier space and integrating over the z co-ordinates we find

$$\langle\phi\rangle(x) = g\int \frac{d^nk_1d^nk_2}{(2\pi)^n\sqrt{E_{k_1}E_{k_2}}}\tilde{\psi}(k_1)\tilde{\psi}^*(k_2)\frac{e^{-ix(k_1-k_2)}}{(k_1-k_2)^2 - m_\phi^2 + i\epsilon(k_1^0-k_2^0)}. \tag{7}$$

For an analytic result we take $\tilde{\psi}(k) = \frac{(2\pi)^{n/2}}{(2\pi\alpha^2)^{n/4}}e^{-\frac{1}{4}(\vec{k})^2/\alpha^2}$ Effectively there is only a significant contribution to the integral when $\vec{k}_1^2 + \vec{k}_2^2 << \alpha^2$. As $\alpha \to 0$ the wavepackets become sharply peaked and this becomes a smaller region. If $m_\psi >> \alpha^2$ then $E_q \approx m_\psi$ through the integral.[2] Therefore we have that

$$\langle\psi|\psi\psi|\psi\rangle(z) \approx \frac{1}{m_\psi}|\int \frac{d^nk}{(2\pi)^{n/2}}\tilde{\psi}(k)e^{-i\vec{z}\vec{k}}|^2 = \frac{1}{m_\psi}|\tilde{\psi}(\vec{z})|^2 \tag{8}$$

Note that where the time is small in comparison to m_ψ, the time dependence drops out of the expression. We are left with a static, stable source over this time interval. Thus

$$\langle\phi(x)\rangle \approx -i\frac{g}{m_\psi}\int d^{n+1}z\, D_R(x-z)|\tilde{\psi}(\vec{z})|^2. \tag{9}$$

[2] This argument will hold true for any peaked wavepacket with width parameterized by α.

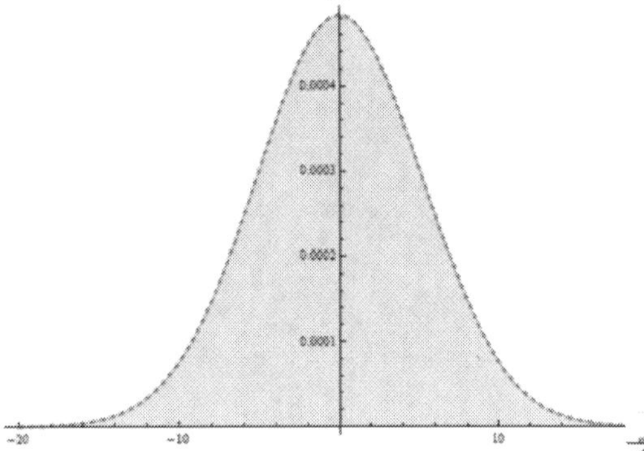

(a) Comparison between numerical and analytic results for $\langle\phi(x)\rangle$ for large m_ψ

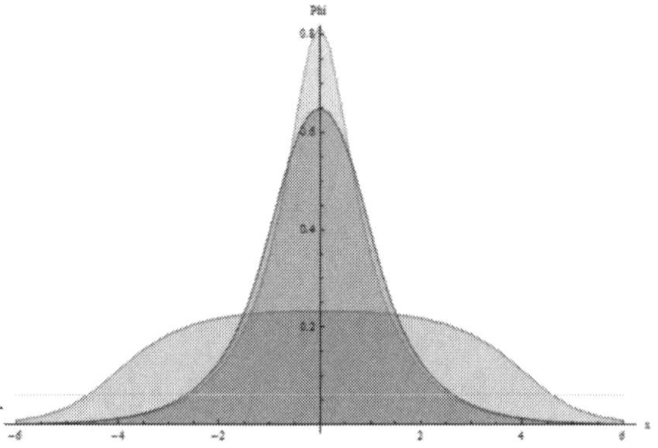

(b) Time evolution of $\langle\phi(x)\rangle$

Figure 2: Plotted expectation values in 1+1 dimensions

For our chosen wavepackets, $|\tilde{\psi}(\vec{z})|^2 = 2^n\sqrt{2\pi\alpha^2}^n e^{-2\alpha^2\vec{z}^2}$. Making the substitution $2\alpha^2 = \frac{1}{4\tau}$, choosing $n = 3$ and integrating the expression over z^0, we will find

$$\langle\phi(x)\rangle \approx (2\pi)^3 \frac{g}{m_\psi} \int d^3z \frac{e^{-m_\phi|\vec{z}|}}{4\pi|\vec{z}|} \frac{e^{-\frac{(\vec{z}-\vec{x})^2}{4\tau}}}{\sqrt{4\pi\tau}^3} \tag{10}$$

We can think of this as the solution of the heat equation with initial condition given by $\frac{e^{-m_\phi|\vec{x}|}}{4\pi|\vec{x}|}$, the Yukawa potential. τ is some function of the width of the asymptotic wavepackets, as we localize this momenta (or spread out the concentration of the source in position space) we smear out the concentration of $\langle\phi(x)\rangle$ given at the origin. For non-zero τ this contribution will be finite. Similar arguments will hold in arbitrary dimensions. Far from the origin the "diffusion" has not yet had a noticable effect, and the result will approach the Yukawa potential (which is exact limit $\tau \to 0$). In $n = 1$ the result can be written in closed form

$$\langle\phi(x)\rangle \approx \frac{g}{m_\phi m_\psi}\frac{\pi}{2}e^{\frac{m_\phi^2}{8\alpha^2}}\left(e^{-m_\phi|\vec{x}|}\text{Erfc}\left(\frac{m_\phi - 4|\vec{x}|\alpha^2}{2\sqrt{2\alpha^2}}\right) + e^{m_\phi|\vec{x}|}\text{Erfc}\left(\frac{m_\phi + 4|\vec{x}|\alpha^2}{2\sqrt{2\alpha^2}}\right)\right). \tag{11}$$

Figure 2 gives the plotted numerical results in $1 + 1$ dimensions.

3. Brief look at the improved energy momentum tensor

In general, expecations of operators in QFT are not finite, and in particular $\langle T_{\mu\nu}\rangle = \langle\sum_i \partial_\mu\phi_i\partial_\nu\phi_i - g_{\mu\nu}\mathcal{L}\rangle$ diverges order by order even after standard renormalization techniques. One must instead compute the improved energy momentum tensor[3] $\langle\Theta_{\mu\nu}\rangle = \langle T_{\mu\nu}\rangle + \langle\frac{1}{4}\frac{n-2}{n-1}\left(k_\mu k_\nu - g_{\mu\nu}k^2\right)\rangle$ for a quantity that is finite to at least one loop order.[3] Notice that this quantity is still satisfies $\partial^\mu\Theta_{\mu\nu} = 0_\nu$ and will produce the same result as $T_{\mu\nu}$ when integrated over all space.

Consider scalar $\lambda\phi^4$ theory with $|\text{in}\rangle = |\phi\rangle$. The divergent contribution (when $\Theta_{\mu\nu}$ is evaluated on the " + " contour) is given exclusively by the time ordered propagators, so we will ignore all other contributions when showing finiteness. Define $M_{\mu\nu}$ as the sum of all time ordered momentum space diagrams contributing to $\langle\Theta_{\mu\nu}\rangle$ at order λ. Then we can find

$$M_{\mu\nu} = 24i\lambda \int \frac{d^dq}{(2\pi)^4} \frac{(q-k)_\mu q_\nu - \frac{1}{2}g_{\mu\nu}((q-k)q - m^2) + \frac{1}{4}\frac{n-2}{n-1}(k_\mu k_\nu - g_{\mu\nu}k^2)}{((q-k)^2 - m^2)(q^2 - m^2)} \tag{12}$$

[3] n here is the number of spatial dimensions.

Journal of Physics: Conference Series **623** (2015) 012019 doi:10.1088/1742-6596/623/1/012019

With $k = k_1 - k_2$ being the difference in momenta of the bra and ket asymptotic states, and m the mass of the ϕ field. Working through this expression we can find the divergent part to be given by

$$\frac{24\lambda}{(4\pi)^{d/2}} \int_0^1 dx \left(\left(x(1-x) - \frac{1}{6} \right) (k_\mu k_\nu - g_{\mu\nu} k^2) \right) \frac{1}{2 - \frac{d}{2}} = 0. \tag{13}$$

Thus we can make sense of the $\Theta_{\mu\nu}$ operator to leading order in λ.

4. Conditional Expectation Value

To understand the improved energy momentum tensor $\langle \Theta_{\mu\nu} \rangle (x)$ for a jet we propose choosing an initial state of two particles. After a two-particle collision, an infinite number of final states are kinematically accessible. Thus $\langle \Theta_{\mu\nu} \rangle$ will be some average value of these possible states. To restrict this expectation value with additional information we define the conditional expectation value. This is the expectation given both initial and final states.

4.1. Derivation

Let r_n be an eigenvalue of the operator $\hat{R}$ and Δ a set of some of these eigenvalues. The probability that a measurement of $\hat{R}$ on some state will yield r_n is 1, what is the probability that it will yield $r_n \in \Delta$? We define

$$M_R(\Delta) = \sum_{r_n \in \Delta} |r_n\rangle\langle r_n| \tag{14}$$

This is the projection operator onto states given by Δ. Define density matrix $\rho = \sum_b \rho_b |b\rangle\langle b|$ with property $\sum_b \rho_b = 1$. The probability that a measurement will yield a result in Δ is the average value of the projection operator. This can be motivated by noting that if Δ_c is a complete set, $M_R(\Delta_c) = 1$. Then $\mathrm{Tr}\{\rho M_R(\Delta_c)\} = 1$. If Δ is a smaller subset, the expectation of this operator now excludes the contribution from the missing states. Thus we say that

$$P(\{R \in \Delta\}|\rho) = \mathrm{Tr}\{\rho M_R(\Delta)\} \tag{15}$$

We now want to consider two observables Q and R represented respectively by operators $\hat{Q}$ and $\hat{R}$ with eigenvectors given by $|q_n\rangle$ and $|r_n\rangle$ with $q_n \in \Gamma$ and $r_n \in \Delta$.

To define the conditional probability, to need to know the probability that a state corresponding to $q_n \in \Gamma$ will be measured, given that a state corresponding to $r_n \in \Delta$ has been measured. This is simply

$$P(\{Q \in \Gamma\}|\tilde{\rho}) = \mathrm{Tr}\{\tilde{\rho} M_Q(\Gamma)\} \tag{16}$$

where $\tilde{\rho}$ is the state of the system after a measurement of $\hat{R}$. How do we find $\tilde{\rho}$? If a state $\tilde{\rho}$ is such that a measurement will yield $r_n \in \Delta$ with certainty, then $\tilde{\rho} \sim M_R(\Delta)\hat{A}M_R(\Delta)$. Enforcing that $\mathrm{Tr}\{\tilde{\rho}\} = 1$ provides a normalization of $\mathrm{Tr}\{M_R(\Delta)\hat{A}M_R(\Delta)\}$.

Suppose we collapsed the state into $\tilde{\rho}(0)$ immediately after preparing the state in $\rho(0)$. Then we would find $\hat{A} = \rho(0)$. Define $\hat{\rho}(t)$ as the time evolved $\hat{\rho}(0)$ state. Then

$$\tilde{\rho}(t) = \frac{M_R(\Delta)\hat{\rho}(t)M_R(\Delta)}{\mathrm{Tr}\{M_R(\Delta)\hat{\rho}(t)M_R(\Delta)\}} = \frac{M_R(\Delta)\hat{\rho}(t)M_R(\Delta)}{\mathrm{Tr}\{\hat{\rho}(t)M_R(\Delta)\}} \tag{17}$$

where we have used the cyclicity of the trace and the fact that $M_R(\Delta)$ is a projection operator. Now we can write

$$P(\{Q \in \Gamma\}|\tilde{\rho}) = \frac{\mathrm{Tr}\{M_R(\Delta)\hat{\rho}(t)M_R(\Delta)M_Q(\Gamma)\}}{\mathrm{Tr}\{\hat{\rho}(t)M_R(\Delta)} = \frac{\mathrm{Tr}\{\hat{\rho}(t)M_R(\Delta)M_Q(\Gamma)M_R(\Delta)\}}{\mathrm{Tr}\{\hat{\rho}(t)M_R(\Delta)\}}. \tag{18}$$

This is the conditional probability that a state corresponding to $q_n \in \Gamma$ will be measured, given that a state corresponding to $r_n \in \Delta$ has been measured.

The conditional expecation value of an operator $\hat{\Theta}$ is defined

$$E(\hat{\Theta}, \{R \in \Delta\}|\rho) = \sum_{q_n \in \Gamma} \Theta_{q_n} P(Q \in \{q_n\}|\tilde{\rho}) \tag{19}$$

the sum of the eigenvalues Θ_{q_n} of $\hat{\Theta}$ weighted by the conditional probability. For $\Theta = 1$ this reduces to $E(1, \{R \in \Delta\}|\rho) = P(Q \in \Gamma|\tilde{\rho})$ as expected. We will take Γ to be a complete set so that any eigenvalue of $\hat{\Theta}$ is accessible.

We can write $\hat{\Theta} = \sum_{q_n \in \Gamma} \Theta_{q_n} |q_n\rangle\langle q_n|$, allowing us to express the conditional probability as

$$E(\hat{\Theta}, \{R \in \Delta\}|\rho) = \sum_{q_n \in \Gamma} \Theta_{q_n} \frac{\mathrm{Tr}\{\hat{\rho}(s) M_R(\Delta)|q_n\rangle\langle q_n|M_R(\Delta)\}}{\mathrm{Tr}\{\hat{\rho}(s) M_R(\Delta)\}} = \frac{\mathrm{Tr}\{\hat{\rho}(s) M_R(\Delta)\hat{\Theta}(t) M_R(\Delta)\}}{\mathrm{Tr}\{\hat{\rho}(s) M_R(\Delta)\}} \tag{20}$$

For the situations in which we are interested, we will set $s = \infty$ (the time at which the state $M_R(\Delta)$ is measured). We can consider $\hat{\Theta}$ to be given at some arbitrary time t because we have used a general eigenbasis given by $|q_n(t)\rangle$. Writing the density matrix in terms of some wavefunction for the system we will have

$$E(\hat{\Theta}, \{R \in \Delta\}|\rho) = \frac{\langle\Psi|M_R(\Delta)\hat{\Theta}(t)M_R(\Delta)|\Psi\rangle(s)}{\langle\Psi|M_R(\Delta)|\Psi\rangle(s)} \tag{21}$$

For an interacting theory, this expectation value can again be related to the asymptotic free field eigenvectors through the interaction picture, and perturbation theory can be done in the usual way. The new contour is given by figure 3.

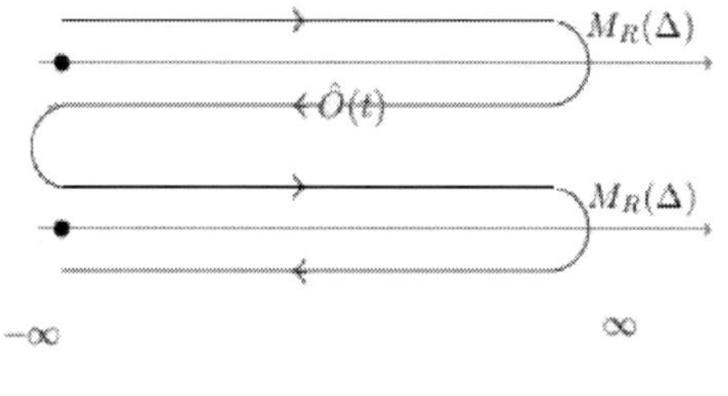

Figure 3: Time evolution indicated by a modified contour with various measurements

5. Conclusion

We have made progress towards building a spacetime description of phenomena in quantum field theory. We have found $\langle\phi(x)\rangle$ to leading order for a simple scalar Yukawa theory, deriving an approximate analytic result. From this an approximate expression for the energy momentum tensor can readily be obtained.

Attempting to calculate $\langle T_{\mu\nu}\rangle$ results in divergences that can not be absorbed by standard counter terms in a general quantum field theory. In order to account for these divergences an improved energy momentum tensor $\Theta_{\mu\nu}$ can be defined. We have briefly demonstrated that one can compute such an object in $\lambda\phi^4$ theory.

We have defined the conditional expectation value within the context of quantum field theory. To extend this work we intend to calculate the conditional expectation value of $\Theta_{\mu\nu}$ given a simple $\psi\psi \to \psi\psi$ scalar particle collision. We believe that this can be used to derive the energy density distribution as a funstion of spacetime for jet-like structures in the perturbative regime.

References

[1] Casalderrey-Solana J, Liu H, Mateos D, Rajagopal K and Wiedemann U A 2011 (*Preprint* 1101.0618)
[2] Morad R and Horowitz W 2014 *JHEP* **1411** 017 (*Preprint* 1409.7545)
[3] Callan Curtis G J, Coleman S R and Jackiw R 1970 *Annals Phys.* **59** 42–73
[4] Peskin M E and Schroeder D V 1995 *An Introduction To Quantum Field Theory (Frontiers in Physics)* (Westview Press) ISBN 0201503972
[5] Kapusta J I and Gale C 2006 *Finite-temperature field theory* 2nd ed (Cambridge University Press) ISBN 0521820820 URL http://www.worldcat.org/isbn/0521820820
[6] Collins J 1984 *Renormalization* (Cambridge: Cambridge University Press)

Search for the Higgs boson in fermionic channels using ATLAS detector

Kazuya Mochizuki, on behalf of the ATLAS collaboration
CPPM, Aix-Marseille Université, CNRS/IN2P3, Marseille, France
E-mail: kazuya.mochizuki@cern.ch

Abstract. The property measurements of the recently discovered boson at a mass around 125 GeV confirm that it is compatible with the Higgs boson predicted by the Standard Model. Since the discovery and the first property measurements were performed using the bosonic decay channels, it is particularly important to observe and measure the Higgs boson using the fermionic decay channels. We present a review of the searches for the Higgs boson decaying to a pair of τ-leptons, muons, and b-quarks with the ATLAS experiment using LHC Run 1 full dataset.

ATL-PHYS-PROC-2015-009
11 March 2015

1. Introduction

In July 2012, the ATLAS [1] and CMS [2] experiments announced the discovery of a new boson at a mass of around 125 GeV [3, 4]. The compatibility of this boson with the Standard Model (SM) prediction was confirmed by measuring properties such as spin, parity, and signal strength at the measured mass using bosonic decay channels, namely $H \to \gamma\gamma$, ZZ^*, and WW^*, thanks to their clean signatures in the final state. In order to fully determine whether this is the SM Higgs boson, it is crucial to observe and measure the Higgs boson using also fermionic channels. In this proceedings article, we report the current status of these searches with the ATLAS experiment. $H \to \tau\tau$, $\mu\mu$, and bb searches are reviewed in section 2, 3, and 4, respectively. These searches are performed using the full ATLAS dataset recorded in LHC Run 1. The integrated luminosity collected in the year 2011 (2012) is 4.6 (20.3) fb^{-1} at a center-of-mass energy of 7 (8) TeV*.

2. $H \to \tau\tau$

In this section, the search for $H \to \tau\tau$ decays is summarized [5]. The branching ratio (BR) of $H \to \tau\tau$ is 6.3% at $m_H = 125$ GeV, due to the relatively high mass of the τ-lepton ($m_\tau = 1.78$ GeV). Because of the presence of undetected neutrinos produced by the τ-decays, the invariant mass of the visible decay components is smaller than the actual $\tau\tau$ invariant mass ($m_{\tau\tau}$). To reconstruct $m_{\tau\tau}$ correctly, the missing mass calculator (MMC) is employed. The MMC scans through the τ-lepton decay probability density functions using the reconstructed $E_{\mathrm{T}}^{\mathrm{miss}}$ and visible τ-lepton mass and then calculates the most likely $\tau\tau$ invariant mass ($m_{\tau\tau}^{\mathrm{MMC}}$).

2.1. Event selection and categorization

Since τ leptons decay either leptonically or hadronically, the analysis is split into three channels, which are $\tau_{\mathrm{lep}}\tau_{\mathrm{lep}}$, $\tau_{\mathrm{lep}}\tau_{\mathrm{had}}$, and $\tau_{\mathrm{had}}\tau_{\mathrm{had}}$ where τ_{lep} is a leptonic τ-decay object ($\tau \to \nu_\tau\nu_\ell\ell$ with $\ell = e, \mu$) and τ_{had} is a hadronic τ-decay object ($\tau \to \nu_\tau$ hadrons).

* The 7 TeV dataset in the $H \to bb$ search corresponds to an integrated luminosity of 4.7 fb^{-1}.

The BRs for $\tau_{\rm lep}\tau_{\rm lep}$, $\tau_{\rm lep}\tau_{\rm had}$, and $\tau_{\rm had}\tau_{\rm had}$ are 12.4%, 45.6%, and 42%, respectively. Across all three channels, the dominant and irreducible background is $(Z \to \tau\tau)$+jets, which is estimated with τ-embedding techniques*. Other background processes, depending on the channels and categories, are top quark production, diboson production, and fakes of leptons or hadronic τ-decays mainly caused by W+jets or multijet events (see Fig 1).

Events are primarily selected by dilepton, single lepton, and di-hadronic tau triggers for $\tau_{\rm lep}\tau_{\rm lep}$, $\tau_{\rm lep}\tau_{\rm had}$, and $\tau_{\rm had}\tau_{\rm had}$ channels, respectively. Since the charge information of the τ-lepton is always inherited by its decay product(s), the analysis requires opposite charges for the decay objects of τ-pairs. For the channels involving $\tau_{\rm lep}$, isolation requirements on electrons or muons are applied, which are dependent on the center-of-mass energy and pileup conditions. To suppress the $t\bar{t}$ background contributions, events with b-tagged jets † with $|\eta^{\rm jet}| < 2.5$ are rejected in the $\tau_{\rm lep}\tau_{\rm lep}$ and $\tau_{\rm had}\tau_{\rm had}$ channels.

$\tau_{\rm lep}\tau_{\rm lep}$: Events with exactly two isolated leptons are selected. No additional $\tau_{\rm had}$ or $\tau_{\rm lep}$ candidates are allowed. The lepton isolation requirement for both leptons reduces the multijet background contribution in this channel.

$\tau_{\rm lep}\tau_{\rm had}$: Events with exactly one isolated lepton and exactly one $\tau_{\rm had}$ candidate are selected. To reduce the W+jets background, a selection on the transverse mass is applied, $m_{\rm T}(\ell, E_{\rm T}^{\rm miss}) < 70$ GeV.

$\tau_{\rm had}\tau_{\rm had}$: In the $\tau_{\rm had}\tau_{\rm had}$ channel, $\tau_{\rm had}$ isolation is required. One isolated medium and one isolated tight $\tau_{\rm had}$ candidates are required in an event with $E_{\rm T}^{\rm miss} > 20$ GeV. Events with electron or muon candidates are rejected. To suppress the main multijet background in this channel, selections are applied on angular separation variables such as $\Delta R(\tau_{\rm had}, \tau_{\rm had})$ and $\Delta\eta(\tau_{\rm had}, \tau_{\rm had})$.

The search is targeted at two production processes of the Higgs boson, which are vector boson fusion (VBF), and gluon fusion (ggH) which is targeted by the Boosted category, where the Higgs boson has $p_{\rm T}^{H} > 100$ GeV. Events are first required to pass the pre-selections before the selections of those VBF and Boosted categories. The VBF category provides the best sensitivity because of the very discriminative event topology in which two high $p_{\rm T}$ jets have a large pseudo-rapidity separation ($\Delta\eta(j_1, j_2)$, see Fig 1). Therefore, events failing to be classified in the VBF category are considered as candidates for the Boosted category.

2.2. *Multivariate analysis and global likelihood fit*

To improve the separation between the signal and background, various kinematic properties are combined together with $m_{\tau\tau}^{\rm MMC}$ by boosted decision trees (BDT). In addition to the simulated samples, some data driven samples are also used for the backgrounds. The BDTs are trained in six categories, which consist of $\tau_{\rm lep}\tau_{\rm lep}$, $\tau_{\rm lep}\tau_{\rm had}$, and $\tau_{\rm had}\tau_{\rm had}$ channels in each of the Booosted and VBF categories. The number of input variables used in these BDTs varies from 5 to 9 depending on the category. These variables describe kinematic properties of the $\tau\tau$ system and jet properties especially in the VBF category. Examples of these input variables are shown in Fig 1. Figure 2 shows the BDT output distributions in the most sensitive channel, $\tau_{\rm lep}\tau_{\rm had}$, in the VBF and Boosted categories separately.

The signal strength μ is extracted from a simultaneous maximum-likelihood fit to the BDT discriminants in the six categories. To constrain the background contributions, this fit is performed also on the event yields in dedicated control regions (CR) for the specific backgrounds in each category. For the $\tau_{\rm lep}\tau_{\rm lep}$ and $\tau_{\rm lep}\tau_{\rm had}$ channels, $Z \to \tau\tau$-enriched, $Z \to \ell\ell$-enriched, fake-enriched, and top CRs are defined. A 'rest' category is defined for the $\tau_{\rm had}\tau_{\rm had}$ channel, where the events pass the pre-selection but not the VBF or Boosted selection. In the $\tau_{\rm had}\tau_{\rm had}$

* The shape of the $(Z \to \tau\tau)$+jets is estimated from $(Z \to \mu\mu)$+jets in data, where muons are replaced with simulated τ-decays.

† Jets identified as originating from b-quark fragmentation by MV1 b-tagging algorithm [6, 7]. The operating point used in this search provides 60–70% efficiency for b-jets in simulated $t\bar{t}$ events.

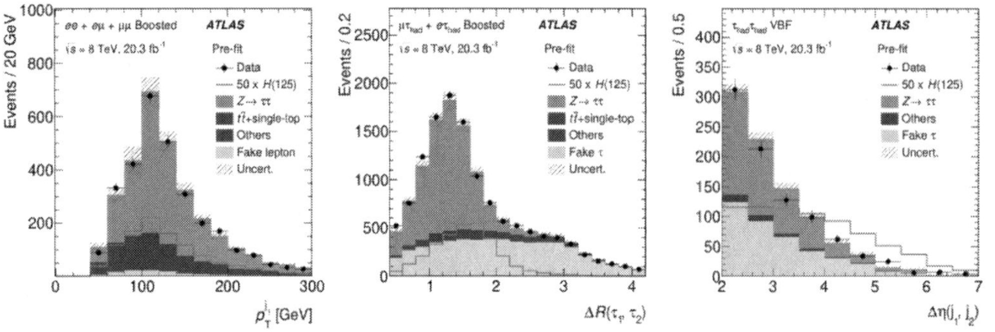

Figure 1. Distributions of important BDT input variables for data collected at a center-of-mass energy of 8 TeV. Leading jet p_T in Boosted $\tau_{lep}\tau_{lep}$ category (left), $\Delta R(\tau_{lep}, \tau_{had})$ in Boosted $\tau_{lep}\tau_{had}$ category (middle) and $\Delta\eta(j_1, j_2)$ in VBF $\tau_{had}\tau_{had}$ category (right). The contributions from the SM Higgs boson with $m_H = 125$ GeV are multiplied by a factor of 50 and superimposed. The error band includes statistical and pre-fit systematic uncertainties [5].

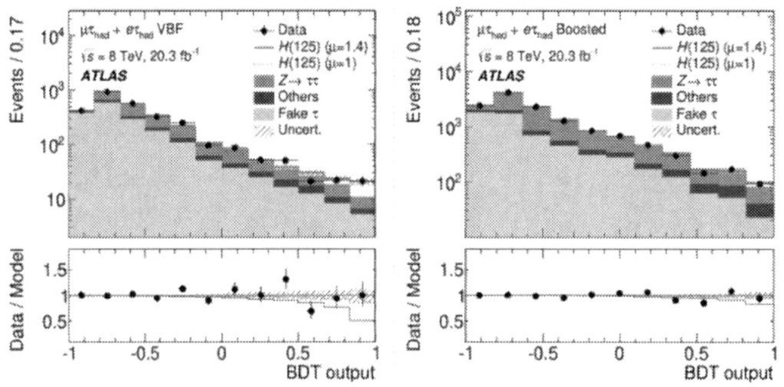

Figure 2. Distributions of the BDT discriminant for data taken at center-of-mass energy of 8 TeV in the signal regions of the VBF (left) and Boosted (right) categories for the $\tau_{lep}\tau_{had}$ channel. The Higgs boson signal ($m_H = 125$ GeV) is shown stacked with a signal strength of $\mu = 1.0$ (dashed red line) and $\mu = 1.4$ (solid red line). The size of the statistical and systematic normalization uncertainties is indicated by the hashed band. The data ratios to the models (background plus Higgs boson contributions with $\mu = 1.4$) are shown in the lower panels. The dashed red and the solid black lines represent the changes in the model when $\mu = 1.0$ or background-only are assumed respectively [5].

channel, $\tau\tau$ invariant mass side-bands ($m_{\tau\tau}^{\text{MMC}} < 110$ GeV or $m_{\tau\tau}^{\text{MMC}} > 150$ GeV) are also used to constrain the backgrounds.

2.3. Results

Figure 3 shows the combined results from all the analysis categories. An excess over the SM background prediction is observed. The extracted signal strength with the global maximum-likelihood fit is $\mu = 1.43^{+0.27}_{-0.26}(\text{stat.})^{+0.32}_{-0.25}(\text{syst.}) \pm 0.09(\text{theo.})$ for $m_H = 125.36$ GeV. The dominant sources of systematic uncertainties are energy scales of the jets and τ_{had}, identification of τ_{had}, and background normalization. The probability p_0 of obtaining a result at least as signal-like as observed in data without signal presence is calculated. For $m_H = 125.36$ GeV, the observed p_0 value is 2.7×10^{-6}, which corresponds to a 4.5σ deviation from the background only hypothesis, while 3.4σ deviation is expected. This provides evidence for $H \to \tau\tau$ at $m_H = 125.36$ GeV.

The two dimensional likelihood contours of the signal strength parameters for production modes of VBF and ggH are shown in Fig 3. The results are compatible with the SM predictions within 1σ.

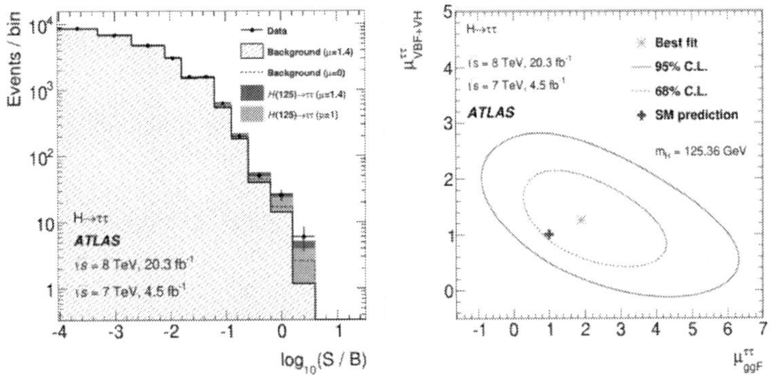

Figure 3. Combined results from $\tau_{\rm lep}\tau_{\rm lep}$, $\tau_{\rm lep}\tau_{\rm had}$, and $\tau_{\rm had}\tau_{\rm had}$ channels in VBF and Boosted categories. The left plot shows event yields as a function of $\log_{10}(S/B)$, where S (signal yield) and B (background yield) are taken from the BDT output bin of each event, assuming $\mu = 1.4$. The predicted background is obtained from the global fit (with $\mu = 1.4$) and signal yields are shown for $m_H = 125$ GeV, at $\mu = 1$ and $\mu = 1.4$ (best-fit value). The background only distribution (dashed line) is obtained from the global fit, but fixing $\mu = 0$. The right plot shows the likelihood contours for the combination of all channels in the ($\mu^{\tau\tau}_{ggF}, \mu^{\tau\tau}_{VBF+VH}$) plane. The 68% (95%) CL contour is shown as dashed (solid) line for $m_H = 125.36$ GeV. The SM expectation is shown by a filled plus symbol. The best fit to data is shown for the case when both the $\mu^{\tau\tau}_{ggF}$ and $\mu^{\tau\tau}_{VBF+VH}$ are unconstrained [5].

3. $H \to \mu\mu$

This section reviews the search for $H \to \mu\mu$ decays [8]. This decay channel is particularly important to measure the Higgs-lepton second generation couplings. In the SM, the BR of $H \to \mu\mu$ is 0.02% at $m_H = 125$ GeV due to the light muon mass. It suffers from the overwhelming and irreducible Drell-Yan background. The analysis focuses on the search for a narrow resonance, exploiting the very high resolution of the $\mu\mu$ invariant mass ($m_{\mu\mu}$). The dominant background in this search is $Z^*/\gamma \to \mu\mu$ production, followed by smaller backgrounds from single and pair production of the top quarks and diboson processes.

3.1. Event selection, categorization and modeling

The trigger used for this search is a combination of single muon and dimuon triggers. The events are required to have two oppositely charged isolated muons with $p_{\rm T} > 15$ GeV, where at least one muon must have $p_{\rm T} > 25$ GeV. All muon candidates are required to be within $|\eta| < 2.5$. To suppress the $t\bar{t}$ and diboson background processes, $E_{\rm T}^{\rm miss} < 80$ GeV is required.

After the pre-selection described above, the categorization is performed for VBF and ggH to maximize the sensitivity. As for the $H \to \tau\tau$ search presented in Sec. 2, events are first checked for the VBF category, then for ggH category. In the VBF category, at least two jets are required with $\Delta\eta(j_1, j_2) > 3$ and $m(j_1, j_2) > 500$ GeV. In the ggH category, the events are further divided into six categories depending on $|\eta^\mu|$ and transverse momentum of the Higgs boson candidate ($p_{\rm T}^{\mu\mu}$). The event is categorized as 'central' category if both muons satisfy

$|\eta^{\mu}| < 1$ and as 'non-central' otherwise. These central and non-central categories are split into three categories based on $p_{\mathrm{T}}^{\mu\mu}$ with boundaries at 15 and 50 GeV. A CR is defined for W+jets background, where one of the jets is misidentified as a muon, by requiring same-sign charges for the muon candidates.

The Higgs signal shape is fit to simulated data with the sum of Crystal Ball and Gaussian functions. Background modeling for the ggH category uses a Breit-Wigner (BW) function convoluted with a Gaussian function while the product of a BW function and an exponential function is taken for the VBF category. To derive the results, a binned maximum likelihood fit is done to the $m_{\mu\mu}$ distribution in the range 110-160 GeV simultaneously in all seven categories with separate distributions in 7 and 8 TeV data. Examples of the fit in 8 TeV data are shown in Fig 4.

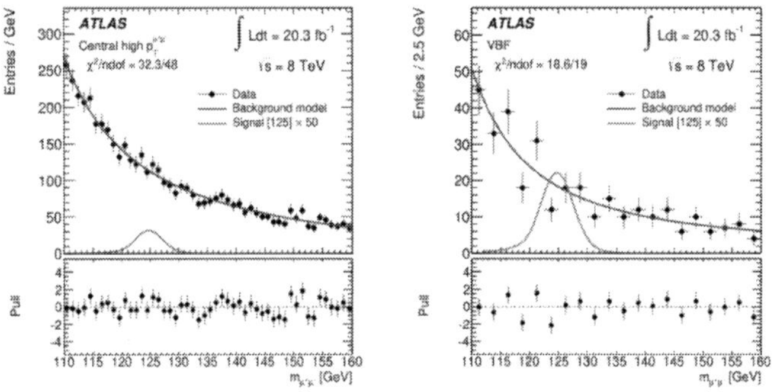

Figure 4. The background model fits to the $m_{\mu\mu}$ distribution for 8 TeV data for the central high $p_{\mathrm{T}}^{\mu\mu}$ category (left) and VBF category (right). The statistical uncertainties are given for the data points. The expected signal is shown for $m_H = 125$ GeV, scaled by a factor of 50 [8].

3.2. Results

The observed data are consistent with the expected backgrounds and no excess is observed. For $m_H = 125.5$ GeV, the observed (expected) limit on the signal strength at 95% CL limit is 7.0 (7.2) times the SM prediction. Under the assumption of $m_H = 125.5$ GeV with the SM production cross section, the 95% CL upper limit on the $H \to \mu\mu$ BR is 1.5×10^{-3}. The search is limited by the statistics.

4. $H \to bb$

This section presents the search for $H \to bb$ decays [9]. $H \to bb$ has the largest BR of 58% at $m_H = 125$ GeV. In this decay channel, VBF and ggH production modes are overwhelmed by b-quark pair production at the LHC. Therefore, the analysis focuses on the VH production ($VH \to Vbb$) mode where the associated vector boson (V) provides a clear tag for the search.

4.1. Event selection and categorization

According to the number of leptons in the final state the search is split into three channels of 0-lepton, 1-lepton, and 2-lepton, targeting $ZH \to \nu\nu bb$, $WH \to \ell\nu bb$, and $ZH \to \ell\ell bb$ processes ($\ell = e, \mu$), respectively. A further categorization is performed in this search based on the p_{T} of the vector boson, defining low and high p_{T}^{V} intervals with a boundary at 120 GeV. The dominant background for this search is $V+$ jets events, especially the $V+$ heavy flavor jets which remain after b-tagging requirements. Diboson (VZ) production generates almost the same event signatures as the VH signal. The VZ production has slightly higher cross-section than VH signal, however smaller than the other backgrounds. Therefore, VZ production is used to validate the search results by measuring the signal strength μ_{VZ}. The other backgrounds are single and pair production of top quarks, and multijet events. Across the three channels, events are required to have two b-tagged jets with $p_T > 20$ GeV, with at least one jet with $p_T > 45$ GeV. Events with forward jets ($|\eta^{jet}| > 2.5$) are discarded for $t\bar{t}$ rejection. Events with a third (non b-tagged) jet are allowed and form a separate category, called the 3-jet bin. This analysis employs MV1c b-tagging algorithm which is a variant of MV1 tagging algorithm with improved c-jet rejection and a modest cost in light flavor jet rejection. Depending on the tightness of the operating point of the MV1c, three categories are defined which are 2 tight tags (TT), 2 medium tags (MM), and 2 light tags (LL).

0-lepton: Events in this channel are selected with an E_T^{miss} trigger which is 100%, 97%, and 80% efficient for $E_T^{miss} > 160$, $E_T^{miss} > 120$ and $E_T^{miss} > 100$ GeV. This channel requires very large E_T^{miss} in the event, $E_T^{miss} > 100$ GeV, caused by a high p_T Z boson decaying to a pair of neutrinos. Many kinematic selections are applied to suppress the multijet and top quark backgrounds. In the 3-jet bin, events with $p_T^V < 120$ are not used.

1-lepton: The main trigger for this channel is single lepton (single electron or single muon) trigger. To recover the events missed by the muon trigger, the E_T^{miss} trigger is also used in the high p_T^V muon channel. The events are required to have exactly one isolated lepton (electron or muon) with $p_T > 20$ GeV, with large H_T* or E_T^{miss} in the low and high p_T^V intervals, respectively. The low p_T^V electron channel is excluded from the analysis.

2-lepton: Events in this channel are selected with a combination of single lepton and dilepton (dielectron or dimuon) triggers. This channel profits from the very clean signature provided by the leptonic decay of the Z boson. Events must have two isolated same flavor leptons with opposite electric charges. The events are required to have dilepton invariant mass in the range 83–99 GeV.

4.2. Multivariate analysis and fit model

Similarly to the $H \rightarrow \tau\tau$ search, BDTs are constructed for 10 categories of signal regions in 0-, 1-, and 2-lepton channels, 2 and 3-jet bins, and low and high p_T^V categories†. Depending on the category, additional kinematic properties and b-jet properties are used to build BDTs together with the most important variable m_{bb}. Figure 5 shows the BDT output distributions in the most sensitive regions of the search.

Multijet backgrounds are estimated using data, since they cannot be estimated with simulated events. In the 0-lepton channel, multijet background is estimated using the so-called "ABCD method" using $\min[\Delta\phi(E_T^{miss}, jet)]$ and $\Delta\phi(E_T^{miss}, p_T^{miss})$. In the 1-lepton channel, the multijet fit is performed on the E_T^{miss} distribution using a CR defined by loosened lepton isolation requirements. In the 2-lepton channel, multijet background is also estimated with the fit to $m_{\ell\ell}$ using a CR defined in similar way to the 1-lepton channel by loosened lepton isolation and identification criteria.

The signal strength μ is extracted by a maximum-likelihood fit to the BDT distributions in the 24 2-tag signal regions defined by three lepton channels, up to two p_T^V intervals, the two number of jet bins, and b-tagging categories. In the 0 and 2-lepton channels, the TT and MM b-tagging categories are merged while they are kept separated in the 1-lepton channel. In the 1-lepton channel, the 2-tag signal regions in the 3-jet bin effectively play the role of CRs because $t\bar{t}$ background dominates these regions. The b-tagging discriminant, $MV1c$, is used in the global fit in the 11 1-tag CRs for three lepton channels, two jet bins, and two p_T^V intervals.

* Scalar sum of E_T^{miss} and p_T of the lepton and leading two jets.

† In low p_T^V 0-lepton channel, the variable used in the global fit is m_{bb} instead of the BDT discriminant.

This particularly improves the normalization of the $V+$ light flavor jets backgrounds.

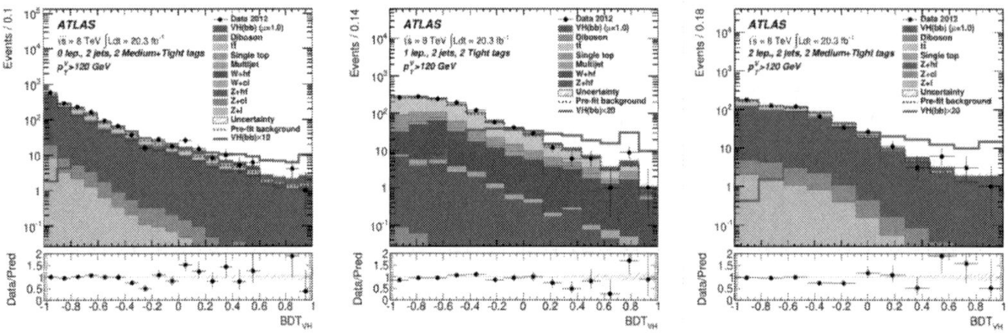

Figure 5. The BDT discriminant distributions in 2-jet high p_T^V categories in MM+TT category 0-lepton channel (left), TT category 1-lepton channel (middle), and MM+TT 2-lepton channel (right). The background contributions after the global fit of the MVA are shown as filled histograms along with the data points. The Higgs boson signal is shown as filled histogram on top of the fitted backgrounds as predicted by the SM ($\mu = 1$), and unstacked open histogram, multiplied by a factor 20. The dashed histogram shows the pre-fit total background. The hatched band shows the statistical and systematic uncertainties. The data ratio to the signal + fitted background is shown in the lower panel [9].

4.3. Results

Figure 6 shows the combined results from all the analysis categories. An excess is observed (expected) at the 1.4σ (2.6σ) level over the background-only hypothesis. The measured signal strength is $\mu_{VH} = 0.52 \pm 0.32(\text{stat.}) \pm 0.24(\text{syst.})$ for $m_H = 125.36$ GeV. The dominant sources of systematic uncertainties are shape and normalization modelings of the $V+$ heavy flavor jets background.

Two cross-checks are performed using the VZ and dijet mass analyses in 8 TeV data. For the cross-check with VZ, a BDT is trained for VZ signal events instead of VH signals. The measured signal strength is $\mu_{VZ} = 0.77 \pm 0.10(\text{stat.}) \pm 0.15(\text{syst.})$, which is consistent with the SM. The cross-check with the dijet mass analysis is done by performing a fit to the m_{bb} distribution rather than to the BDT discriminant. In this cross-check, the selections are tighter, with finer p_T^V intervals (see Fig 6). The measured signal strength of the Higgs boson signal is $\mu_{VH} = 1.23 \pm 0.44(\text{stat.}) \pm 0.41(\text{syst.})$. The observed (expected) significance of an excess over the background-only hypothesis is 2.2σ (1.9σ) in the dijet mass cross-check.

Figure 7 shows the signal strength measured separately in 7 and 8 TeV data as well as in the three lepton channels for the combined sample.

5. Summary

Higgs boson searches in decay channels to $\tau\tau$, $\mu\mu$ and bb, performed in the ATLAS experiment, are presented. Direct evidence of $H \to \tau\tau$ is observed. The $H \to \mu\mu$ search is not sensitive yet because of the limited data statistics. However, the result confirms non-universal couplings of $H \to \ell\ell$ as predicted by the SM. An excess with 1.4σ significance is observed in the $VH \to Vbb$ search. Overall, all the results are compatible with the SM predictions. The LHC Run 2 will provide data with an increased center-of-mass energy of 13 TeV in 2015 and at higher luminosity, which will allow for more precise studies of the Higgs boson properties in fermionic decay channels.

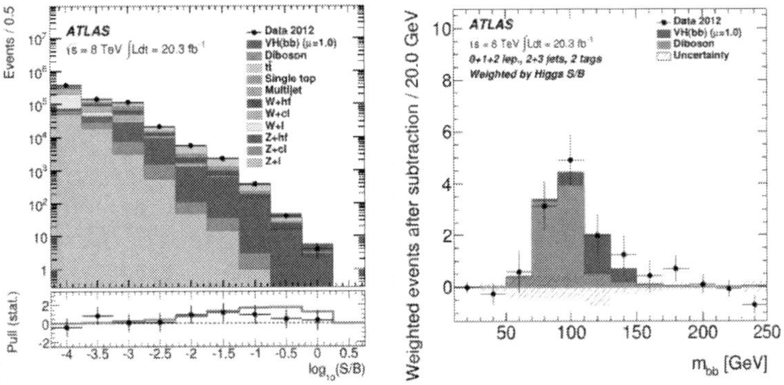

Figure 6. The left plot shows event yields as a function of $\log(S/B)$ for data, background and Higgs boson signal with $m_H = 125$ GeV for the 8 TeV data in 0-, 1- and 2-lepton channels of the $H \rightarrow bb$ analysis. The Higgs boson signal contribution is stacked on top of background with $\mu = 1.0$. The pull of data with respect to the background-only prediction is also shown with statistical uncertainties only. The red line shows the pull of the background + signal ($\mu = 1.0$) with respect to the background-only prediction. The right plot shows the m_{bb} distribution after subtraction of all the backgrounds except the VZ processes, for the dijet mass cross-check performed for the 8 TeV data. The contributions from all lepton channels, p_T^V intervals, number of jet bins, and 2-tag b-tagging categories are summed with weights defined by their respective S/B ratio after the global fit. The Higgs boson signal is stacked over the VZ contribution with $\mu = 1.0$. The hatched band shows the size of the statistical and systematic uncertainty on the fitted background [9].

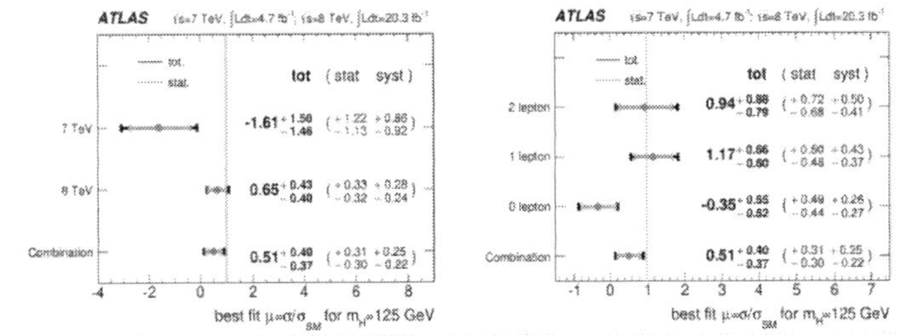

Figure 7. The signal strength measured separately in 7 and 8 TeV data (left) and in the 0-, 1-, and 2-lepton channels in 7 and 8 TeV data (right) for $m_H = 125$ GeV. The individual μ values for the three lepton channels are obtained from a simultaneous fit with the signal strength for each lepton channel floating independently [9].

References

[1] ATLAS Collaboration, JINST **3** (2008) S08003.
[2] CMS Collaboration, JINST **3** (2008) S08004.
[3] ATLAS Collaboration, Phys.Lett. **B716** (2012) 1–29, `arXiv:1207.7214 [hep-ex]`.
[4] CMS Collaboration, Phys.Lett. **B716** (2012) 30–61, `arXiv:1207.7235 [hep-ex]`.
[5] ATLAS Collaboration, CERN-PH-EP-2014-262 (2015), `arXiv:1501.04943 [hep-ex]`.
[6] ATLAS Collaboration, ATLAS-CONF-2014-004 (2014). `http://cds.cern.ch/record/1664335`.
[7] ATLAS Collaboration, ATLAS-CONF-2014-046 (2014). `http://cds.cern.ch/record/1741020`.
[8] ATLAS Collaboration, Phys.Lett. **B738** (2014) 68–86, `arXiv:1406.7663 [hep-ex]`.
[9] ATLAS Collaboration, JHEP **1501** (2015) 069, `arXiv:1409.6212 [hep-ex]`.

Journal of Physics: Conference Series **623** (2015) 012021 doi:10.1088/1742-6596/623/1/012021

Jet Nuclear Modification Factor from the AdS/CFT Correspondence

R Morad and W A Horowitz

Department of Physics, University of Cape Town, Private Bag X3, Rondebosch 7701, SA

E-mail: `razieh.morad@uct.ac.za` and `wa.horowitz@uct.ac.za`

Abstract.
We define a novel definition of the holographic light hadron jet by a separation of scales from plasma to jet, which leads to the re-emergence of the late-time Bragg peak in the instantaneous jet energy loss rate. We use the holographic energy loss of light quark and our new jet prescription in AdS/CFT to calculate the nuclear modification factor of jet for a brick of plasma (both static and expanding plasma) and compare the AdS/CFT results with the experimental data for most central Pb-Pb collision at LHC at 2.76 TeV center-of-mass energy. Defining a renormalized AdS/CFT jets that we argue better reflect QCD physics, we find a surprisingly good agreement between our toy model which is the first fully strongly coupled calculations, and preliminary jet suppression data from heavy ion collisions at LHC.

1. Introduction

The spectacular measurements from the Relativistic Heavy Ion Collider (RHIC) and the Large Hadron Collider (LHC) provide compelling evidence that the matter produced in heavy ion collision is a deconfined state of QCD, Quark-Gluon Plasma (QGP) [1, 2, 3, 4], at temperatures above ~160 MeV which appears to be nearly perfect, with an extremely low viscosity-to-entropy ratio $\eta/s \sim 1/4\pi$ [5, 6]. While lattice QCD is the proper tool for understanding the static equilibrium thermodynamics of such strongly coupled plasma, it does not allow us to calculate its dynamics evolution on heavy-ion collision.

Recently, a novel tool called "the AdS/CFT correspondence" [7, 8, 9, 10, 11] provide valuable insight into the strongly coupled plasma. This conjecture state that the $\mathcal{N} = 4$ SYM theory in the large N_c and large 't Hooft coupling is dual to classical supergravity on ten-dimensional $AdS_5 \times S^5$ geometry [7]. In order to study the theory at finite temperature, one can add black hole (BH) to the geometry [8] which yields to the AdS-Sch metric. Fundamental quarks are described by open strings moving in the 10d geometry. In the large λ limit, the quantum fluctuation of string world sheet are suppressed and the dynamics of string are described by the classical string theory.

Jets are produced within the expanding fireball and probe the QGP. Analyses the energy loss of these energetic partons as they travel throw QGP may reveal extremely valuable information about the dynamics of the plasma and exhibit distinctive properties such as jet-quenching which can clearly be observed at RHIC [1, 2, 3, 4] and more recently LHC [12, 13, 14].

In this paper, we study the light quark jet energy loss in both gravitational background dual to static and expanding plasma. We propose a new prescription of jets in string theory based on the separation of hard and soft sectors. We demonstrate that the light quark jet energy loss reveal the "Bragg peak" at late times. Then we compute the nuclear modification factor of jets R_{AA}^{jet}, renormalise the quantity, and compare the results with the preliminary CMS data [15].

2. Light quark jet energy loss

According to the AdS/CFT correspondence [10], the $\mathcal{N}=4$ SYM theory at finite temperature is dual to a $10d$ black hole geometry with the AdS-Schwarzschild (AdS-Sch) metric as follows,

$$ds^2 = \frac{L^2}{u^2}\left[-f(u)\,dt^2 + d\mathbf{x}^2 + \frac{du^2}{f(u)}\right], \tag{1}$$

where $f(u) \equiv 1-(u/u_h)^4$ is the blackening factor and L is the AdS curvature radius. Four dimensional Minkowski coordinates are denoted by x_μ and the coordinate u is an inverse radial coordinate. So, the boundary of the AdS-Sch spacetime is at $u=0$ and the event horizon is located at $u=u_h$. The temperature of the equilibrium SYM plasma relates to the event horizon as $T \equiv \frac{1}{(\pi u_h)}$. World sheet coordinates are as σ^a where $\tau \equiv \sigma^0$ is denoted as the timelike world sheet coordinate, while the spatial coordinate is $\sigma \equiv \sigma^1$.

Fundamental representation quarks added to the $\mathcal{N}=4$ SYM theory are dual to open strings moving in the $10d$ geometry. Addition of a $\mathcal{N}=2$ hypermultiplet to the $\mathcal{N}=4$ SYM theory is performed by adding D7 branes to the $10d$ geometry [16]. These branes extend along the radial coordinate from the boundary at $u=0$ down to maximal coordinate at $u=u_m$ as well as fill the whole 4d Minkowski space. Also, they wrap on S^3 from the S^5 sphere. The bare mass M of quark is proportional to $1/u_m$ [17], so for massless quarks the D7 brane should fill the whole radial direction. Open strings that are attached to the D7 brane are dual to the quark-anti quark pairs on the field theory side. In the $5d$ geometry these strings can fall unimpeded toward the event horizon until their end points reach the radial coordinate u_m where the D7 brane ends. Since for sufficiently light or massless quarks $u_m > u_h$, open string end points can fall into the horizon.

We are interested in studying the back-to-back jets so we consider the configurations in which the two endpoints of string move away from each other as the total spatial momentum of the string vanishes. By choosing the appropriate frame, one half of the string has a large spatial momentum in x direction, and the other half of the string carries a large negative spatial momentum. We will limit our attention to strings which move in one direction in the R^3 space, x direction, so the embedding function of string $X^\mu(\tau,\sigma)$ will be a map to $(t(\tau,\sigma), x(\tau,\sigma), u(\tau,\sigma))$. So, the profile of an open string which is created at a point in space at time $t=t_c$ is given by

$$t(0,\sigma) = t_c\,,\quad x(0,\sigma) = 0\,,\quad u(0,\sigma) = u_c\,. \tag{2}$$

By these conditions, the string created at time t_c and by time evolution, the string evolves from a point into an extended object and the string endpoints fall toward the horizon.

The dynamics of string is governed by the Polyakov action

$$S_P = -\frac{T_0}{2}\int d^2\sigma\,\sqrt{-\eta}\,\eta^{ab}\,\partial_a X^\mu \partial_b X^\nu\, G_{\mu\nu}\,. \tag{3}$$

Variation of the Polyakov action with respect to the embedding functions X^μ lead to the equation of motion as

$$\partial_a\left[\sqrt{-\eta}\,\Pi^a_\mu\right] = \frac{1}{2}\sqrt{-\eta}\,\eta^{ab}\frac{\partial G_{\nu\rho}}{\partial X^\mu}\,\partial_a X^\nu \partial_b X^\rho, \tag{4}$$

where Π^a_μ are the canonical momentum densities associated to the string obtained from varying the action with respect to the derivatives of the embedding functions.

In order to study jets in AdS/CFT we need to define the proper objects in the dual string theory that corresponds to a jet, a slippery object even in field theory; jets are truly only defined by the algorithm used to measure them. Presumably the ideal way to compute jet observables in the dual theory is to compute the energy momentum tensor associated with a high-momentum probe and "run" a jet finding algorithm on the result. However, one can define a jet prescription in the AdS/CFT and calculate the rate of energy loss from the string profile itself.

The authors of [18] are motivated by the localization of the baryon density on the boundary which is of scale of order $\Delta x \sim 1/\pi T$ and defined jet as a part of string which is in the Δx spatial distance from the endpoint. We called this as the "$\Delta x - prescription$" of jet [18].

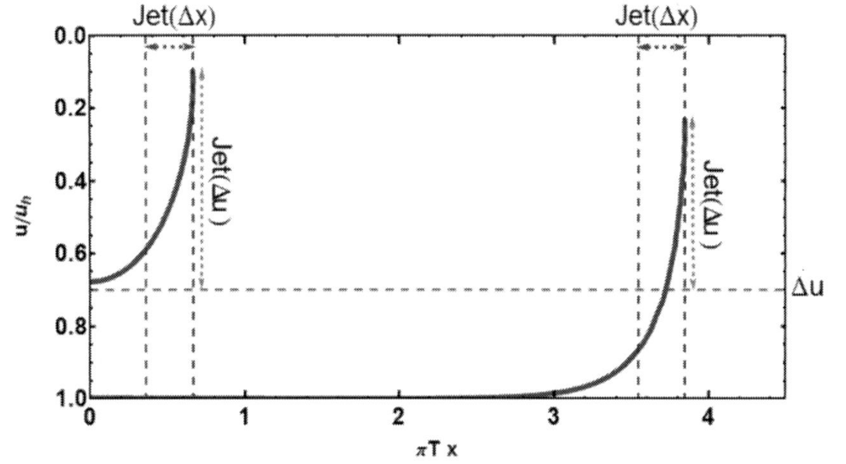

Figure 1. Illustration of the Δx and Δu prescriptions of a jet in the string theory; see text for details.

In this paper, motivated by the separation of energy scales in, e.g., thermal field theory, we propose rather a Δu prescription which we believe will ultimately provide a closer approximation to the result of a more complete calculation, figure 1. Since the radial coordinate in the string theory sets an energy scale in the field theory, in our Δu prescription the portion of the string above some cutoff $u = u^*$ in the radial direction is considered part of the jet; the portion of the string below the cutoff is considered part of the thermalized medium. By choosing any value of u above the black hole horizon as the cutoff, we regain the natural result that a jet that is thermalized no longer has detectable energy or momentum.

We evaluate the energy loss rate of jet in the radial $\sigma = u$ parametrization for both prescriptions of jet and plot in figure 2(a,b). In order to define jet using the $\Delta x - prescription$, we choose $\sigma_\kappa(t)$ as $\Delta x = 0.3/\pi T$. Our result on the energy loss rate of light quark using the $\Delta u - prescription$ shows a Bragg peak at late times which means the explosive transfer of quark energy to the plasma at late times and is consistence with the previous works [18].

Since the quark-gluon plasma produced at ultra-relativistic heavy ion collision is an expanding and cooling medium, we study the light quark jet energy loss in a time dependent gravity dual to the boost invariant flow [19]. This geometry is similar to the static black hole geometry, but the location of the horizon moves in the bulk as $\tau^{1/3}$ where τ is the proper time. Also the temperature of plasma cools as $T(\tau)\ \tau^{-1/3}$.

We calculate the energy loss rate of light quark using both $\Delta x - prescription$ and $\Delta u - prescription$ of jet. We consider the initial temperature of plasma in JP metric T_c the same

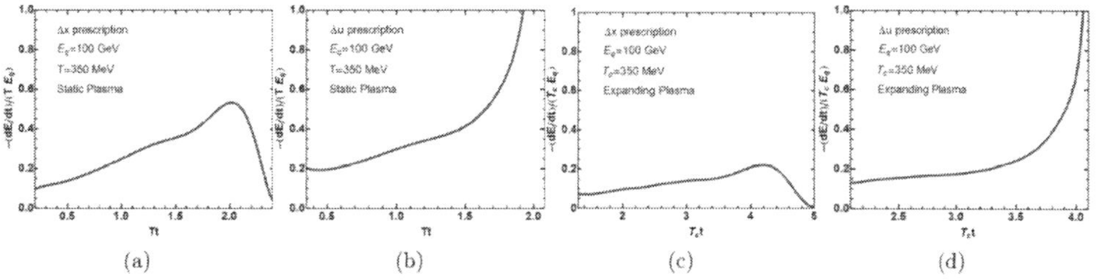

Figure 2. The instantaneous energy loss of a light quark jet as a function of time in the AdS-Sch (a,b) and JP metrics (c,d) in the Δx prescription and Δu prescription. The normalization constant $E_q = 100$ GeV is the initial energy of the jet, which has a virtuality of 175 GeV^2, and $T = 350$ MeV is the temperature of the plasma.

as the temperature of plasma in AdS-Sch metric. The result is shown in figure 2(c,d) which demonstrates that the behavior of light quark energy loss in the JP metric is same as the AdS-Sch metric, but the distance that quark traveled before thermalizing increases.

3. Nuclear Modification Factor of Jet

We calculate an approximation of the nuclear modification factor R_{AA} for jet using our energy loss model of jet. We consider the contributions of both quark and gluon jets. We assume that the produced parton with initial energy p_T^i loses a fraction of its energy ϵ with probability $P\left(\epsilon|p_T^i, L, T\right)$ as the final energy of parton is given by $p_T^f = (1-\epsilon)p_T^i$. Also, we suppose that the the AdS energy loss is approximately independent of the initial energy [20], and the gluons loss their energy by the factor of 2 in the large N_c limit respect to the quarks.

We suppose that the production spectrum can be approximately by a power law [20], with slowly varying with respect to p_T, then we may find a simple equation for the jet nuclear modification factor as follows,

$$R_{AA}^{R\to jet}(p_T) = \left\langle \int d\epsilon\, P(\epsilon|p_T, L, T)\left(1-\epsilon^R\right)^{n_R(p_T)-1}\right\rangle. \tag{5}$$

For an absolutely uniform nucleus that is a 1D line, the geometric average is carried out as an integral over a line of production points with a parton that propagates through the line. In this case, $R_{AA}^{R\to jet}(p_T)$ is obtained from the below line integral [20]

$$R_{AA}^{R\to jet}(p_T) = \int_0^{L_{max}} \frac{dl}{L_{max}}\left(1-\epsilon^R(p_T, l, T)\right)^{n_R(p_T)-1}. \tag{6}$$

We calculate R_{AA} of jet by using our results of jet energy loss in both AdS-Sch and JP metric. Our results are shown in figure 3 (a). The purple curve shows the R_{AA} obtained from the AdS-Sch metric, while the blue curve is the R_{AA} obtained from the JP metric. The AdS/CFT results of jet energy loss show an over suppression of jets in both static and expanding plasma.

The point-like initial condition falling string that we consider here is dual to creation of a pair of quark-antiquark which fly away each other in the strongly coupled plasma, interact and loss their energy. We expect that jets produced in the pp collision do not loss their energy. So, we consider the falling string with the same initial conditions in AdS_5 metric. Our results show that the string falls in the empty AdS_5. So, jet loses its energy in the vacuum! We calculate the R_{AA} for a falling string in AdS_5 metric and show in figure 3 (a) in red curve.

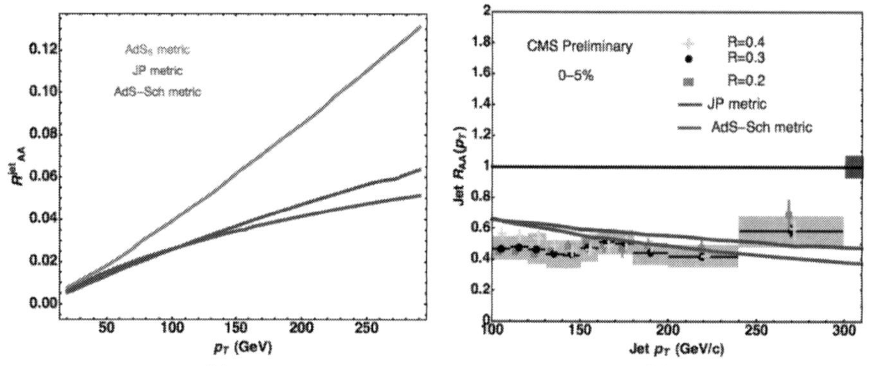

(a) (b)

Figure 3. (a) Jet R_{AA} as a function of p_T in the most central Pb-Pb collision obtained via AdS/CFT in AdS_5 (red), JP (blue) and AdS-Sch (purple) metrics. (b) AdS/CFT jet R_{AA} as a function of p_T compared with the preliminary CMS data in different effective cone sizes for anti-k_T jets using the Bayesian unfolding method for most central Pb-Pb collision at the LHC with $\sqrt{s} = 2.76$ TeV per nucleon [15].

Since the definition of R_{AA} measured the difference between the medium and vacuum effects on jet, we define a renormalized R_{AA} in AdS/CFT as

$$R_{AA}^{renorm} = \frac{R_{AA}^{medium}}{R_{AA}^{AdS_5}} \tag{7}$$

We plot the renormalized R_{AA}^{renorm} for jets in both AdS-Sch and JP metric in figure 3 (b) and compare with the CMS preliminary data for the most central Pb-Pb collision at $\sqrt{s_{NN}} = 2.76\,GeV$ [15]. A suppression factor of 0.5 for high p_T jets is observed in central Pb-Pb collision in comparison to the pp collisions. Our results also show surprisingly agreement with CMS preliminary results on jet R_{AA}.

4. Conclusions

In this paper we have purposed a novel prescription of jet in the context of string theory and AdS/CFT correspondence. We have defined jet as a part of a falling string which lies above the radial coordinate $u^\star$ near the event horizon. Our motivation is defining the jet as hard partons. The instantaneous energy loss of light quark is identified with the energy flux from the point at $u(\tau, \sigma_\kappa) = u^\star$. We have shown that using the $\Delta u - prescription$ of jet, the light quark energy loss exhibit the Bragg peak again at late times in both static and expanding plasmas. This late time behavior of jet energy lost implies that after traveling substantial distances through the plasma, the thermalization of light quark ends with a large amount of energy transferring to the plasma which is similar to the energy loss rate of a fast charge particle moving through ordinary matter.

We considered a brick of plasma and calculated the nuclear modification factor of jet in both AdS-Sch and JP metric. We assumed that the temperature of plasma around 350 MeV at AdS-Sch metric and at initial time in JP metric. Our results show an aver suppression of jet of order of ten respect to the data. We investigated that it is because of the falling string setup at AdS space. In fact, R_{AA} of jet using the falling string in AdS_5 which is dual to jets in vacuum is not one, even though it is less than one. We introduced a renormalized R_{AA}^{renorm} by dividing the

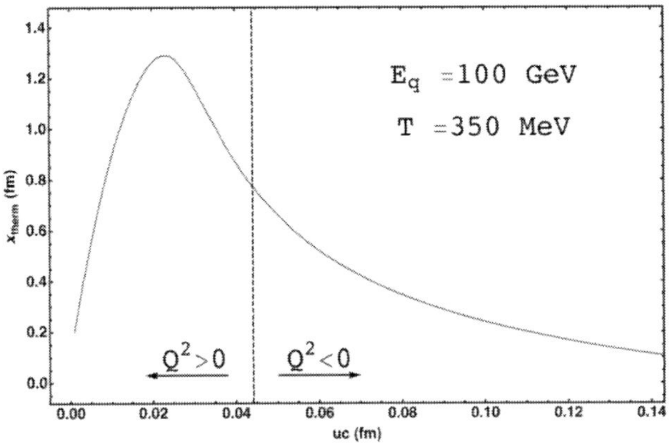

Figure 4. The maximum stopping distance of jet versus the virtuality of quark Q^2. Our result show a nontrivial dependency of the thermalization time to the initial condition of the string.

R_{AA} in the medium to the R_{AA} in the vacuum. Surprisingly, our ratio shows good agreement with the experimental data on the R^{jet}_{AA} of most central Pb-Pb collision at LHC figure 3 (b).

On the other hand, the light quark energy loss is highly depends on the initial conditions of falling string figure 4 . The only way to determine the energy loss of a jet precisely in strongly coupled regime is solving the gravitational bulk-to-boundary problem. One can solve Einstein's equations for the perturbation in the $5d$ geometry due to the presence of the string and according to the bulk to boundary map interpret the near boundary behaviour of the metric perturbation as the perturbation in the SYM energy-momentum tensor by the presence of jet which will be left for the future work.

References

[1] Adams J *et al.* (STAR Collaboration) 2005 *Nucl.Phys.* **A757** 102–183 (*Preprint* nucl-ex/0501009)
[2] Adcox K *et al.* (PHENIX Collaboration) 2005 *Nucl.Phys.* **A757** 184–283 (*Preprint* nucl-ex/0410003)
[3] Arsene I *et al.* (BRAHMS Collaboration) 2005 *Nucl.Phys.* **A757** 1–27 (*Preprint* nucl-ex/0410020)
[4] Back B, Baker M, Ballintijn M, Barton D, Becker B *et al.* 2005 *Nucl.Phys.* **A757** 28–101 (*Preprint* nucl-ex/0410022)
[5] Policastro G, Son D T and Starinets A O 2001 *Phys.Rev.Lett.* **87** 081601 (*Preprint* hep-th/0104066)
[6] Kovtun P, Son D T and Starinets A O 2005 *Phys.Rev.Lett.* **94** 111601 (*Preprint* hep-th/0405231)
[7] Maldacena J M 1999 *Int.J.Theor.Phys.* **38** 1113–1133 (*Preprint* hep-th/9711200)
[8] Witten E 1998 *Adv.Theor.Math.Phys.* **2** 253–291 (*Preprint* hep-th/9802150)
[9] Gubser S, Klebanov I R and Polyakov A M 1998 *Phys.Lett.* **B428** 105–114 (*Preprint* hep-th/9802109)
[10] Aharony O, Gubser S S, Maldacena J M, Ooguri H and Oz Y 2000 *Phys.Rept.* **323** 183–386 (*Preprint* hep-th/9905111)
[11] Casalderrey-Solana J, Liu H, Mateos D, Rajagopal K and Wiedemann U A 2011 (*Preprint* 1101.0618)
[12] Yin Z B (ALICE Collaboration) 2013 *Acta Phys.Polon.Supp.* **6** 479–484
[13] Aad G *et al.* (ATLAS Collaboration) 2010 *Phys.Rev.Lett.* **105** 252303 (*Preprint* 1011.6182)
[14] Chatrchyan S *et al.* (CMS Collaboration) 2011 *Phys.Rev.* **C84** 024906 (*Preprint* 1102.1957)
[15] Collaboration C (CMS Collaboration) 2012
[16] Karch A and Katz E 2002 *JHEP* **0206** 043 (*Preprint* hep-th/0205236)
[17] Herzog C, Karch A, Kovtun P, Kozcaz C and Yaffe L 2006 *JHEP* **0607** 013 (*Preprint* hep-th/0605158)
[18] Chesler P M, Jensen K, Karch A and Yaffe L G 2009 *Phys.Rev.* **D79** 125015 (*Preprint* 0810.1985)
[19] Janik R A and Peschanski R B 2006 *Phys.Rev.* **D73** 045013 (*Preprint* hep-th/0512162)
[20] Horowitz W A 2010 (*Preprint* 1011.4316)

Central exclusive quarkonia production in the forward region at LHCb

Katharina Müller
Winterthurerstr. 190, CH-8057 Zurich, Switzerland
E-mail: kmueller@physik.uzh.ch

Abstract. The LHCb detector and LHC running conditions are ideally suited to measure central exclusive production. Recent results of central exclusive production of J/Ψ, $\Psi(2S)$ and double charmonium are presented. Results are consistent with theoretical expectations. Prospects for measurements of central exclusive production with a new detector installed for the next running period are discussed.

1. Introduction

Central exclusive production (CEP), $pp \to pXp$, in which the protons remain intact and the system X is produced with a rapidity gap on either side, proceeds via the exchange of colourless, neutral particles, either photons or combinations of gluons, for example pomerons. Experimentally, this leads to a unique signature with a small number of particles in the detector, either produced directly or as decay products, and two rapidity gaps that extend to the outgoing protons. The outgoing protons are not detected but escape through the beam-pipe. CEP allows the study of quantum chromodynamics (QCD) and the pomeron. Furthermore, it allows searches for exotic states in a low-background experimental environment and to probe the gluon distribution of the proton.

2. LHCb detector and running conditions

The LHCb detector [1] is a single-arm forward spectrometer, fully instrumented in the pseudorapidity range $2 < \eta < 5$ ('forward region'). The detector includes a high-precision tracking system consisting of a silicon-strip vertex detector (VELO) surrounding the pp interaction region. The VELO has sensitivity to charged particles with momenta above 100 MeV/c in the range $3.5 < \eta < 1.5$ ('backward region'), while extending the sensitivity of the forward region to $1.5 < \eta < 5$. Though tracks in the backward region have no momentum measurement, the backward acceptance provides an efficient veto for charged particles.

The running conditions at LHCb are well suited for CEP measurements because the experiment is operated at a low number of interactions per beam crossing. The data used in the analyses presented below correspond to an integrated luminosity of about 1 fb^{-1} collected in 2011 at a centre-of-mass energy $\sqrt{s} = 7$ TeV and 2 fb^{-1} collected in 2012 at $\sqrt{s} = 8$ TeV.

3. Exclusive J/Ψ and Ψ(2S) production

Exclusive J/Ψ and Ψ(2S) meson production in hadron collisions proceed at leading order via the exchange of a photon and a pomeron, which at sufficiently hard scales can be described by two

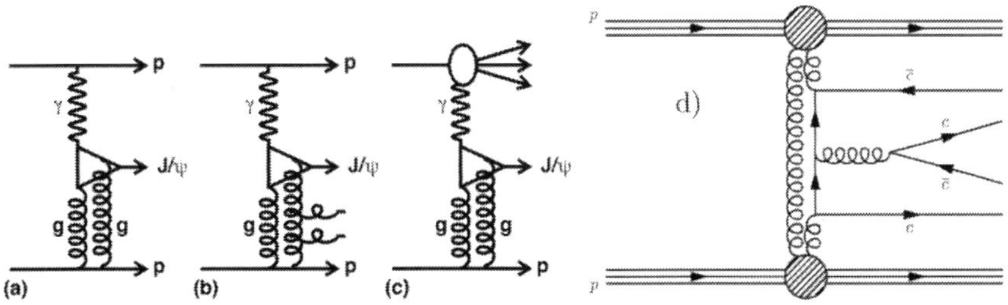

Figure 1. Feynman diagrams for the central exclusive production of a) charmonium through photon-pomeron fusion and d) double charmonium through double pomeron exchange. Typical diagrams for inelastic background from b) additional gluon radiation or c) proton dissociation.

gluons as shown in Fig. 1 a). Measurements of exclusive J/Ψ and Ψ(2S) production thus provide a test of QCD and help to understand the pomeron, which plays a critical role in the description of diffraction and soft processes. The measurements are sensitive to saturation effects [2] and probe x, the fractional momentum of the parton, down to $5 \cdot 10^{-6}$ in the pseudorapidity range of LHCb.

The analysis is based on about 1 fb^{-1} of data collected at $\sqrt{s} = 7$ TeV [3]. Events are selected with two muons and no other track in the VELO. Furthermore, it is required that there is no reconstructed photon. This ensures two rapidity gaps which sum to 3.5 units in the forward region and one to two units, depending on the vertex position, in the backward region. Muon pairs are combined to form meson candidates. Their transverse momentum squared is required to be $p_T^2 < 0.8$ GeV^2/c^2 to reduce inelastic background, and their invariant masses must lie within 65 MeV/c^2 of their PDG values [4]. 55,985 J/Ψ and 1565 Ψ(2S) candidates are found.

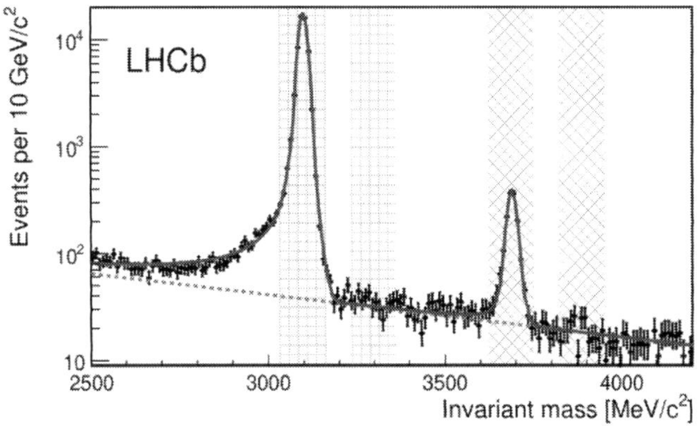

Figure 2. Invariant mass distribution of muon pairs after the selection requirements. The horizontally (diagonally) hatched regions show the J/Ψ (Ψ(2S)) signal and side-band regions. The data are fitted (solid curve) with Crystal Ball functions for the signals and an exponential function for the non-resonant background (dashed curve) [3].

Three background components are considered: non-resonant background, feed-down from exclusive production of other mesons, and inelastic production of mesons with additional gluon radiation or proton dissociation shown in Fig. 1b) and c). The di-muon invariant mass distribution is shown in Fig. 2. It is fitted with Crystal Ball functions [5] to describe the resonant contributions and an exponential for the non-resonant background. Within the mass range of the analysis the non-resonant background is estimated to amount to $(0.8 \pm 0.1)\%$ for J/Ψ and $(17.0 \pm 0.3)\%$ for $\Psi(2S)$ candidates.

Exclusively produced χ_c or $\Psi(2S)$ mesons can feed down to mimic an exclusive J/Ψ decay when the particles produced in association with the J/Ψ are not detected. Their contribution is estimated using simulated events normalised to an enriched background sample in the data. The feed-down from χ_c and $\Psi(2S)$ is estimated to account for $(7.6 \pm 0.9)\%$ and $(2.5 \pm 0.2)\%$ of the J/Ψ candidates respectively. Feed-down into the $\Psi(2S)$ selection is expected to be very small, and mainly due to $\chi_c(2P)$ or X(3872) decays. It is estimated from data, by relaxing the requirement on the number of photons, to be $(2 \pm 2)\%$.

The largest background is due to inelastic production of J/Ψ and $\Psi(2S)$ mesons with additional gluon radiation or proton dissociation where the additional particles are not detected in LHCb. Since the inelastic background has a higher p_T than the signal the background contribution is determined by a fit to the p_T^2 distribution of the meson. In Regge theory, it is assumed that elastic J/Ψ and $\Psi(2S)$ meson production follows an exponential $d\sigma/dt \propto \exp(bst)$, where $t \simeq p_T^2$ is the four-momentum transfer squared at the proton-pomeron vertex and b a constant for a given process. The assumption that the signal and inelastic background can be parametrised by an exponential was tested by the H1 collaboration and found to hold for low p_T [6]. Fitting the p_T^2 distribution with three components, an exponential for elastic CEP, an exponential for the inelastic background, and the feed-down component from data, allows the signal fraction to be extracted (Fig. 3). The slopes found from the fit are in good agreement with expectations from HERA using Regge theory for the extrapolation in energy. The signal fraction is measured to be $(59 \pm 1)\%$ for J/Ψ and $(52 \pm 7)\%$ for $\Psi(2S)$ with $p_T^2 < 0.8\text{GeV}^2/\text{c}^2$.

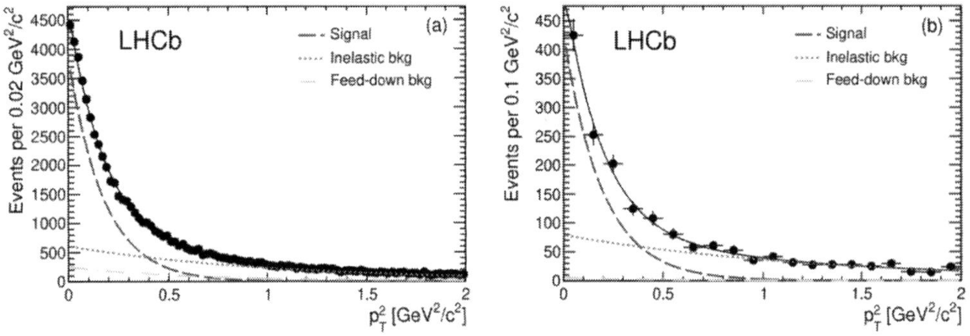

Figure 3. Transverse momentum squared distributions for (a) J/Ψ and (b) $\Psi(2S)$ candidates, where the non-resonant background contribution has been subtracted using side-bands. The points are data, the solid curve is the total fit, the other curves describe the signal and the backgrounds from feed-down and inelastic events [3].

The cross-section times branching fraction to two muons, with pseudorapidities between 2.0 and 4.5, is determined in ten bins of meson rapidity, y. The efficiencies for tracking, muon identification and trigger are determined from data, the selection efficiency from simulation. The number of visible proton-proton interactions per beam crossing, n, is assumed to follow a Poisson distribution, $P(n) = \mu^n \exp(\mu)/n!$, where μ is the average number of visible interactions. Averaged over the data-taking period, the efficiency for a single interaction is then found to be

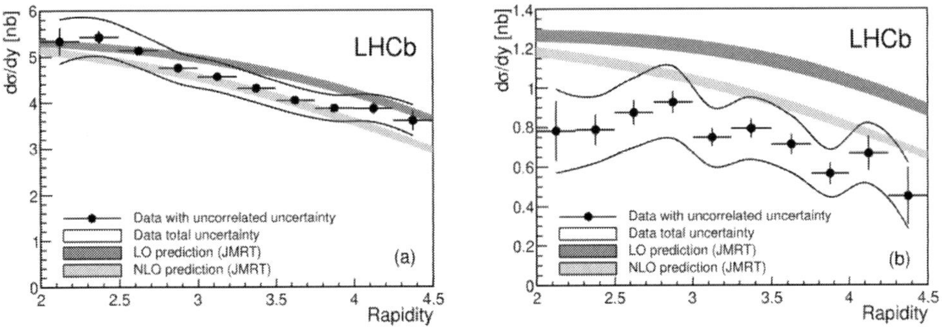

Figure 4. Differential cross-section for (a) J/Ψ and (b) Ψ(2S) production compared to LO and NLO predictions of [7]. The band indicates the total uncertainty, most of which is correlated between bins [3].

$\epsilon_{single} = (24.1 \pm 0.3)\%$.

The cross-sections times branching fraction to two muons having pseudorapidities between 2.0 and 4.5 are measured to be:

$$\sigma(J/\Psi \rightarrow \mu\mu) = 291 \pm 7(\text{stat}) \pm 19(\text{syst})\text{pb}$$
$$\sigma(\Psi(2S) \rightarrow \mu\mu) = 6.5 \pm 0.9(\text{stat}) \pm 0.4(\text{syst})\text{pb}.$$

The differential distributions are presented in Fig. 4, where the error bars indicate the uncorrelated statistical uncertainties and the band is the total uncertainty. The systematic uncertainties are mostly correlated between bins. The measurements are compared to leading and next-to-leading order predictions [7]. Better agreement is observed between data and the NLO prediction than between data and the LO prediction.

Exclusive production of J/Ψ in pp collisions is related to photoproduction through

$$\frac{\mathrm{d}\sigma(pp)}{\mathrm{d}y} = r_+ k_+ \frac{\mathrm{d}n}{\mathrm{d}k_+}\sigma(\gamma p)(W_+) + r_- k_- \frac{\mathrm{d}n}{\mathrm{d}k_-}\sigma(\gamma p)(W_-)$$

where $\mathrm{d}n/\mathrm{d}k$ are photon fluxes for photons of energy $k \simeq (M(\mathrm{J}/\Psi/2)\exp(\pm|y|)$, $W_\pm$ the γp centre of mass energy, $(W^\pm)^2 = 2k_\pm s$, and $r_\pm$ are absorptive corrections. The LHCb results cannot unambiguously determine the photoproduction cross-section due to contributions from both W_+ and W_- , corresponding to the photon being either an emitter or a target, respectively. A model-dependent determination of $\sigma(\gamma p \rightarrow \mathrm{J}/\Psi p(W_{+(-)})$ is obtained from the LHCb differential cross-section measurement, assuming the power-law result from a fit to the HERA data [6] for $\sigma(\gamma p \rightarrow \mathrm{J}/\Psi p(W_{-(+)})$ as shown in Fig. 5. Here, the LHCb data is shown together with HERA and fixed target photoproduction results and the power-law fit from [6]. There are two correlated points for each LHCb measurement. Deviations from the power law dependence are observed at the energies probed by the LHCb measurements. These can be explained by higher order corrections or saturation effects.

4. Double charmonium production

Cross-sections and invariant mass spectra of charmonium pairs are sensitive to the presence of additional particles in the decay chain such as glueballs or tetraquarks [9]. LHCb has measured the inclusive production of J/Ψ pairs [8] to be in broad agreement with the QCD predictions,

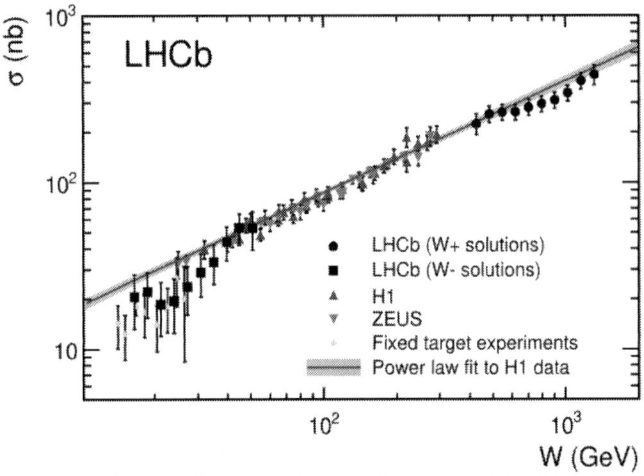

Figure 5. Photoproduction cross-section as a function of the centre-of-mass of the photon-proton system with the power-law fit from superimposed [6]. The LHCb data points for $W_+(W_-)$ are derived assuming the power-law fit for $W_-(W_+)$ and are correlated. The uncertainties are correlated between bins [8].

although the invariant mass distribution of the di-meson system is shifted to higher values in data. In the inclusive case, this shift could be an indication of double parton scattering (DPS) effects [10]. The principal production mechanism for the CEP of two charmonia is through double pomeron exchange (DPE) as shown in Fig. 1 d).

CEP of J/ΨJ/Ψ, J/ΨΨ(2S) and Ψ(2S)Ψ(2S) are examined using a data sample corresponding to an integrated luminosity of about 3 fb^{-1} [11]. The J/Ψ and Ψ(2S) mesons are identified through their decays to two muons. The selection of J/Ψ or Ψ(2S) candidates requires four reconstructed tracks, three of them identified as muons, and no other activity in the event. The invariant masses of oppositely charged muon candidates is shown in Fig. 6 a). Accumulations of events are apparent around the J/Ψ and Ψ(2S) masses.

Figure 6 b) shows the higher mass combination when asking that the lower mass combination is consistent with the J/Ψ mass. There are 37 J/ΨJ/Ψ, 5 J/ΨΨ(2S) and no Ψ(2S)Ψ(2S) candidates. The only substantial feed-down background to the J/ΨJ/Ψ signal comes from J/ΨΨ(2S) where Ψ(2S)→J/ΨX with X unreconstructed.

The cross-sections, at an average energy of 7.6 TeV, for the di-meson system to be in the rapidity range $2.0 < y < 4.5$ with no other charged or neutral energy inside the LHCb acceptance are measured to be

$$\sigma(\mathrm{J}/\Psi\mathrm{J}/\Psi) = 58 \pm 10(\mathrm{stat}) \pm 6(\mathrm{syst})\mathrm{pb}$$
$$\sigma(\Psi(2S)\Psi(2S)) = 63^{+27}_{-18}(\mathrm{stat}) \pm 10(\mathrm{syst})\mathrm{pb}$$
$$\sigma(\mathrm{J}/\Psi\Psi(2S)) < 237(\mathrm{syst})\mathrm{pb}.$$

The upper bound for the cross-section for J/ΨΨ(2S) production is given at a 90% CL.

Similarly to the previous analysis, the elastic fraction is estimated from the p_T^2 distribution. It is shown in the left plot of Fig. 7a) for the J/ΨJ/Ψ candidates. Central exclusive events peak below 1 GeV^2/c^2 while the tail to higher values is characteristic of inelastic production. A maximum likelihood fit is performed to the sum of two exponentials, $f_{el}b_s\exp(-b_sp_T^2) + (1 -$

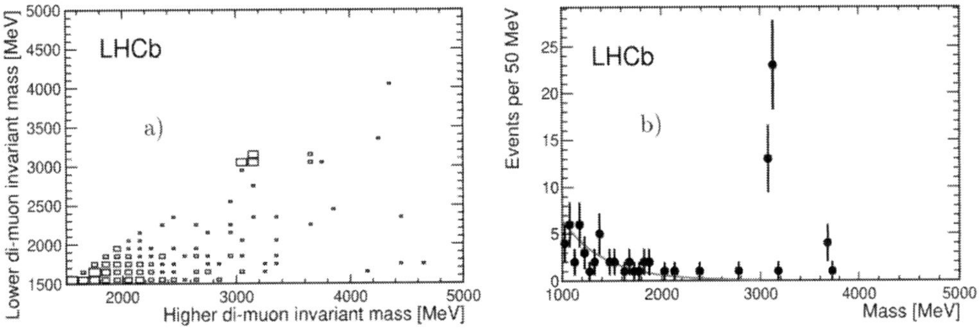

Figure 6. Invariant masses (a) of pairs of oppositely charged muons in events with exactly four tracks and invariant mass of the second pair of tracks (b) where the first pair has a mass consistent with the J/Ψ or Ψ(2S) meson [11].

$f_{el})b_b\exp(-b_b p_T^2)$, where b_s, b_b are the slopes for the signal and background and f_{el} is the fraction of elastic events. Due to the small sample size, the value of b_b is constrained using the distribution for exclusive dimuon candidates whose invariant mass lies in the range 6 to 9 GeV/c^2. Their p_T^2 distribution is shown in Fig. 7b). It has a prominent peak in the first bin corresponding to the electromagnetic two-photon exchange process, $pp \to p\mu^+\mu^- p$. The tail to larger values is characteristic of events with proton dissociation. Only the region $1.5 < p_T^2 < 10$ GeV^2/c^2 is used in the fit to determine the slope $b_b = 0.29 \pm 0.02$ GeV^{-2}c^2. With b_b fixed the fit to the signal distribution yields $b_s = 2.9 \pm 1.3$ GeV^{-2}c^2 and $f_{el} = 0.42 \pm 0.13$. Using this signal fraction the cross-section for exclusive production of J/ΨJ/Ψ is found to be 24 ± 9 pb. This is in agreement with theoretical calculations [12] which predict cross-sections between 6 and 34 pb.

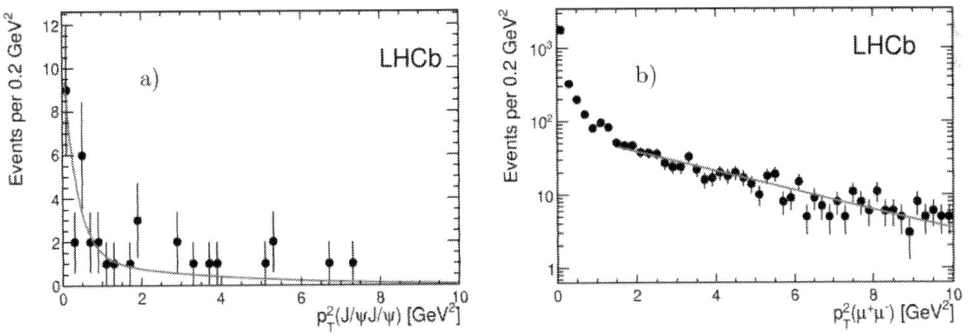

Figure 7. Transverse momentum squared distribution of a) candidates for exclusively produced J/Ψ J/Ψ and b) dimuons with invariant mass between 6 and 9 GeV/c^2. The curves are fits to the data as described in the text [11].

5. Shower counters

The coverage of the LHCb detector is the limiting factor for the precision of CEP measurements. During the LHC shutdown new forward shower counters (FSC) have been installed, significantly extending the rapidity coverage in the very forward and backward region $5 < |\eta| < 8$. The FSCs consist of five stations with four retractable plastic scintillator plates placed on either side of the interaction point at 7.5 m, 19 m and 114 m from the interaction region. A schematic of one station in the closed and open configuration is shown in Fig. 8. Information from this new detector will greatly improve LHCbs ability to separate elastic from inelastic events and help to understand the shape of the background. Furthermore, a veto on the activity in the FSCs will be used in the trigger allowing a more efficient triggering mechanism for CEP events.

Figure 8. Drawings of one plane of the forward shower counters which were installed around the beam-pipe, in the LHC tunnel, on either side of the interaction point. The picture shows the support structure, scintillators and PMTs, which can be retracted from the beam-line.

6. Conclusions

The design of the LHCb detector makes it well suited to the study of central exclusive production. It has excellent particle identification and the ability to trigger on low multiplicity events and low transverse momentum objects. Measurements of differential cross-sections for exclusive JΨ and Ψ(2S) have been discussed. They agree with the shapes predicted by next-to-leading order predictions. A clear signal for double charmonium production in the absence of other activity in the LHCb acceptance is observed. It is estimated that $(42 \pm 13)\%$ of these events are elastically produced.

The additional instrumentation of the LHCb detector with forward shower counters will significantly increase the coverage of the detector and will allow more precise exclusive measurements in the next running period. The signal of these counters can be used as a veto in the trigger and help to further reduce the inelastic background. The uncertainty on this background is presently the dominant systematic uncertainty.

References

[1] A. A. Alves Jr. et al. The LHCb detector at the LHC. *JINST*, 3:S08005, 2008.

[2] L. Motyka and G. Watt. Exclusive photoproduction at the Tevatron and CERN LHC within the dipole picture. *Phys.Rev.*, D78:014023, 2008.

[3] Roel Aaij et al. Updated measurements of exclusive J/ψ and $\psi(2S)$ production cross-sections in pp collisions at $\sqrt{s} = 7$ TeV. *J.Phys.*, G41:055002, 2014.

[4] K. Nakamura et al. Review of particle physics. *J. Phys.*, G37:075021, 2010.

[5] Tomasz Skwarnicki. *A study of the radiative cascade transitions between the Υ' and Υ resonances.* PhD thesis, Institute of Nuclear Physics, Krakow, 1986. http://inspirehep.net/record/230779/DESY-F31-86-02.

[6] C. Alexa et al. Elastic and Proton-Dissociative Photoproduction of J/psi Mesons at HERA. *Eur.Phys.J.*, C73(6):2466, 2013.

[7] S.P. Jones, A.D. Martin, M.G. Ryskin, and T. Teubner. Probes of the small x gluon via exclusive J/ψ and Υ production at HERA and the LHC. *JHEP*, 1311:085, 2013.

[8] R Aaij et al. Exclusive J/ψ and $\psi(2S)$ production in pp collisions at $\sqrt{s} = 7$ TeV. *J.Phys.*, G40:045001, 2013.

[9] A.V. Berezhnoy, A.K. Likhoded, A.V. Luchinsky, and A.A. Novoselov. Double J/psi-meson Production at LHC and 4c-tetraquark state. *Phys.Rev.*, D84:094023, 2011.

[10] R. Aaij et al. Observation of J/ψ pair production in pp collisions at $\sqrt{s} = 7TeV$. *Phys.Lett.*, B707:52–59, 2012.

[11] R. Aaij et al. Observation of charmonium pairs produced exclusively in pp collisions. *J.Phys.*, G41(11):115002, 2014.

[12] L.A. Harland-Lang, V.A. Khoze, and M.G. Ryskin. Exclusive production of double J/ψ mesons in hadronic collisions. 2014.

LHCb results in proton-nucleus collisions at the LHC

Katharina Müller
Winterthurerstr. 190, CH-8057 Zurich, Switzerland
E-mail: kmueller@physik.uzh.ch

Abstract. The forward acceptance of the LHCb detector allows it to probe proton-ion collision in a unique kinematic range, complementary to the other LHC experiments. The production of J/Ψ and Υ-mesons decaying into two muons is studied at the LHCb experiment in proton-lead collisions at a proton-nucleon centre-of-mass energy $\sqrt{s_{NN}} = 5$ TeV. The analysis is based on a data sample corresponding to an integrated luminosity of 1.6 nb^{-1}. The nuclear modification factor and the forward-backward production ratio are determined for J/Ψ and Υ(1S) mesons. Clear suppression of prompt J/Ψ production is observed with respect to the production in *pp* collisions at large rapidity, while the suppression of J/Ψ from *b*-hadron decays is less pronounced. The nuclear modification factor for Υ(1S) mesons in the forward region is found to be similar to those for J/Ψ from *b*-hadron decays. Furthermore, a first observation of *Z* bosons in proton-lead collisions is reported.

1. Introduction

In ultra-relativistic heavy-ion collisions, the production of heavy quarkonia or electroweak bosons is expected to be suppressed with respect to proton-proton collisions, if a quark-gluon plasma, QGP, is created [1]. Such a suppression of heavy quarkonia and *Z* boson production with respect to *pp* collisions can also take place in proton-nucleus (*pA*) collisions, where a quark-gluon plasma is not expected to be created but only cold nuclear matter effects, such as nuclear absorption, parton shadowing and parton energy loss in initial and final states can occur [2, 3, 4]. The study of *pA* collisions therefore provides important input to disentangle the QGP effects from cold nuclear effects, probe nuclear parton distribution functions, which are poorly constrained, and provide a reference sample for nucleus-nucleus collisions.

In early 2013, the LHCb detector [5] collected two data samples corresponding to 1.6 nb^{-1} of proton-lead collisions at a centre-of-mass energy per proton-nucleon pair of $\sqrt{s_{NN}} = 5$ TeV. The two data samples correspond to two different beam configurations, with either the proton or the lead beam into the direction of LHCb. These are referred to as forward and backward configuration respectively. Owing to the asymmetric beam configuration the LHCb acceptance corresponds to $1.5 < y < 4.0$ ($-5.0 < y < -2.5$) for the forward (backward) configuration. Results on J/Ψ [6], Υ [7] and *Z* [8] production are reported below.

2. J/Ψ and Υ production

J/Ψ [6] and Υ [7] mesons are reconstructed in the di-muon final state with the transverse momentum, p_T, of the di-muon system restricted to $p_T < 14$ GeV/c ($p_T < 15$ GeV/c) for J/Ψ (Υ). The excellent vertexing capability of LHCb allows a separation of prompt J/Ψ mesons and J/Ψ mesons from *b*-hadron decays (J/Ψ from *b*). The number of prompt J/Ψ and J/Ψ from *b*

Journal of Physics: Conference Series **623** (2015) 012023 doi:10.1088/1742-6596/623/1/012023

candidates is determined by a combined fit to the di-muon invariant mass and pseudo-proper time distributions. The pseudo-proper time is defined as

$$t_z = \frac{(z_{J/\Psi} - z_{PV}) \times M_{J/\Psi}}{p_z},$$

where $z_{J/\Psi}$ is the z position of the J/Ψ decay vertex, z_{PV} that of the primary vertex, p_z the z component of the measured J/Ψ momentum, and $M_{J/\Psi}$ the mass of the J/Ψ. The signal di-muon invariant mass distribution in each p_T and rapidity (y) bin is modelled with a Crystal Ball function [9], and the combinatorial background with an exponential function. The t_z signal distribution is described by the sum of a δ-function at $t_z = 0$ for prompt J/Ψ production and an exponential decay function for J/Ψ from b, both convolved with a double-Gaussian resolution function whose parameters are free in the fit. Figure 1 shows the projections of the combined fit in two rapidity bins in the forward and backward configuration.

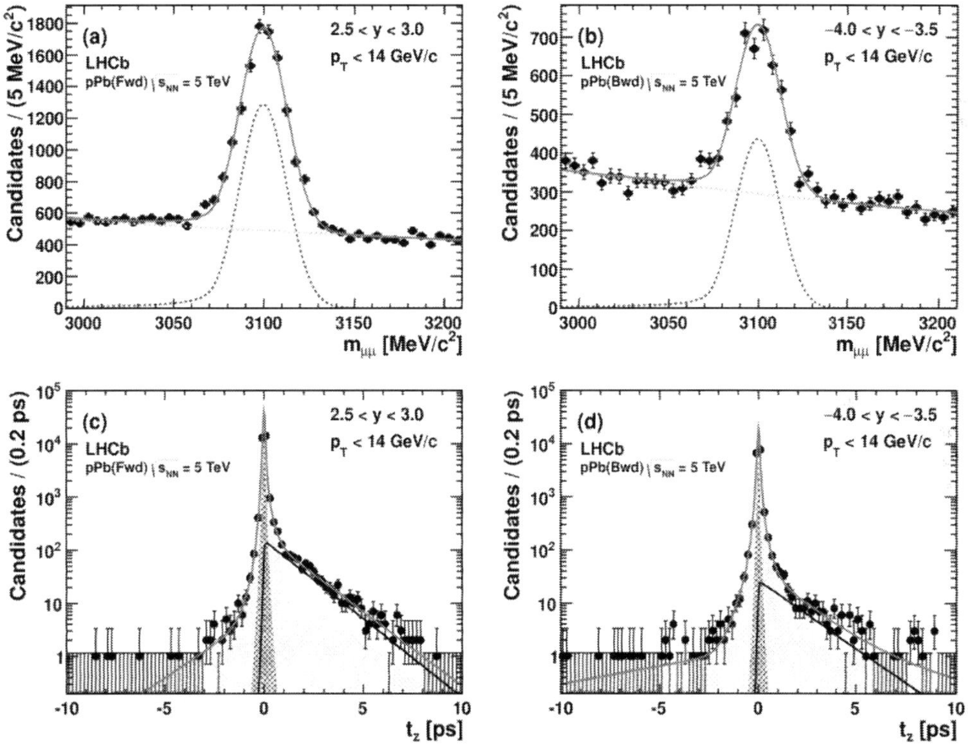

Figure 1. Projections of the combined fit on (a, b) di-muon invariant mass and (c, d) t_z in two representative bins in the (a, c) forward and (b, d) backward samples. For the mass projections the (red solid curve) total fitted function is shown together with the (blue dotted curve) J/Ψ signal and (green dotted curve) background contributions. For the t_z projections the total fitted function is indicated by the solid red curve, the background by the green hatched area, the prompt signal by the blue area and J/Ψ from b by the solid black curve [6].

The di-muon invariant mass distribution for the Υ candidates in the two samples are shown in Fig. 2. While all three Υ resonances are observed in the forward direction, only Υ(1S) gives a significant signal in the backward configuration. An unbinned extended maximum likelihood fit to the invariant mass distribution of the selected candidates is performed to determine the signal yields of $\Upsilon(1S)$, $\Upsilon(2S)$ and $\Upsilon(3S)$ mesons in a fit range $8400 < m_{\mu^+\mu} < 11400$ MeV/c^2. A sum of three Crystal Ball functions is used to describe the $\Upsilon(1S)$, $\Upsilon(2S)$ and $\Upsilon(3S)$ signal components, while the combinatorial background is modelled with an exponential function.

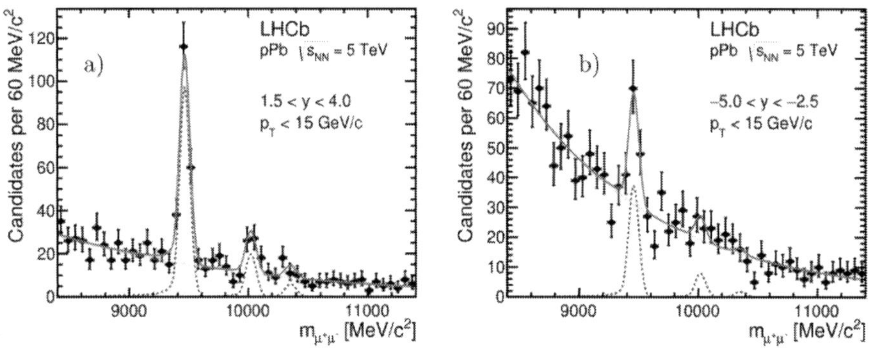

Figure 2. Invariant mass distribution of di-muon pairs in the (a) forward and (b) backward samples of pPb collisions. The transverse momentum is restricted to $p_T < 15$ GeV. The rapidity range is $1.5 < y < 4.0(5.0 < y < 2.5)$ for the forward (backward) sample. The black dots are the data points, the blue dashed curve indicates the signal component, the green dotted curve represents the combinatorial background, and the red solid curve is the sum of the signal and background components [7].

Higher combinatorial background in the backward configuration is observed for J/Ψ and Υ production due to the larger multiplicity in lead-proton collisions.

Figure 3 shows the single differential production cross-sections for prompt J/Ψ and J/Ψ from b as functions of p_T in the forward and backward regions. The cross-section for J/Ψ from b is about a factor of ten smaller than for prompt J/Ψ with a very similar p_T dependence similar to what was observed by LHCb in pp collisions [10].

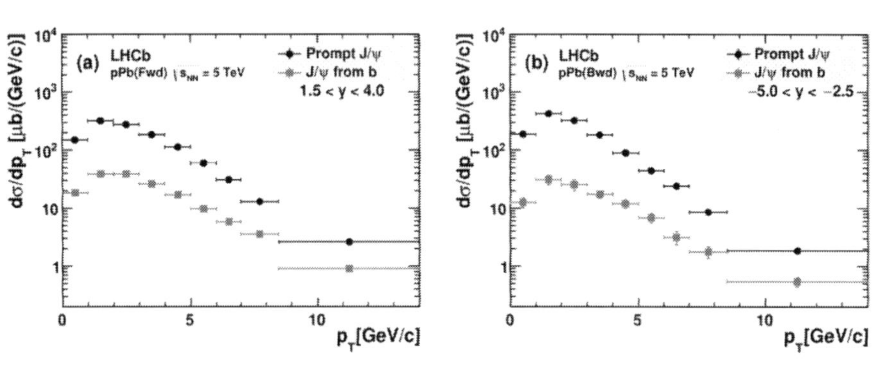

Figure 3. Single differential production cross-sections for (black dots) prompt J/Ψ and (red squares) J/Ψ from b as functions of p_T in the (a) forward and (b) backward region [6].

3. Cold nuclear effects

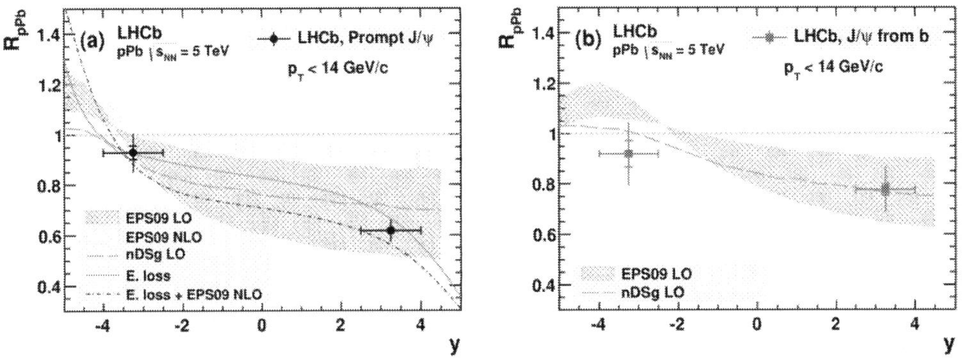

Figure 4. Forward-backward production ratios (R_{FB}) for (a) prompt J/Ψ and (b) J/Ψ from b as functions of rapidity [6] together with theoretical predictions from (yellow dashed line and brown band) [2, 11], (blue band) [3], and (green solid and blue dash-dotted lines) [4].

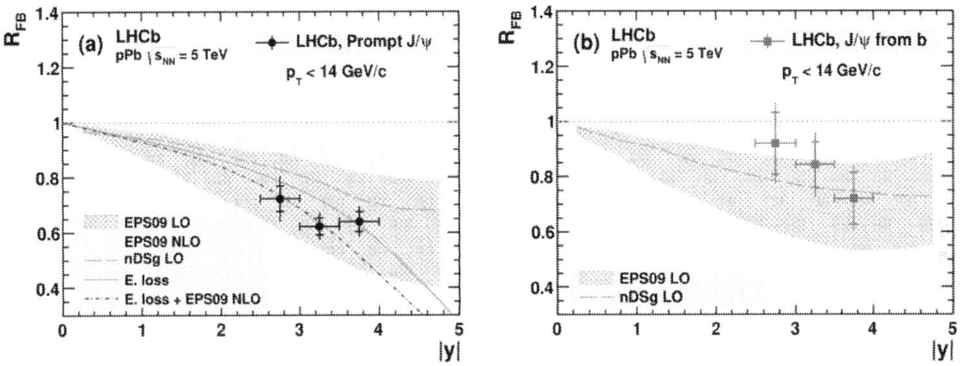

Figure 5. Nuclear modification factor (R_{pA}) for (a) prompt J/Ψ and (b) J/Ψ from b as functions of rapidity [6] together with theoretical predictions from (yellow dashed line and brown band) [2, 11], (blue band) [3], and (green solid and blue dash-dotted lines) [4].

Nuclear effects are usually characterised by the nuclear modification factor R_{pA} and the forward-backward production ratio R_{FB},

$$R_{pA} = \frac{d\sigma_{pA}/dy}{A d\sigma_{pp}/dy}, \qquad R_{FB} = \frac{d\sigma_{pA(y>0)}/dy}{d\sigma_{pA(y<0)}/dy},$$

which depend on the production cross-section of a given particle in pA collisions and for R_{pA} also on the cross-section in pp collisions at the same centre-of-mass energy as well as the atomic number A. The advantage of measuring the R_{FB} is that it does not rely on the knowledge of the production cross-section in pp collisions and that experimental systematic uncertainties and theoretical scale uncertainties cancel partially.

To determine the nuclear modification factor R_{pA}, the reference cross-sections in pp collisions at $\sqrt{s_{NN}} = 5$ TeV are needed. Since there is no direct measurement, these are obtained by a power-law fit to the previous LHCb measurements of J/Ψ and Υ production at 2.76 TeV, 7 TeV and 8 TeV [12, 13]. Cold nuclear effects are studied in three rapidity bins for J/Ψ; the low statistics of the Υ sample only allow a measurement for Υ(1S) in one rapidity bin.

Figure 4 and 5 show the nuclear modification factors and the forward-backward production ratios, for prompt J/Ψ mesons and J/Ψ from b as functions of rapidity [6], compared to different theoretical predictions [2, 11, 3, 4]. A clear suppression of about 40% at large rapidity is observed for prompt J/Ψ production. The measurements agree with most predictions. The data show a modest suppression of J/Ψ from b production in the forward region, with respect to that in pp collisions. This is the first indication of the suppression of b hadron production in proton-lead collisions. The nuclear modification factor and forward-backward production ratio for J/Ψ from b reflect that cold nuclear matter effects on b hadrons are less pronounced than for J/Ψ.

Figure 6 shows R_{pA} and R_{FB} for Υ(1S) [7] together with the LHCb results of prompt J/Ψ and J/Ψ from b with theoretical predictions. The data are consistent with a suppression in the forward region and a possible enhancement in the backward region. In the forward region, the suppression of $\Upsilon(1S)$ mesons is smaller than that of prompt J/Ψ mesons and similar to J/Ψ from b. Data and theoretical predictions [3], which include coherent energy loss and nuclear shadowing as parametrised with EPOS09 [4], agree within the large experimental uncertainties.

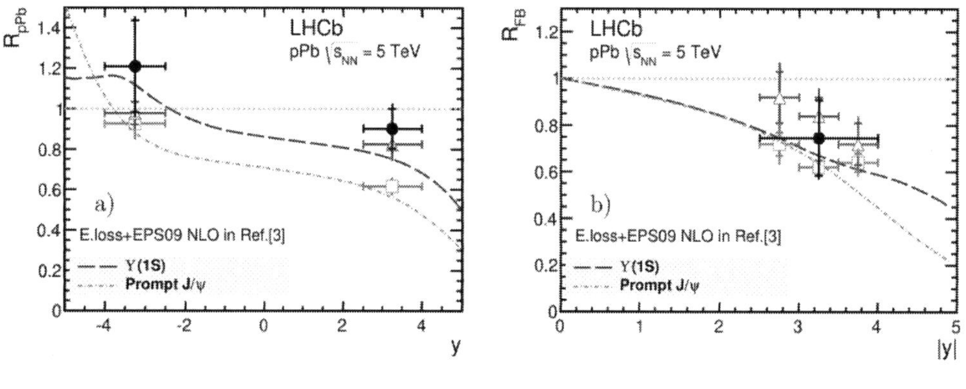

Figure 6. (a) R_{FB} and (b) R_{pA} for Υ, prompt J/Ψ and J/Ψ from b as functions of rapidity [7] with theoretical predictions including energy loss and nuclear shadowing [3].

4. Inclusive Z boson production in proton-lead collisions

The Z candidates are reconstructed in the di-muon final state. Background contributions from muon mis-identification and the decay of heavy flavour mesons are determined from data. A total of 15 candidates are selected with a purity of above 99%, corresponding to a significance of 10.4σ (6.8σ) for the Z signal in the forward (backward) direction. Figure 7 shows the di-muon invariant mass of the Z candidates in the forward and backward direction. The inclusive Z boson production cross-section is measured to be $\sigma(Z \rightarrow \mu\mu) = 13.5^{+5.4}_{-4.0} \pm 1.2$ nb in the forward and $\sigma(Z \rightarrow \mu\mu) = 10.7^{+8.4}_{-5.1} \pm 1.0$ nb in the backward configuration. Here, the first uncertainty is statistical and the second systematic. The measurements are compared to theoretical predictions calculated at NNLO using the FEWZ generator [14] and computed with and without considering nuclear effects based on the EPS09 nuclear PDF set [15] in Fig. 8. The statistical precision of

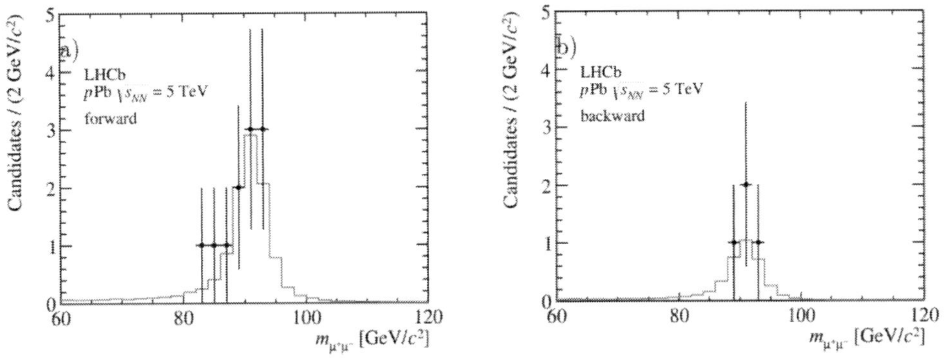

Figure 7. Di-muon invariant mass for the Z candidates in the (a) forward and (b) backward sample [8].

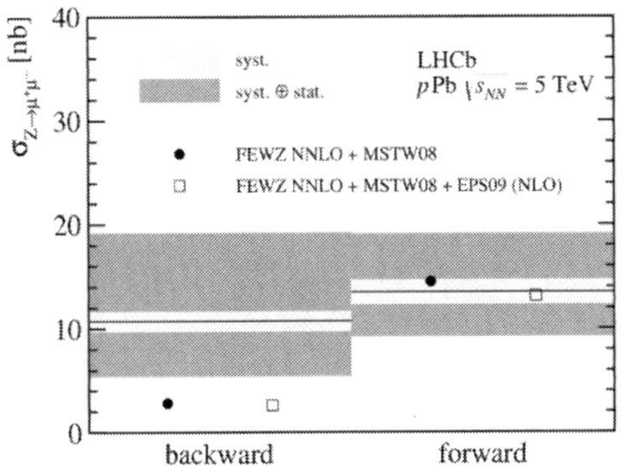

Figure 8. Experimental results and theoretical predictions for the Z production cross-section in proton-lead collisions [8].

the measured cross-sections prevents conclusions on the presence of nuclear effects. However, the observation of Z production in proton-nucleus collisions demonstrates the excellent potential of the study of electroweak bosons in proton-lead collisions at LHCb.

5. Conclusions

The production of prompt J/Ψ mesons, J/Ψ from b-hadron decays and Υ is studied in pPb collisions with the LHCb detector at the nucleon-nucleon centre-of-mass energy $\sqrt{s_{NN}} = 5$ TeV. The measurement is performed as a function of the transverse momentum and rapidity of the J/Ψ (Υ) in the region $p_T < 14$ (15) GeV/c and $1.5 < y < 4.0$ (forward) and $5.0 < y < 2.5$ (backward). The measurements indicate that cold nuclear matter effects are less pronounced for J/Ψ from b-hadron decays and Υ, than for prompt J/Ψ. The results show good agreement with the available theoretical predictions and provide useful constraints to the parameterisation of theoretical models. The measured nuclear modification factor for prompt J/Ψ mesons shows that it is necessary to include cold nuclear matter effects in the interpretation of quark-gluon plasma signatures in heavy-ion collisions. The limited statistics of the Z production cross-section does not allow a conclusion on the presence of cold nuclear effects.

References

[1] T. Matsui and H. Satz. J/ψ Suppression by Quark-Gluon Plasma Formation. *Phys.Lett.*, B178:416, 1986.
[2] E.G. Ferreiro, F. Fleuret, J.P. Lansberg, and A. Rakotozafindrabe. Impact of the Nuclear Modification of the Gluon Densities on J/ψ production in pPb collisions at $\sqrt{s_{NN}} = 5$ TeV. *Phys.Rev.*, C88(4):047901, 2013.
[3] J.L. Albacete, N. Armesto, R. Baier, G.G. Barnafoldi, J. Barrette, et al. Predictions for $p + Pb$ Collisions at $\sqrt{s_{NN}} = 5$ TeV. *Int.J.Mod.Phys.*, E22:1330007, 2013.
[4] Francois Arleo and Stephane Peigne. Heavy-quarkonium suppression in p-A collisions from parton energy loss in cold QCD matter. *JHEP*, 1303:122, 2013.
[5] Jr. Alves, A. Augusto et al. The LHCb Detector at the LHC. *JINST*, 3:S08005, 2008.
[6] R. Aaij et al. Study of J/ψ production and cold nuclear matter effects in pPb collisions at $\sqrt{s_{NN}} = 5$ TeV. *JHEP*, 1402:072, 2014.
[7] R. Aaij et al. Study of Υ production and cold nuclear matter effects in pPb collisions at $\sqrt{s_{NN}} = 5$ TeV. *JHEP*, 1407:094, 2014.
[8] R. Aaij et al. Observation of Z production in proton-lead collisions at LHCb. *JHEP*, 1409:030, 2014.
[9] Tomasz Skwarnicki. *A study of the radiative cascade transitions between the Upsilon-prime and Upsilon resonances.* PhD thesis, Institute of Nuclear Physics, Krakow, 1986. http://inspirehep.net/record/230779/DESY-F31-86-02.
[10] R. Aaij et al. Measurement of J/ψ production in pp collisions at $\sqrt{s} = 7$ TeV. *Eur.Phys.J.*, C71:1645, 2011.
[11] Z. Conesa del Valle, E.G. Ferreiro, F. Fleuret, J.P. Lansberg, and A. Rakotozafindrabe. Open-beauty production in pPb collisions at $\sqrt{s_{NN}}$=5 TeV: effect of the gluon nuclear densities. *Nucl.Phys.*, A926:236–241, 2014.
[12] ALICE and LHCb Collaborations. Reference pp cross-sections for J/ψ studies in proton-lead collisions at $\sqrt{s_{NN}} = 5.02$ TeV and comparisons between ALICE and LHCb results. 2013.
[13] The LHCb Collaboration. Reference pp cross-sections for $\Upsilon(1S)$ studies in proton-lead collisions at $\sqrt{s_{NN}} = 5.02$ TeV and comparisons between ALICE and LHCb results. 2014.
[14] Ye Li and Frank Petriello. Combining QCD and electroweak corrections to dilepton production in FEWZ. *Phys.Rev.*, D86:094034, 2012.
[15] K.J. Eskola, H. Paukkunen, and C.A. Salgado. EPS09: A New Generation of NLO and LO Nuclear Parton Distribution Functions. *JHEP*, 0904:065, 2009.

International Workshop on Discovery Physics at the LHC (Kruger2014) IOP Publishing
Journal of Physics: Conference Series **623** (2015) 012024 doi:10.1088/1742-6596/623/1/012024

The Upgrade of the ATLAS Tile Calorimeter Readout Electronics for Phase II

Robert Graham Reed
On behalf of the ATLAS Tile Calorimeter System.
School of Physics, University of the Witwatersrand, Johannesburg, South Africa
E-mail: `robert.reed@cern.ch`

Abstract. The Large Hadron Collider at CERN is scheduled to undergo a major upgrade, called the Phase II Upgrade, in 2022. The ATLAS Tile Calorimeter (TileCal) community will do major modifications to the sub-detector to account for the increased luminosity. More specifically, a large proportion of the current front and back-end electronics will be upgraded in order to digitize all signals generated in the Calorimeters. A Demonstrator program has been established, which combines the current and future architectures, as a proof of principle. The insertion of the first demonstrator is planned for the end of 2015.

1. Introduction

The ATLAS detector [1], one of two general purpose detectors at the LHC, consists of multiple sub-detectors which are designed to detect interesting physics in 40 million bunch-crossings per second. The TileCal, the central region in Fig. 1, is used to measure energies and directions of hadrons, jets, τ and leptons. Particles pass through plastic scintillating tiles which emit light that is guided to photo-multiplier tubes (PMTs). This provides the means to measure energy and direction of hadrons. These light signals are directed to front-end electronics where they are digitised and processed for the first level of triggering. Data from the front-end electronics is transferred off the detector to the back-end electronics where further processing occurs. The ATLAS detector is scheduled to undergo a major upgrade in the year 2022 in order to handle an increased luminosity.

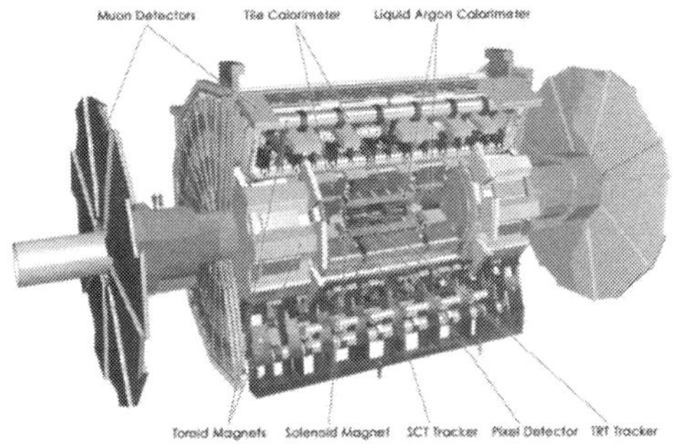

Figure 1: Schematic of the ATLAS detector.

Table 1 illustrates the increase in bandwidth for the TileCAL from the current to future luminosities, which will increase by a factor of 5, providing up to 41 Tbps. As a result the entire front and back-end electronics will be upgraded for the Phase II upgrade [2].

Table 1: Tile Calorimeter Bandwidth Upgrade.

Phase	Present	Upgrade
Number of fibers	256	4096
Fiber Bandwidth	800 Mbps	10 Gbps
Total Bandwidth	205 Gbps	41 Tbps

2. Upgrade of the Readout Electronics

2.1. Current Architecture

The current front-end electronics are located in a super-drawer situated on the outer edge of the Tile Calorimeter. Fig. 2 shows the current flow of information. A pulse of light caused by charged particles passing through the scintillating plastic is directed to a PMT using wavelength shifting fibres. The signal is directed to the 3-in-1 cards where two different gains are applied. It is then digitized by an ADC and stored in a pipeline, to wait for a level 1 accept (L1A), on the Digitizer Board. If a L1A is received the data is sent to the interface board where it is formatted and sent at a rate of 100 kHz to the Read of Driver (ROD) located on the back-end. As luminosity increases so does pileup and the amount of data generated. The larger pileup and

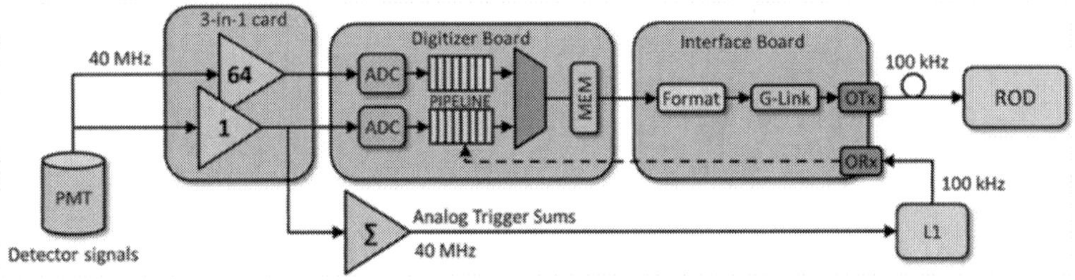

Figure 2: Current read out electronics.

data mean that a better precision and granularity is needed. The radiation levels are expected to increase which can cause errors in sensitive electronics hence radiation hard components are needed. The upgrade allows for the simplification and reduction of maintenance needs by moving a large portion of the front-end electronics off the detector to the back-end. This also reduces the radiation induced errors and provides an opportunity to replace obsolete components.

2.2. Upgrade Architecture

The upgraded front-end electronics are shown in Fig. 3. The components are discussed in more detail below.

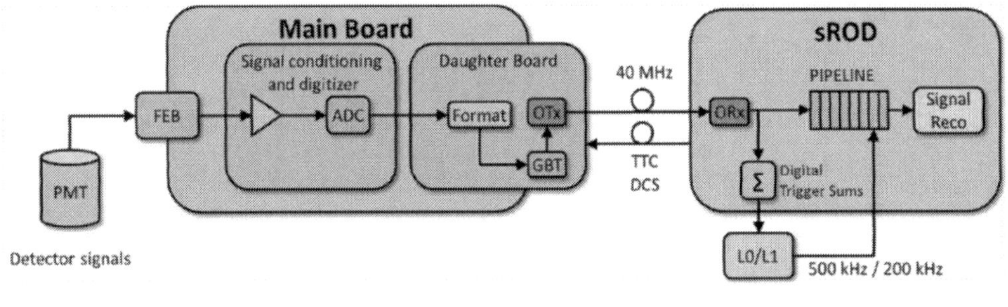

Figure 3: Upgraded read out electronics.

2.2.1. Front-End Boards. There are currently three different front end boards (FEBs) that are being evaluated.

(i) Modified 3-in-1 Boards - It is based on the current design with improvements in radiation hardness, linearity and noise. Shown in Fig. 4a). This board is being designed by the Enrico Fermi Institute, University of Chicago.

(ii) QIE ASIC - This board is a custom Application-Specific Integrated Circuit (ASIC). Shown in Fig. 4b). It is a completely different operating design compared to the 3-in-1 board. It uses a Charge Integrator and Encode (QIE) chip that splits the current into four different ranges with no shaping. This is useful for pile-up as it provides a way to measure every 25 ns. It is a joint effort between the Argonne National Laboratory (ANL) and Fermilab.

(iii) FATALIC - This stands for the Front-end for Atlas TilecAL Inegrated Circuit. Shown in Fig. 4c). This board combines two ASIC solutions (TACTIC and FATALIC). It provides shaping with three different ranges using a 12-bit pipelined ADC. It is also possible to operate at 25 ns. This board is designed at the Laboratoire de Physique Corpusculaire in Clermont-Ferrand (LPC).

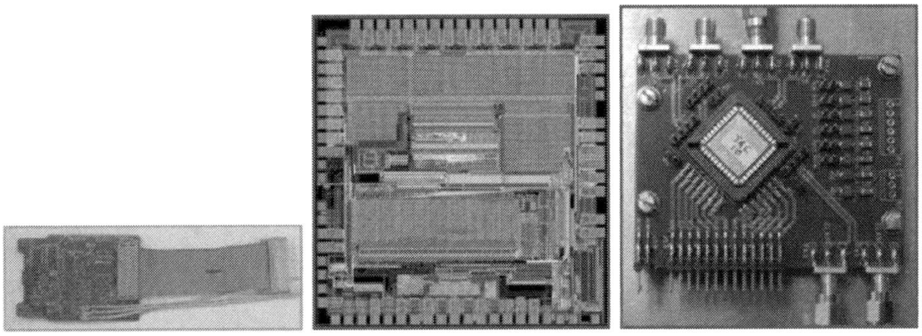

(a) (b) (c)

Figure 4: a) Modified 3-in-1 board b) QIE ASIC and c) FATALIC Board

2.2.2. Main Board. The Main board, shown in Fig. 5, is the motherboard for the different components in the drawer. It provides auxiliary functions such as mechanical support and power and is an interface between the different devices. It also allows the monitoring and control of the chosen FEB as well as the low voltage power supplies. The board is divided in the middle along its length in a mirror configuration to provide redundancy.

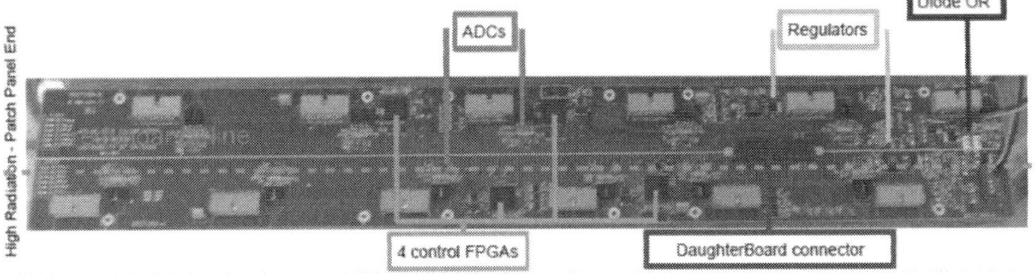

Figure 5: Main board for use with modified 3-in-1 board.

2.2.3. Daughter Board. The Daughter Board (DB), shown in Fig. 7, is connected to the Main Board along the mirror line. The DB provides a high speed communication platform between the detector and the back-end electronics. It is also mirrored down the length of the DB to provide redundancy. The board receives the digitized signals from the main board, packages and sends it to the back-end infrastructure using a QSFP+ optical link at a rate of 40 Gbps (Each link provides 10 Gbps). The board has a bit error rate of less than 10^{-18} which is approximately 1 error in 1000 days.

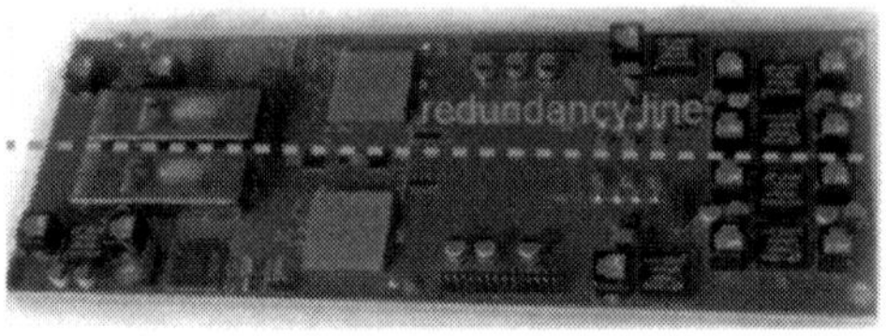

Figure 6: Daughter Board with redundancy line shown.

2.2.4. Super Read Out Driver. The main back-end Read Out Driver (ROD) will be upgraded to receive the increased amount of data. This Super Read Out Driver (sROD) will be the interface between all front and back-end electronics. It will send all Trigger, Timing and Control commands as well as the Detector Control System (DCS) commands to the electronics situated on the detector. The sROD receives and pre-processes the data while storing it in a digital pipeline to wait for the L0/L1 triggering decisions. A prototype has already been built [3]. In order to handle this much data the sROD uses a Virtex7 Field Programmable Gate Array.

Figure 7: Upgraded Read out Driver for the back end electronics.

2.2.5. Back-end Infrastructure. The sROD will be housed in an Advanced Telecommunications Computing Architecture (ATCA) chassis. This is a replacement of the current Versa Module Europa Crates. The ATCA standard will be adopted in ATLAS for the Phase II upgrade in the year 2022 [4]. The ATCA chassis allows intelligent monitoring and control while offering high speed connectivity via a 40 Gbps backplane. Monitoring and Control is achieved via the Simple Network Management Protocol to the Chassis Shelf Manager. The Chassis and all components can be integrated into the existing Detector Control System [5] using a new ATCA-DCS Integration Framework Tool shown in Fig. 8.

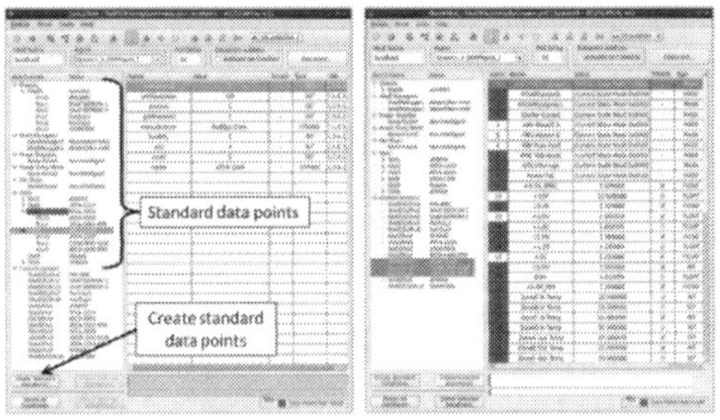

Figure 8: Integration tool to incorporate ATCA Chassis into the Detector Control System.

3. Demonstrator Project

A Demonstrator project [6] has been established in order to test the feasibility and performance of the proposed upgraded electronics. In order to perform the needed tests a prototype hybrid demonstrator will be installed in the detector at the end of 2015. A hybrid between the new and old read out architectures is needed in order to be backward compatible with the existing system. Fig. 9 shows the hybrid readout architecture. The major difference with the hybrid architecture is the analog triggering and the 100 kHz L1A to the sROD. It also makes use of the modified 3-in-1 card but all three will be investigated during beam tests for the final upgrade.

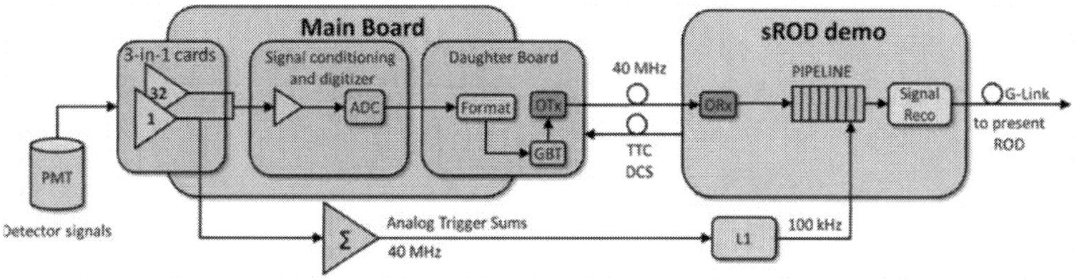

Figure 9: Demonstrator read out electronics.

4. Conclusions

The Phase II upgrade at the LHC will result in an increase in the luminosity by a factor of 5. The associated conditions with such a high luminosity calls for a redesign of the current Tile Calorimeter Readout electronics. The upgrade plan involves the design and insertion of a demonstrator that will evaluate and qualify the proposed upgrade architecture. The demonstrator will completely digitize the signals and transmit them to the back-end at the same rate as bunch crossings (40 MHz). The insertion of the demonstrator will be in the winter of 2015.

References

[1] The ATLAS Collaboration 2008 *Journal of Instrumentation* **3** S08003–S08003 URL `http://stacks.iop.org/1748-0221/3/i=08/a=S08003`

[2] The ATLAS Collaboration 2012 Letter of Intent for the Phase-II Upgrade of the ATLAS Experiment URL `http://cds.cern.ch/record/1502664?ln=en`

[3] Carrió F, Castillo V, Ferrer A, Fiorini L, Hernández Y, Higón E, Mellado B, March L, Moreno P, Reed R, Solans C, Valero A and Valls J A 2014 *Journal of Instrumentation* **9** C02019–C02019 URL `http://iopscience.iop.org/1748-0221/9/02/C02019`

[4] Ballestreroa S, Farthouatb P, Gorinib B, Hillc J, Hufferd M, Joosb M, Miottob G L, Levinsone L, Lichardb P, Morettinif P, Perrotg G, Sankeyh D, Schlenkerb S, Spiwoksb R, Bioti A V, Varij R, Vermeulenk J and Zhangl J 2014 ATCA in ATLAS Tech. rep. URL `https://edms.cern.ch/file/1304001/1/ATCA-Backupdoc-rev2.pdf`

[5] Lantzsch K, Arfaoui S, Franz S, Gutzwiller O, Schlenker S, Tsarouchas C A, Mindur B, Hartert J, Zimmermann S, Talyshev A, Oliveira Damazio D, Poblaguev A, Braun H, Hirschbuehl D, Kersten S, Martin T, Thompson P D, Caforio D, Sbarra C, Hoffmann D, Nemecek S, Robichaud-Veronneau A, Wynne B, Banas E, Hajduk Z, Olszowska J, Stanecka E, Bindi M, Polini A, Deliyergiyev M, Mandic I, Ertel E, Marques Vinagre F, Ribeiro G, Santos H F, Barillari T, Habring J, Huber J, Arabidze G, Boterenbrood H, Hart R, Iakovidis G, Karakostas K, Leontsinis S, Mountricha E, Ntekas K, Filimonov V, Khomutnikov V, Kovalenko S, Grassi V, Mitrevski J, Phillips P, Chekulaev S, D'Auria S, Nagai K, Tartarelli G F, Aielli G, Marchese F, Lafarguette P and Brenner R 2012 *Journal of Physics: Conference Series* **396** 012028 URL `http://stacks.iop.org/1742-6596/396/i=1/a=012028`

[6] C Bohm on behalf of the ATLAS Tile Calorimeter System 2013 *2013 IEEE Nuclear Science Symposium and Medical Imaging Conference* URL `https://cds.cern.ch/record/1494585/files/ATL-TILECAL-PROC-2012-014.pdf`

$B^0_{(s)} \to \mu^+\mu^-$ combination at the LHC

Maximilian Schlupp[1]

Department of Physics, Technische Universität Dortmund, Otto-Hahn-Str 4, 44227 Dortmund, Germany

E-mail: maximilian.schlupp@cern.ch

Abstract. The combined analysis of the B^0_s and B^0 meson decays into two muons based on the full CMS and LHCb datasets is presented. The simultaneous fit to the datasets yields the most precise measurement of the branching fractions to date with a total uncertainty of about 23% for the B^0_s meson and 38% for the B^0 decay into two muons. The measured branching fractions are compatible with the Standard Model predictions within 1.2σ and 2.2σ for the B^0_s and B^0 signals, respectively. The statistical significance for the B^0_s signal is 6.2σ and 3.2σ for the B^0 signal. This results in the first observation of B^0_s meson decays into two muons and the first evidence for the B^0 sister decay.

1. Introduction

Rare decays of beauty mesons provide a unique environment to search for deviations in the Standard Model of particle physics (SM). These highly suppressed processes are sensitive to expansions of the SM as small deviations from the SM predictions can have a significant effect. Two of the most promising examples able to test for "New Physics" models are the decays $B^0_{(s)} \to \mu^+\mu^-$, where either a B^0 or a B^0_s meson decays into two muons. The unique role of these decays arise from the suppression mechanisms. Firstly, the $B^0_{(s)} \to \mu^+\mu^-$ decays proceed through flavour changing neutral currents which are forbidden at lowest order pertubation theory. Hence, the dominant SM contributions come from weak penguin or box-diagrams. Secondly, due to the spin configuration and the special chiral nature of the weak interaction $B^0_{(s)} \to \mu^+\mu^-$ decays are helicity suppressed. The SM predictions [1] for the branching fractions for $B^0_s \to \mu^+\mu^-$ and $B^0 \to \mu^+\mu^-$ incorporating the latest combined value of the top quark mass from the LHC and Tevatron experiments [2] are

$$\mathcal{B}(B^0_s \to \mu^+\mu^-) = (3.66 \pm 0.23) \times 10^{-9}$$
$$\mathcal{B}(B^0 \to \mu^+\mu^-) = (1.06 \pm 0.09) \times 10^{-10}\,.$$

This picture changes if beyond SM theories are realised in nature. Especially additional (pseudo-)scalar particle that break the helicity suppression can enhance the branching fraction by a significant amount. Consequently, $B^0_{(s)} \to \mu^+\mu^-$ decays are highly sensitive to extension of the SM with an enlarged scalar sector, e.g. supersymmetric or two Higgs doublet models.
Additionally the ratio of the branching fractions of $B^0 \to \mu^+\mu^-$ and $B^0_s \to \mu^+\mu^-$ provides

[1] on behalf of the LHCb and CMS collaborations

an exceptional observable of the underlaying flavour structure. This quantity can precisely be predicted within the SM as

$$\mathcal{R} \equiv \frac{\mathcal{B}(B^0 \to \mu^+\mu^-)}{\mathcal{B}(B^0_s \to \mu^+\mu^-)} = 0.0295^{+0.0028}_{-0.0025} .$$

In particular it discriminates the SM and the so-called minimal flavour violating (MFV) extensions from new theories with a different flavour sector.
After thirty years of searches for $B^0_{(s)} \to \mu^+\mu^-$ decays in 2012 the LHCb collaboration published the first evidence for the $B^0_s \to \mu^+\mu^-$ decay with 3.5σ significance [3] using $2\,\text{fb}^{-1}$ of proton-proton collisions at center of mass energies of $\sqrt{s} = 7$ and $8\,\text{TeV}$. In 2013 the LHCb and CMS experiments simultaneously reported their analyses on the complete LHC-RunI dataset of $3\,\text{fb}^{-1}$ and $25\,\text{fb}^{-1}$, respectively [4, 5]. Both collaborations reached approximately the same sensitivity for both decay channels, however neither could claim an observation of $B^0_s \to \mu^+\mu^-$ nor an evidence for $B^0 \to \mu^+\mu^-$. In order to calculate the combined significance of the $B^0_{(s)} \to \mu^+\mu^-$ signals at the LHC, the datasets of the LHCb and CMS experiment are combined and a joint analysis is performed. The results are presented in this proceeding and are published in Ref. [6]. More details on the individual publications can be found in Ref. [4, 5].

2. Joint CMS and LHCb analysis

This analysis presents the full combination of the individual publications realised by merging the two datasets and describing the data with a common probability density function. Hence, the two different fit models need to be harmonised. The $\Lambda^0_b \to p\mu^-\bar{\nu}$ background is added to the default background description including randomly selected muon pairs (combinatorial background), semileptonic B meson decays such as $B^0_{(s)} \to h^-\mu^+\nu$, partially reconstructed decays $B^{+,0} \to \pi^{+,0}\mu^+\mu^-$, and the peaking component of $B \to hh'$, where h could be a kaon or pion. Additionally the branching fraction prediction for $\Lambda^0_b \to p\mu^-\bar{\nu}$ is updated to the latest calculation [7, 8] and the kinematic properties of the simulated decays are changed in order to get more realistic momentum distributions for the final state muons. Finally, the lifetime bias correction [9, 10, 11] is applied to the $B^0_s \to \mu^+\mu^-$ signal descriptions.
Besides these harmonisations, the analysis strategy is unchanged. Both experiments select signal candidates with a soft preselection followed by a multivariate classifier, a boosted decision tree (BDT), to further separate signal from background. Then the invariant mass distribution of the muon pair is fit in categories of the multivariate classifier. Due to the approximately constant running conditions, the full LHCb dataset is only split into eight BDT bins. In case of CMS, the data is also categorised dependent on the year of data-taking and the detector region. The invariant mass distributions of the muon pairs are split in up to three bins of the BDT classifiers resulting in twelve total categories for the CMS dataset.
The twenty combined CMS and LHCb categories of invariant mass distributions are fit using the simultaneous extended maximum likelihood method. The shared parameters between all the categories are the signals branching fractions $\mathcal{B}(B^0_s \to \mu^+\mu^-)$ and $\mathcal{B}(B^0 \to \mu^+\mu^-)$, the branching fraction of the normalisation channel $B^+ \to J/\psi K^+$, and the ratio of hadronisation fractions f_s/f_d.
The combined fit yields

$$\mathcal{B}(B^0_s \to \mu^+\mu^-) = (2.8^{+0.7}_{-0.6}) \times 10^{-9} \text{ and}$$
$$\mathcal{B}(B^0 \to \mu^+\mu^-) = (3.9^{+1.6}_{-1.4}) \times 10^{-10},$$

where the errors combine the statistical and systematic uncertainties. The latter accounts for 35% and 18% of the total uncertainty for $B^0_s \to \mu^+\mu^-$ and $B^0 \to \mu^+\mu^-$, respectively. Using

Wilks' theorem [12], the statistical significances for $B_s^0 \to \mu^+\mu^-$ and $B^0 \to \mu^+\mu^-$ are 6.2σ and 3.2σ, respectively. The expected significances assuming the SM values are 7.4σ for the B_s^0 and 0.8σ for the B^0 signal. Consequently, the decay of $B_s^0 \to \mu^+\mu^-$ is discovered for the first time and the first evidence for $B^0 \to \mu^+\mu^-$ decays is found.
The total dataset together with the different signal and background components in the six bins with the highest expected sensitivity is shown in Fig. 1. The bins are ranked by $S/\sqrt{S+B}$, where S and B are the number of expected signal events assuming the SM branching fractions and the number of background events under the B_s peak in each category. Two simultaneous

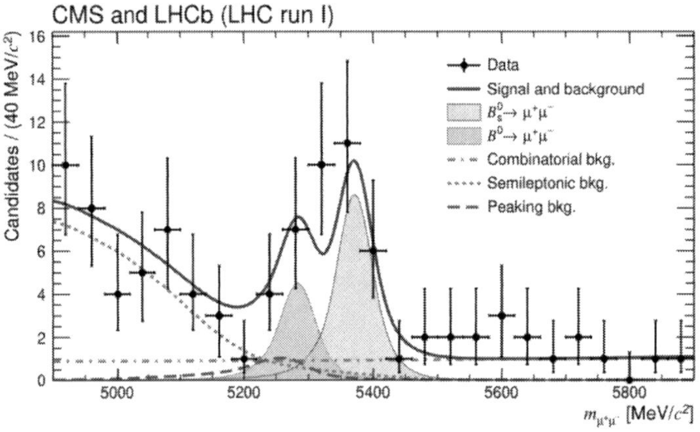

Figure 1. Dimuon invariant mass distributions of the six bins with the highest expected sensitivity according to $S/\sqrt{S+B}$. The data is shown as black points, the total signal and background model is the solid blue line, where the signal components are light blue for $B_s^0 \to \mu^+\mu^-$ and yellow shaded for $B^0 \to \mu^+\mu^-$. Additionally, the different background components are the combinatorial (green dash-dotted), the semileptonic (salmon dotted) and the peaking background (violet dashed).

fits are performed to measure the signal branching fractions relative to their SM prediction $\mathcal{S}_{\rm SM}^{B^0_{(s)}}$ and the ratio $\mathcal{R}$ of the branching fractions of B^0 and B_s^0. The results are

$$\mathcal{S}_{\rm SM}^{B_s^0} = 0.76^{+0.20}_{-0.18} \quad \text{and}$$
$$\mathcal{S}_{\rm SM}^{B^0} = 3.7^{+1.6}_{-1.4} \, ,$$

which are compatible with the SM within 1.2σ and 2.2σ for B_s^0 and B^0, respectively. The ratio of branching fractions is measured to be $\mathcal{R} = 0.14^{+0.08}_{-0.06}$ and compatible with the SM within 2.3σ. The likelihood profile for the signal branching fractions and their ratio are shown in Fig. 2 and Fig. 3, respectively.

3. Confidence regions evaluation

In order to confirm that the approximations of Wilks' theorem hold, the confidence interval of the $B^0 \to \mu^+\mu^-$ signal is calculated using the unified frequentist method proposed by Feldman and Cousins [13]. Instead of assuming a χ^2-distribution with one degree of freedom for the test statistic $q = -2\Delta \ln L$, for each $\mathcal{B}(B^0 \to \mu^+\mu^-)$ hypothesis the actual test statistic distribution is constructed using pseudo-experiments. To generate the simulated experiments the

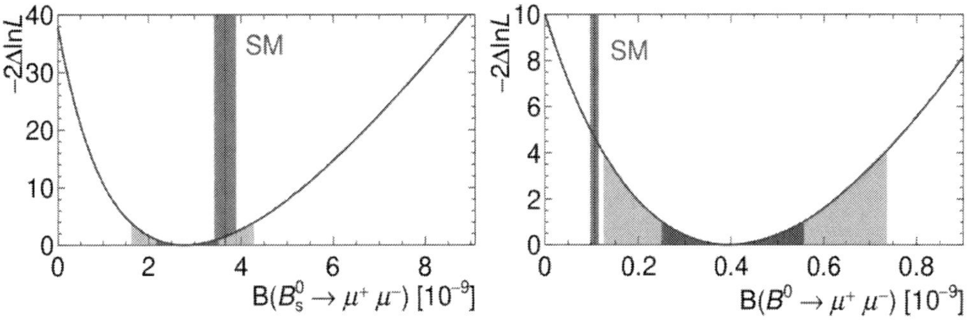

Figure 2. Likelihood profile for the branching fractions of $B_s^0 \to \mu^+\mu^-$ (left) and $B^0 \to \mu^+\mu^-$ (right). The $\pm 1\sigma$ and $\pm 2\sigma$ confidence regions are marked in dark and light cyan, while the SM predictions are denoted by the red bands.

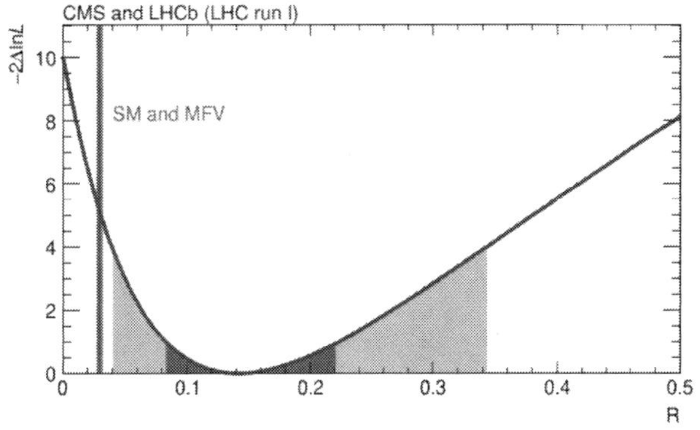

Figure 3. Likelihood profile for $\mathcal{R} \equiv \mathcal{B}(B^0 \to \mu^+\mu^-)/\mathcal{B}(B_s^0 \to \mu^+\mu^-)$ with the $\pm 1\sigma$ and $\pm 2\sigma$ confidence regions highlighted in dark and light cyan. The red bands defines the predictions of the SM and MFV models.

nuisance parameters are fixed to their best value given by the result of a fit to the data with fixed $\mathcal{B}(B^0 \to \mu^+\mu^-)$. The test statistic distribution is calculated from two maximum likelihood fits, where the $\mathcal{B}(B^0 \to \mu^+\mu^-)$ parameter is either fixed or freely floating. The mean values of the constraints on the nuisance parameters are randomised around the best-fit value used to generate each pseudo-experiment. It is found that $\mathcal{B}(B^0 \to \mu^+\mu^-)$ is in $[2.5, 5.6] \times 10^{-10}$ with 68.27% confidence level and in $[1.4, 7.4] \times 10^{-10}$ with 94.45% confidence level (CL). The outcome of this study is displayed as a 1−CL curve in Fig. 4. The black points together with a spline interpolation in solid blue are the results of the pseudo-experiments and the approximation using Wilks' theorem is drawn as a dashed grey line. The corresponding $\pm 1\sigma$ and $\pm 2\sigma$ confidence intervals are indicated in dark and light cyan. In addition the significance is calculated from the same pseudo-experiments as it is defined as the 1−CL value of the $\mathcal{B}(B^0 \to \mu^+\mu^-) = 0$ hypothesis translated into units of standard gaussian widths. For this reason $O(10^5)$ pseudo-experiments are simulated assuming no $B^0 \to \mu^+\mu^-$ signal yielding a statistical significance of 3.0σ. This confirms that the approximations in Wilks' theorem hold.

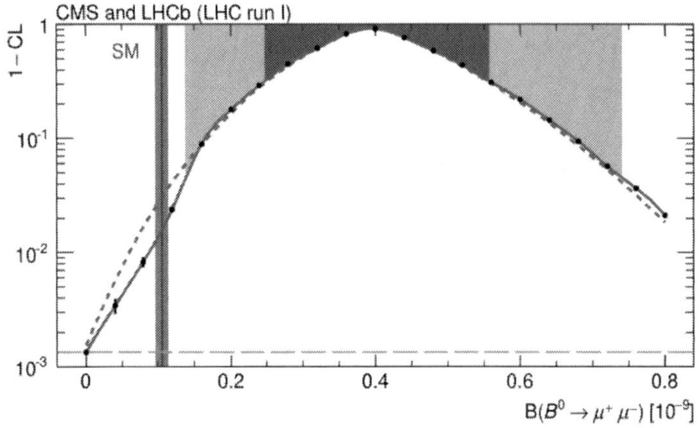

Figure 4. 1−CL curve in dependence of $\mathcal{B}(B^0 \to \mu^+\mu^-)$ hypothesis. The black points mark the 1−CL values obtained from the Feldman-Cousins procedure, while the solid blue line is their spline interpolation. The dashed grey line is the result from the evaluation of 1−CL using Wilks' theorem. The SM prediction for $\mathcal{B}(B^0 \to \mu^+\mu^-)$ is shown as a red band and the $\pm 1\sigma$ and $\pm 2\sigma$ confidence intervals are indicated in dark and light cyan.

Finally, confidence regions are evaluated in the two-dimensional $\mathcal{B}(B^0 \to \mu^+\mu^-)$–$\mathcal{B}(B^0_s \to \mu^+\mu^-)$ plane. Therefore, the test statistic $q = -2\Delta \ln L$ is calculated from the nominal fit and a maximum likelihood fit to the data, where both signal branching fractions are fixed to the corresponding hypotheses. The two-dimensional confidence regions are defined to have the same coverage as the corresponding one-dimensional gaussian σ intervals. The result is shown in Fig. 5.

4. Conclusions

The joint search for $B^0_{(s)} \to \mu^+\mu^-$ decays of the combined LHC-RunI dataset of the CMS and LHCb experiments is presented. For the first time the decay $B^0_s \to \mu^+\mu^-$ is observed with a statistical significance of 6.2σ. Its branching fraction is fully compatible with the SM prediction at 1.2σ.

In case of the $B^0 \to \mu^+\mu^-$ decay, an excess of events is seen over the background-only hypothesis resulting in a statistical significance of 3.2σ using Wilks' theorem and 3.0σ using pseudo-experiments. This signal is compatible with the SM within 2.2σ.

The ratio of the branching fractions of $B^0 \to \mu^+\mu^-$ and $B^0_s \to \mu^+\mu^-$ is measured as $\mathcal{R} = 0.14^{+0.08}_{-0.06}$ and therefore compatible with the SM within 2.3σ.

Measuring all observables consistently with the SM predictions significantly reduces the allowed parameters space for theories beyond the SM. However, the measurement is statistically limited. During the upcoming RunII of the Large Hadron Collider both experiments will significantly increase the amount of collected B mesons, which will allow to study $B^0_{(s)} \to \mu^+\mu^-$ decays with greater precision.

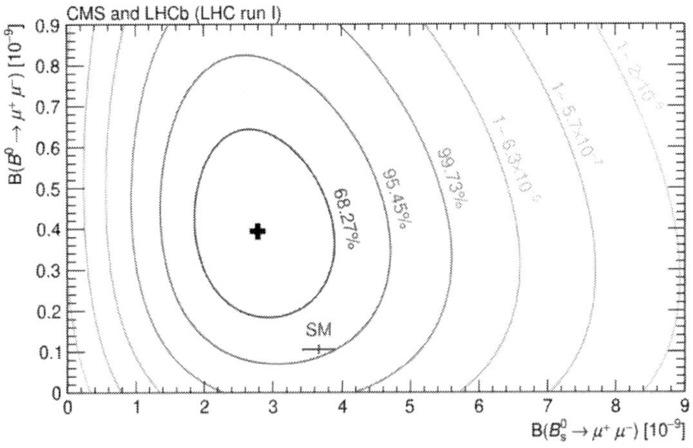

Figure 5. Two-dimensional confidence regions in the $\mathcal{B}(B^0 \to \mu^+\mu^-)$–$\mathcal{B}(B_s^0 \to \mu^+\mu^-)$ plane. The black cross marks the global best-fit point. The contours indicate the region with the corresponding coverage given on the plot and the SM prediction with its uncertainties is shown as a red cross.

References

[1] Bobeth C, Gorbahn M, Hermann T, Misiak M, Stamou E *et al.* 2014 *Phys.Rev.Lett.* **112** 101801 (*Preprint* 1311.0903)
[2] 2014 First combination of Tevatron and LHC measurements of the top-quark mass ATLAS-CONF-2014-008, CDF-NOTE-11071, CMS-PAS-TOP-13-014, D0-NOTE-6416 (*Preprint* 1403.4427)
[3] Aaij R *et al.* (LHCb Collaboration) 2013 *Phys.Rev.Lett.* **110** 021801 (*Preprint* 1211.2674)
[4] Aaij R *et al.* (LHCb Collaboration) 2013 *Phys.Rev.Lett.* **111** 101805 (*Preprint* 1307.5024)
[5] Chatrchyan S *et al.* (CMS Collaboration) 2013 *Phys.Rev.Lett.* **111** 101804 (*Preprint* 1307.5025)
[6] Khachatryan V *et al.* (CMS Collaboration, LHCb Collaboration) 2014 (*Preprint* 1411.4413)
[7] Khodjamirian A, Klein C, Mannel T and Wang Y M 2011 *JHEP* **1109** 106 (*Preprint* 1108.2971)
[8] Aaij R *et al.* (LHCb Collaboration) 2014 *Phys.Lett.* **B734** 122 (*Preprint* 1402.6242)
[9] De Bruyn K, Fleischer R, Knegjens R, Koppenburg P, Merk M *et al.* 2012 *Phys.Rev.Lett.* **109** 041801 (*Preprint* 1204.1737)
[10] DeBruyn K, Fleischer R, Knegjens R, Koppenburg P, Merk M *et al.* 2012 *Phys.Rev.* **D86** 014027 (*Preprint* 1204.1735)
[11] Dunietz I, Fleischer R and Nierste U 2001 *Phys.Rev.* **D63** 114015 (*Preprint* hep-ph/0012219)
[12] Wilks S S 1938 *The Annals of Mathematical Statistics* **9** 60–62 (URL http://www.jstor.org/stable/2957648)
[13] Feldman G J and Cousins R D 1998 *Phys.Rev.* **D57** 3873–3889 (*Preprint* physics/9711021)

International Workshop on Discovery Physics at the LHC (Kruger2014) IOP Publishing
Journal of Physics: Conference Series **623** (2015) 012026 doi:10.1088/1742-6596/623/1/012026

FAIR - Cosmic Matter in the Laboratory

Horst Stöcker[1,3,5], **Thomas Stöhlker**[1,2,6] **and Christian Sturm**[1,4]

[1] GSI Helmholtzzentrum für Schwerionenforschung GmbH, Planckstr. 1, D-64291 Darmstadt, Germany
[2] Helmholtz Institut Jena, Fröbelstieg 3, D-07743 Jena, Germany
[3] Judah M. Eisenberg endowed chair for Theoretical Physics, Institut für Theoretische Physik, J.W. Goethe Universität Frankfurt, Max-von-Laue Str. 1, D-60438 Frankfurt Main, Germany
[4] Institut für Kernphysik, J.W. Goethe Universität Frankfurt, Max-von-Laue Str. 1, D-60438 Frankfurt Main, Germany
[5] Frankfurt Institute for Advanced Studies, Ruth-Moufang-Str. 1, D-60438 Frankfurt am Main, Germany
[6] Friedrich-Schiller-Universität Jena, 07737 Jena, Germany

E-mail: h.stoecker@gsi.de

Abstract. To explore cosmic matter in the laboratory - this fascinating research prospect becomes available at the Facility for Antiproton and Ion Research, FAIR. The new facility is being constructed within the next five years adjacent to the existing accelerator complex of the GSI Helmholtz Centre for Heavy Ion Research at Darmstadt/Germany, expanding the research goals and technical possibilities substantially. This includes new insights into the dynamics of supernovae depending on the properties of short-lived neutron-rich nuclei which will be investigated with intense rare isotope beams. New insights will be provided into the interior of stars by exploring dense plasmas with intense heavy-ion beams combined with a high-performance laser - or into neutron star cores by probing the highest baryon densities in relativistic nucleus-nucleus collisions at unprecedented collision rates. To the latter, the properties of hadrons play an important part which will be systematically studied by high precision hadron spectroscopy with antiproton beams at unmatched intensities. The worldwide unique accelerator and experimental facilities of FAIR will open the way for a broad spectrum of unprecedented fore-front research supplying a large variety of experiments in hadron, nuclear, atomic and plasma physics as well as biomedical and material science which will be briefly described in this article. This article is based on the FAIR Green Paper [4] and gives an update of former publications [5] - [12].

1. Introduction

The new international facility FAIR (Facility for Antiproton and Ion Research in Europe [1]-[3]) is being constructed adjacent to the site of the GSI Helmholtz Centre for Heavy Ion Research at Darmstadt, Germany (see Fig. 1). It will substantially expand research goals by providing worldwide unique accelerator and detector facilities for a large variety of unprecedented fore-front science. The main thrust of FAIR research focuses on the structure and evolution of cosmic matter on a microscopic scale. This will deepen our understanding of fundamental questions such as:

- How does the complex structure of matter at all levels arise from the basic constituents and the fundamental interactions ?

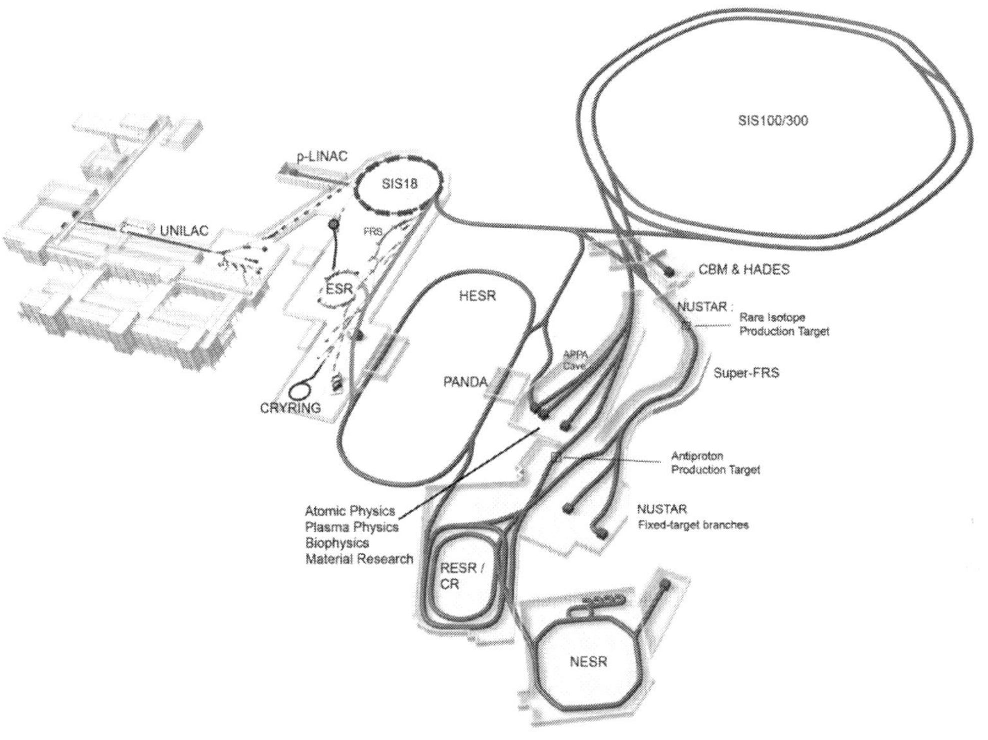

Figure 1. The existing GSI facility is shown on the left, in dark-blue the existing GSI accelerator facilities and ion sources. On the right the new FAIR complex is displayed in red with its 100 and 300 Tm super conducting double-ring synchrotrons SIS100 and SIS300, detector areas (i) for atomic and plasma physics, biomedical and material sciences (APPA), (ii) for the relativistic nucleus-nucleus collision experiments CBM and HADES, (iii) for the radioactive beam facility (NUSTAR), the and (iv) for the hadron physics detector system PANDA as well as the rare-isotope and antiproton production targets with storage rings. Proposed beamlines for transport of protons and ions from SIS18 directly to the High Energy Storage Ring HESR [13] and of antiprotons and exotic ions from HESR to the ESR/CRYRING are shown in addition (light blue). After the start phase the facility will be completed by the experimental storage rings NESR (New Experimental Storage Ring) and RESR (Recuperated Experimental Storage Ring) enhancing capabilities of secondary beams and by the superconducting synchrotron SIS300 providing true parallel operation of experimental programs as well as particle energies twenty-fold higher compared to those achieved so far at GSI [4].

- How can the structure of hadronic matter be deduced from the strong interaction ? In particular, what is the origin of hadron masses ?
- What is the structure of matter under the extreme conditions of temperature and density found in astrophysical objects ?
- What was the evolution and the composition of matter in the early Universe ?
- What is the origin of the elements in the Universe ?

To address these fundamental questions FAIR will provide an extensive range of particle beams from protons and their antimatter partners, antiprotons to ion beams of all chemical elements up to the heaviest (stable) one, uranium, with world record intensities. As a joint effort of more than 50 countries the new facility builds, and substantially expands, on the present accelerator system at GSI Helmholtz Centre for Heavy Ion Research, both in its research goals and its technical possibilities. Compared to the present GSI facility, the beam intensities will increase by a factor of 100 for primary beams and up to a factor of 10000 for secondary radioactive beams with an excellent beam brilliance of primary as well as secondary beams. This will be achieved through innovative beam handling techniques, many aspects of which have been developed at GSI over recent years with the present accelerator complex. This includes in particular stochastic, and electron-beam cooling of high-energy, high-charge state ion beams in storage rings and bunch compression techniques. Even laser cooling has been demonstrated to be powerful tool for certain ion species. To realize parallel operation of the research programs the design of FAIR includes the superconducting double-ring synchrotrons SIS100 and SIS300. Both have a circumference of 1100 meters each and a magnetic rigidity of 100 and 300 Tm, respectively. Since a key feature of the new facility will be the generation of intense, high-quality secondary beams, the facility design contains a system of associated storage rings for beam collection, cooling, phase space optimization and experimentation (see Fig. 1).

The *Modularized Start Version* (MSV) [3, 4] of FAIR consists of the superconducting synchrotron SIS100 which feeds primary beams for fixed-target experiments and production targets for secondary beams of rare isotopes and antiprotons with the two cooler-storage rings HESR (High Energy Storage Ring) and CR (Collector Ring). Detector areas for all scientific pillars are included. Following an upgrade for high intensities, the existing GSI accelerators UNILAC and SIS18 will serve as injectors. The facility will be completed by experimental storage rings enhancing capabilities of secondary beams and by the superconducting synchrotron SIS300 providing parallel operation of experimental programs as well as particle energies twenty-fold higher compared to those achieved so far at GSI.

To increase even further the flexibility in beam handling and operation, currently detailed investigations of short additional beamline sections are getting pursued (light blue colored in Fig. 1) [13]. These beamlines will allow to transport stable as well as exotic ions from SIS18 (as well as from ESR or FRS) directly to the High Energy Storage Ring HESR and of antiprotons and exotic ions from HESR to the ESR.

2. The Experimental Program of FAIR

The approved FAIR research program embraces 14 experiments. These form the four scientific pillars of FAIR and offer a large variety of unprecedented forefront research in hadron, nuclear, atomic and plasma physics as well as applied sciences. Already today, over 2500 scientists and engineers are involved in the design and preparation of the FAIR experiments. The four scientific pillars are named

- APPA – Atomic Physics, Plasma Physics and Applications,
- CBM and HADES – Compressed Baryonic Matter,
- NUSTAR – Nuclear Structure, Astrophysics and Reactions,
- PANDA – AntiProton ANnihilation in DArmstadt .

2.1. APPA – A look into the interior of giant planets and stars

In the Universe at large, matter exists predominantly either as hot dense plasma in the interior of stars or in stellar atmospheres, as medium- or low-temperature dense plasma in the interior of giant planets, or as hot plasma of very low density in interstellar space. In particular the regime of dense plasmas is so far only scarcely explored, since it is difficult to approach experimentally.

Journal of Physics: Conference Series **623** (2015) 012026 doi:10.1088/1742-6596/623/1/012026

Intense heavy-ion beams provide an efficient tool to create large samples of this high-energy-density matter with volumes of the order of a few mm^3 or even cm^3 and lifetimes of the order of several tenths of nanoseconds by isochoric and uniform heating of solid matter. The uniform physical conditions and comparatively long life times facilitate application of a broad spectrum of the diagnostic tools for studying these states of matter. For Plasma physics the availability of high-energy, high-intensity ion-beams in unique combination with a petawatt-laser enables the investigation of high energy density matter in regimes of temperature, density and pressure not accessible so far [16]. It will allow probing new areas in the phase diagram and long-standing open questions of basic equation-of-state research can be addressed. The efficient generation of these states in macroscopic amounts will be boosted by the implementation of new, efficient time and space resolved diagnostics methods based on high-energy proton microscopy and intense lasers. Around these topics, two plasma physics collaborations were built: the High Energy Density Matter generate with intense Heavy Ion Beams (HEDgeHOB) [16] and the Warm Dense Matter (WDM).

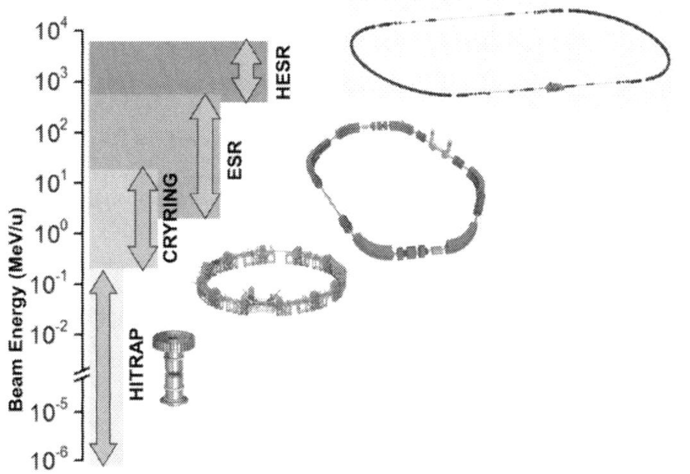

Figure 2. Portfolio of storage and trapping facilities at FAIR. These facilities cover continuously a kinetic beam-energy range of more than 10 orders of magnitude, ranging from rest in the laboratory up to large γ values.

Atomic physics at FAIR [14, 15, 17] will focus on the exploration of matter in extreme electromagnetic fields by exploiting the storage and trapping facilities. The unrivaled combination of storage and trapping facilities for antiprotons and heavy ions are a further unique feature that distinguish FAIR from all other particle accelerator facilities worldwide (planned or in operation). As depicted in Fig. 2, these facilities cover a kinetic beam-energy range of more than 10 orders of magnitude, ranging from rest in the laboratory up to large γ values. For atomic physics, FLAIR [17] will exploit these research opportunities for experiments with low-energy anti-protons whereas SPARC [14, 15] will pursue experiments with stable and exotic nuclei at high atomic charge-states over the whole beam-energy range available. This will allow the extension of atomic-physics research across virtually the full range of atomic matter, i.e. concerning the accessible ionic charge states as well as beam energies.

In SPARC experiments two major research areas are planed: collision dynamics in strong electromagnetic fields and fundamental interactions between electrons and heavy nuclei up to bare uranium. In the first area we will use the relativistic heavy ions for a wide range of collision studies. In the extremely short, relativistically enhanced field pulses, the critical field limit (Schwinger limit) for lepton-pair production can be surpassed by orders of magnitudes and breakdowns of perturbative approximations for pair production are expected. The detection methods of reaction microscopes will give the momentum of all fragments when atoms or molecules are disintegrating in strong field pulses of the ions. This allows exploring regimes of multi-photon processes not accessible by other means. In particular, the storage ring HESR will be exploited for collision studies and fundamental atomic processes will be investigated for cooled heavy-ions at well-defined charge states interacting with photons, electrons and atoms. One may note, that the HESR can store cooled beams at energies of up to a few GeV/u and can thus enable unique precision experiments with cooled relativistic energies (γ values ranging from 2 to 6.

For SPARC, the experiments in the high energy regime will be complemented by the CRYRING at ESR (see Fig. 1) [18]. Here, at the low-beam energies (between 1 and 14 MeV/u) atomic interactions for highly-charged ions are governed by strong perturbations and quasi-molecular effects. The other class of experiments will focus on the determination of properties of stable and unstable nuclei by atomic physics techniques on the one hand, and precision tests of quantum electrodynamics (QED) and fundamental interactions in extremely strong electromagnetic fields on the other hand. A further important scenario for this class of experiments will be the slowing-down, trapping and cooling of particles in the ion trap facility HITRAP [19]. This scenario will enable high-accuracy experiments in the realm of atomic and nuclear physics, as well as the determination of fundamental constants or even highly-sensitive tests of the Standard Model.

Variants of civil construction sequences are presently investigated to enable an earlier start of the scientific program of FAIR, among which we would like to note the direct coupling of SIS18 to HESR and of HESR to ESR and CRYRING [13]. Both are outside of the Modularized Start Version of FAIR, but both APPA and the low-energy antiproton and Ion (Flair) community see this as a cost-effective model for an early start of the science of module 5 as outlined in the Green Paper of FAIR [4].

Applied research benefits from the large range of beam energies and intensities for various activities, in particular in biophysics and material science. The biophysics research has two main topics: space radiation effects on humans (astronauts) and spacecraft instrumentation (microelectronics); and particle therapy using high-energy protons and heavy ions for cancer and non-cancer diseases. Research in material science is dedicated to heavy-ion induced modifications in solids under extreme conditions such as short ion pulses, high fluences, and the simultaneous application of pressure and temperature. A multipurpose irradiation facility, able to uniformly scan large targets and equipped with various instrumentations for in-situ and on-line monitoring, has been designed. This facility will be built and exploited by the BIOMAT collaboration consisting of international collaborations on heavy-ion biophysics (BIO) and material science (MAT).

2.2. CBM and HADES – A look into the interior of neutron stars

The stability of neutron stars is guaranteed by the Pauli principle of nucleons together with the repulsive part of the nucleon-nucleon interaction. Therefore the structure of a neutron star is dictated by the strong interaction with the nuclear equation-of-state as key ingredient. The transition from the outer to the inner core of a neutron star is specified by the appearance of exotic phases, e.g. matter containing additional particles species carrying strangeness such as hyperons, pion or kaon condensed matter or deconfined quark matter. The question whether the interior of a neutron star is already in a (superconducting) quark phase is an active field of research. To explore nuclear matter densities beyond saturation density in the laboratory nucleus-nucleus reactions provide the only possibility.

The mission of high-energy nucleus-nucleus collision experiments worldwide is to investigate the properties of strongly interacting matter under these extreme conditions. At very high collision energies, as available at RHIC and LHC, the measurements concentrate on the study of the properties of deconfined QCD matter at very high temperatures and almost zero net baryon densities. Results from lattice QCD indicate that the transition from confined to deconfined matter at vanishing net baryon density is a smooth crossover, whereas in the region of high net baryon densities, accessible with heavy-ion reactions at lower beam energies, a first-order phase transition is expected [20]. Its experimental confirmation would be a substantial progress in the understanding of the properties of strongly interacting matter.

Complementary to high-energy nucleus-nucleus collision experiments at RHIC and LHC, the CBM experiment [21, 22, 23] as well as HADES [25] - [30] at SIS100/300 will explore the QCD phase diagram in the region of very high baryon densities and moderate temperatures by investigating heavy-ion collision in the beam energy range 2 - 44 AGeV. This approach includes the study of the nuclear matter equation-of-state, the search for new forms of matter, the search for the predicted first order phase transition to the deconfinement phase at high baryon densities, the QCD critical endpoint, and the chiral phase transition, which is related to the origin of hadron masses.

In the case of the predicted first order phase transition, basically one has to search for non-monotonic behavior of observables as function of collision energy and system size. The CBM experiment at FAIR is being designed to perform this search with a large range of observables, including very rare probes at these energy regime like charmed hadrons. Produced near threshold, their measurement might be well suited to discriminate hadronic from partonic production scenarios. The former requires pairwise creation of charmed hadrons, the latter the recombination of c-quarks created in first chance collisions of the nucleus-nucleus reaction. Ratios of hadrons containing charm quarks as a function of the available energy may provide direct evidence for a deconfinement phase.

The properties of hadrons are expected to be modified in a dense hadronic environment which is eventually linked to the onset of chiral symmetry restoration at high baryon densities and/or high temperatures. The experimental verification of this theoretical prediction is one of the most challenging questions in modern strongly interacting matter physics. The dileptonic decays of light vector mesons (ρ,ω,ϕ) provide the tool to study such modifications since the lepton daughters do not undergo strong interactions and can therefore leave the dense hadronic medium essentially undistorted by final-state interaction. For these investigations the ρ meson plays an important role since it has a short lifetime and through this a large probability to decay inside the reaction zone when created in a nucleus-nucleus collision. As a detector system dedicated to high-precision di-electron spectroscopy at beam energies of 1 – 2 AGeV, the modified HADES detector at SIS100 will measure e^+e^- decay channels as well as hadrons [28, 29, 30] in collisions of light nuclei up to 10 AGeV beam energy. In addition and complementary, the CBM experiment will cover the complete FAIR energy range by measuring both the e^+e^- and the $\mu^+\mu^-$ decay channels.

Most of the rare probes like lepton pairs, multi-strange hyperons and charm will be measured for the first time in the FAIR energy range. The goal of the CBM experiment as well as HADES is to study rare and bulk particles including their phase-space distributions, correlations and fluctuations with unprecedented precision and statistics. These measurements will be performed in nucleus–nucleus, proton–nucleus, and proton–proton collisions at various beam energies. The unprecedented beam intensities will allow to study extremely rare probes with high precision which have not been accessible by previous nucleus-nucleus experiments at the AGS, SPS and the beam energy scan program (BES) at RHIC.

Complementary to the CBM program at FAIR and at about the same time as FAIR the Nuclotron-based Ion Collider facility NICA will come into operation at the Joint Institute for Nuclear Research (JINR) in Dubna [24]. The main goal of the project is to study hot and dense strongly interacting matter in nucleus-nucleus collisions at center-of-mass energies $\sqrt{s_{NN}} = 4$-11 GeV with the Multi-Purpose Detector (MPD). As a collider, NICA achieves higher collision energies while CBM will perform experiments with up to 4 orders of magnitude higher collision rates.

2.3. NUSTAR – Rapid nuclear processes in supernovae and the elemental abundances in the universe

Reactions between nuclei play a decisive role in many astrophysical processes in the universe. Nuclear structure effects and the dynamics of nuclear reactions are directly reflected in the various evolutionary stages of stars, in the light curves of stellar explosions, and in the elemental abundance distributions in the Universe. Unstable nuclei, far away from stability, are involved and their properties determine the fate of the relevant astrophysical processes. The radioactive beam facility at FAIR offers world-wide unique experimental opportunities for this area of research.

The main scientific thrusts in the study of nuclei far from stability are aimed at three areas of research: (i) the structure of nuclei, the quantal many-body systems built by protons and neutrons and governed by the strong force, towards the limits of stability, where nuclei become unbound, (ii) nuclear astrophysics delineating the detailed paths of element formation in stars and explosive nucleosynthesis that involve short-lived nuclei, (iii) and the study of fundamental interactions and symmetries exploiting the properties of specific radioactive nuclei.

The central part of the NUSTAR programme at FAIR [31, 32] is the high acceptance Super-FRS with its multi-stage separation that will provide high intensity mono-isotopic radioactive ion beams of bare and highly-ionized exotic nuclei at and close to the driplines. This separator, in conjunction with high intensity primary beams with energies up to 1.5 AGeV, is the keystone for a competitive NUSTAR physics programme. This opens the unique opportunity to study the evolution of nuclear structure into the yet unexplored territory of the nuclear chart and to determine the properties of many short-lived nuclei which are produced in explosive astrophysical events and crucially influence their dynamics and associated nucleosynthesis processes.

2.4. PANDA – The structure of hadrons and the generation of mass in the visible universe

In the evolution of the universe, microseconds after the big bang, a coalescence of quarks and gluons to hadrons occurred associated with the generation of hadron masses. The elementary light quarks, the up and down quarks that make up the nucleon, have very small masses amounting to only a few percent of the total mass of the nucleon. Most of the baryon's masses, and hence of the visible universe stems from the QCD interaction. The generation of mass is thus associated with the confinement of quarks and the spontaneous breaking of chiral symmetry, one of the fundamental symmetries of QCD in the limit of massless quarks.

The big challenge in hadron physics is to achieve a quantitative understanding of strongly interacting complex systems at the level of quarks and gluons. In $p\bar{p}$-annihilation, particles with gluonic degrees of freedom as well as particle-antiparticle pairs are copiously produced, allowing spectroscopic studies with unprecedented statistics and precision. The PANDA experiment at FAIR [33, 34, 35] will bring new fundamental knowledge in hadron physics by pushing the precision barrier towards new limits. The charmonium ($c\bar{c}$) spectroscopy will take advantage of high-precision measurements of mass, width and decay branches. Particular emphasis is placed on mesons with open and hidden charm, which extends ongoing studies in the light quark sector to heavy quarks, and adds information on contributions of the gluon dynamics to hadron masses. The search for exotic hadronic matter such as hybrid mesons or heavy glueballs will profit by precise scanning of resonance curves of narrow states as well. Additionally, the precision gamma-ray spectroscopy of single and double hypernuclei will allow extracting information on their structure and on the hyperon-nucleon and hyperon-hyperon interaction.

Recently, much attention has been attracted by the observation at electron-positron colliders of the so-called X, Y and Z states with masses around 4 GeV as measured by experiments e.g. Belle at KEKB [36], BESIII at BEPC [37] and others. These heavy mesons show very unusual properties, whose theoretical interpretation is entirely open.

An extension of the FAIR antiproton facility to a $\bar{p}p$-collider with a maximum center-of-mass energy $\sqrt{s}^{max} = 30\,\mathrm{GeV}$ [39] and luminosities of about $3 \cdot 10^{30} cm^{-2} s^{-1}$ [38] would allow to analyze these exotic objects and their beauty complements at quantal states not accessible to electron-positron colliders. FAIR would thus open the way to a rich spectrum of Beauty Physics by formation of beauty baryon-antibaryon pairs with a large variety of beauty, charm, strange and light quarks, as well as beauty mesons with spin $\neq 1$. Technically, the antiproton and proton beam would be stored and accelerated in opposite directions inside the HESR which would require a second injection point as well as a second electron cooler. Furthermore the PANDA detector can serve as the mid-rapidity detector system. This fascinating opportunity will enrich the PANDA physics program into new, unprecedented realms and will push the physics program of the High Energy Storage Ring HESR for decades to come.

3. Summary

After the official launch of the project in 2007, nine countries (in alphabetical order: Finland, France, Germany, India, Poland, Romania, Russia, Slovenia and Sweden) signed the international agreement on the construction of the Facility for Antiproton and Ion Research, FAIR, on October 4th, 2010. The construction has started in 2013, so that the first beams will be delivered by the super conducting synchrotron SIS100 in 2019. The scientific program of FAIR cover

- the investigation of bulk matter in the dense and hot plasma state, atomic physics of ultra-high electro-magnetic fields, material and biological research using intense, highly-stripped ions and high-intensity laser fields (APPA);
- the exploration of the QCD phase-diagram and the phase transition to the deconfinement phase at high baryon densities in high-intensity nucleus-nucleus collisions (CBM and HADES);
- the investigation of the nuclear chart and nuclear structure far from stability and of nuclear astrophysics with radio-active beams (NUSTAR);
- the exploration of the hadron structure, QCD vacuum, and the nature of the strong force in the non-perturbative regime of quantum-chromo-dynamics with high-energy beams of antiprotons (PANDA),

forming together the four scientific pillars. The design of FAIR includes the superconducting double-ring synchrotrons SIS100 and SIS300 to realize parallel operation of the research programs. Since a key feature of the new facility will be the generation of intense, high-quality secondary beams, the facility design contains a system of associated storage rings for beam collection, cooling, phase space optimization and experimentation.

The first phase of the FAIR project comprises the superconducting synchrotron SIS100 as well as experimental areas to perform experiments for all research pillars. It will allow to carry out an outstanding and world-leading research program in hadron, nuclear, atomic and plasma physics as well as applied sciences. Due to the high luminosity which exceeds current facilities by orders of magnitude, experiments will be feasible that could not be done elsewhere. FAIR will expand the knowledge in various scientific fields beyond current frontiers. Moreover, exciting synergy effects between the various research fields and instruments at FAIR in five parallel and crossed operation modes will prove what has been the basic idea of the FAIR communities: *four* scientific pillars with *one* machine.

References

[1] H.H. Gutbrod et al. (Eds.), *FAIR Baseline Technical Report*, ISBN 3-9811298-0-6 Nov. 2006 .
[2] W.F. Henning, *FAIR: Recent developments and status*, Nucl. Phys. A 805 (2008) 502c .
[3] H. Stöcker, *FAIR: Challenges Overcome and Still to be Met*, published in Conf.Proc.C0806233:moycgm01, 2008 .
[4] Green Paper of FAIR: *The Modularized Start Version* , https://www.gsi.de/documents/DOC-2009-Nov-124-1.pdf , October 2009
[5] C. Sturm, B. Sharkov and H. Stöcker, *1,2,3 ... FAIR !* , Nucl. Phys. A 834 (2010) 682c .
[6] H. Stoecker, C. Sturm, *The FAIR Start* , Nucl. Phys. A855 (2011) 506 .
[7] H. Stoecker, C. Sturm, *The facility for antiproton and ion research FAIR: cosmic matter in the Laboratory* , Nucl. Phys. A862-863 (2011) 92 .
[8] C. Sturm, H. Stoecker, *The facility for antiproton and ion research FAIR* , Phys.Part.Nucl.Lett. 8 (2011) 865 .
[9] H. Stoecker, C. Sturm, *The facility for antiproton and ion research FAIR* , Phys.J.Ro. Vol. 58, Number 9 - 10 (2013) 1023 .
[10] H. Stoecker, C. Sturm, *FAIR - Cosmic Matter in the Laboratory*, Proceedings of Nuclear Physics: Presence and Future, FIAS Interdisciplinary Science Series, Vol. 2 (Springer Verlag, 2014) .
[11] H. Stoecker, C. Sturm, *FAIR - Cosmic Matter in the Laboratory* , Astron. Nachr., Vol. 335 (2014) 581 .
[12] H. Stoecker, T. Stoehlker, C. Sturm, *FAIR - Cosmic Matter in the Laboratory* , accepted for publication in the Proceedings of the J-PARC Symposium 2014, JPS Conference Proceedings 2015 .
[13] N. Angert, B. Franzke, Y. Litvinov, R. Maier, F. Nolden, D. Prasuhn, L. Schmitt, H. Simon, M. Steck, Th. Stöhlker, priv. communication .
[14] Th. Stöhlker et al. , *Atomic physics with highly-charged ions at the future FAIR facility: A status report* , Nuclear Instruments and Methods in Physics Research B 261 (2007) 234 .
[15] Th. Stöhlker et al. , *SPARC: The Stored Particle Atomic Research Collaboration At FAIR* , AIP Conf. Proc. 1336 (2011) 132 .
[16] I.V. Lomonosov and N.A. Tahir, *Prospects of High-Energy Density Matter Research at the Future FAIR Facility at Darmstadt* , Nucl. Phys. News 16 1 (2006) 29 .
[17] E. Widmann, *FLAIR, a Facility for Low-energy Antiproton and Ion Research* , *EXA05 International Conference on Exotic Atoms and Related Topics*, DOI 10.1553/exa05s459 (2005) S. 459 .
[18] Th. Stöhlker et al. , *SPARC collaboration: new strategy for storage ring physics at FAIR* , Hyperfine Interact., DOI 10.1007/s10751-014-1047-2 (2014) .
[19] F. Herfurth et al., *HITRAP - heavy, highly-charged ions and antiprotons at rest*, Acta Physica Polonica B 41 (2010) 457 .
[20] B. Friman, C. Höhne, J. Knoll, S. Leupold, J. Randrup, R. Rapp, P. Senger (Editors) , *The CBM Physics Book*, Lecture Notes in Physics 814 (2011) 1 .
[21] P. Senger et al. [CBM Collaboration] , *Compressed baryonic matter: Experiments at GSI and FAIR*, Phys. Part. Nucl. 39 (2008) 1055 .
[22] P. Senger et al. [CBM Collaboration] , *Probing dense baryonic matter*, Prog. Part. Nucl. Phys. 62 (2009) 375 .
[23] J. Heuser et al. [CBM Collaboration] , *The compressed baryonic matter experiment at FAIR*, Nucl.Phys. A904-905 (2013) 941c .
[24] V.D. Kekelidze et al. , *Status of the NICA Project*, PoS (EPS-HEP 2013) 106 .
[25] G. Agakishiev et al. [HADES Collaboration] , *The High-Acceptance Dielectron Spectrometer HADES* , Eur. Phys. J. A 41 (2009) 243 .
[26] G. Agakishiev et al. [HADES Collaboration] , *Origin of the low-mass electron pair excess in light nucleus-nucleus collisions* , Phys.Lett.B690 (2010) 118 - 122 .
[27] G. Agakishiev et al. [HADES Collaboration] , *Dielectron production in Ar+KCl collisions at 1.76A GeV* , Phys.Rev. C84 (2011) 014902 .
[28] I. Fröhlich for the HADES Collaboration , *Future perspectives at SIS-100 with HADES-at-FAIR*, arXiv:0906.0091v2 [nucl-ex] (2009).
[29] P. Tlusty et al. [HADES Collaboration] , *Future experiments with HADES at FAIR*, AIP Conf.Proc. 1322 (2010) 116 .
[30] K. Lapidus et al. [HADES Collaboration] , *The HADES-at-FAIR project* , Phys.Atom.Nucl. 75 (2012) 589.

Journal of Physics: Conference Series **623** (2015) 012026 doi:10.1088/1742-6596/623/1/012026

[31] R. Krücken et al. [NUSTAR Collaboration] , *The Nustar Facility At FAIR,* J. Phys. G 31 (2005) S1807 .

[32] B. Rubio and T. Nilsson, *NUSTAR,* Nucl. Phys. News 16 (2006) 9 .

[33] Physics Performance Report for PANDA: *Strong Interaction Studies with Antiprotons,* arXiv:0903.3905v1 [hep-ex].

[34] K. Fohl et al. [PANDA Collaboration] , *The Panda Detector At The Future FAIR Laboratory,* Eur. Phys. J. ST 162 (2008) 213 .

[35] J.S. Lange for the PANDA Collaboration , *The Panda Experiment: Hadron Physics With Antiprotons At FAIR,* Int. J. Mod. Phys. A 24 (2009) 369 .

[36] Z. Q. Liu et al. (Belle Collaboration) , *Study of $e^+e^?$ $\rightarrow$ $\pi^+\pi^?J/\psi$ and Observation of a Charged Charmoniumlike State at Belle,* Phys. Rev. Lett. 110 (2013) 252002 .

[37] M. Ablikim et al. (BESIII Collaboration) , *Observation of a charged $(D\bar{D}^*)$-mass peak in $e^+e^- \rightarrow \pi^+(D\bar{D}^*)$ - at E_{cm}=4.26 GeV,* Phys. Rev. Lett 112 (2014) 022001 .

[38] R. Maier, private communication .

[39] M. Strikmann, H. Stoecker, F. Maas and R. Maier , *The authors have suggested to extend the present FAIR design by moderate changes into a full-fledged $\bar{p}p$ collider with $\sqrt{s}^{max}$ = 30 GeV accessing the Beauty Baryon and Meson Physics at FAIR* . 2013 , manuscript in preparation .

Search for the critical point of strongly interacting matter at the CERN SPS NA61/SHINE experiment[1]

Ludwik Turko
for the NA61/SHINE Collaboration

Institute of Theoretical Physics, University of Wroclaw, pl. M. Borna 9, 50-205 Wroclaw, Poland

E-mail: ludwik.turko@ift.uni.wroc.pl

Abstract. The NA61/SHINE experiment performs a detailed study of the onset of deconfinement and search for critical point of hadronic matter by colliding nuclei of different sizes at various beam momenta from 13A to 158A GeV/c. Experimental setup and results on the theoretically expected signatures are discussed.

1. Introduction

The NA61/SHINE, understood as The **S**uper Proton Synchrotron (SPS) **H**eavy **I**on and **N**eutrino **E**xperiment, is the continuation and the extension of the NA49 experiment [1, 2]. It uses the same experimental fix target set up as NA49 but the physics programme has been extended to measure the production of charged pions and kaons out of a thin carbon target and a replica of the T2K target what is necessary to test accelerator neutrino beams [3]. Data taking began in 2007. These $p - C$ data allow also for better understanding of nuclear cascades in the cosmic-air showers - necessary in the in the Pierre Auger and KASCADE experiments [4, 5]. The main aim of the NA61/SHINE, however, is a study of the onset of deconfinement and search for critical point of hadronic matter. The NA49 experiment studied hadron production in Pb+Pb interactions while the NA61/SHINE collect data at varying collision energy (13A-158A GeV) and size of the colliding systems. This is equivalent to the two dimensional scan of the hadronic phase diagram in the T, μ_B plane. The ion collisions programme was initiated in 2009 with the $p - p$ collisions used later on as reference data.

The programme allows to measure also fluctuations of various physical quantities which are sensitive to the phase transition. These fluctuations create experimental signature for the critical point. An analysis of fluctuations of various observables, particularly in a range of energies around 8 GeV at the center of mass in interactions of light nuclei per colliding nucleon pair (it corresponds to the beam energy of 30 GeV in the frame of a stationary target), is the main goal of the NA61/SHINE experiment. This is the kinematical region where NA49 data indicate that the onset of deconfinement in central Pb+Pb collisions. It is mainly based on the observation of structures in the energy dependence of hadron production in central Pb+Pb collisions which are not observed in elementary interactions [6, 7]

[1] Talk presented at the International Workshop: Discovery Physics at the LHC – Kruger 2014, RPA, December 1 6 , 2014

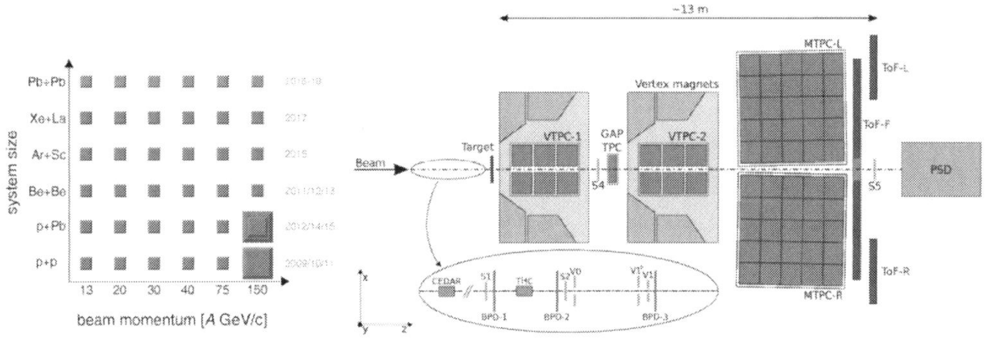

Figure 1. Left: For the programme on strong interactions NA61 / SHINE scans in the system size and beam momentum. In the plot the recorded data are indicated in green, the approved future data taking in red, whereas the proposed extension for the period 20152018 in gray.
Right: The NA61 / SHINE detector consists of a large acceptance hadron spectrometer followed by a set of six Time Projection Chambers as well as Time-of-Flight detectors. The high resolution forward calorimeter, the Projectile Spectator Detector, measures energy flow around the beam direction. For hadron-nucleus interactions, the collision volume is determined by counting low momentum particles emitted from the nuclear target with the LMPD detector (a small TPC) surrounding the target. An array of beam detectors identifies beam particles, secondary hadrons and nuclei as well as primary nuclei, and measures precisely their trajectories.

The collaboration consists of about 150 physicists from 15 countries and 30 institutions. It is the second largest non-LHC experiment at CERN.

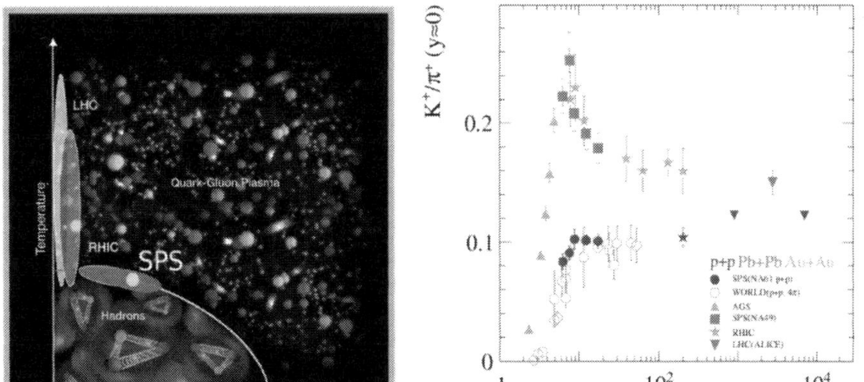

Figure 2. Left: Phase diagram of strongly interacting matter in the (T, μ_B) plane
Right: Horn - a strong maximum of the ratio of K^+/pi^+ multiplicities. A reduced shadow of the horn structure is visible in $p + p$ reactions.

Journal of Physics: Conference Series **623** (2015) 012027 doi:10.1088/1742-6596/623/1/012027

2. Some new NA61/SHINE results

NA61 measurements established energy dependence of the inelastic cross section $^{7}Be + {}^{9}Be$

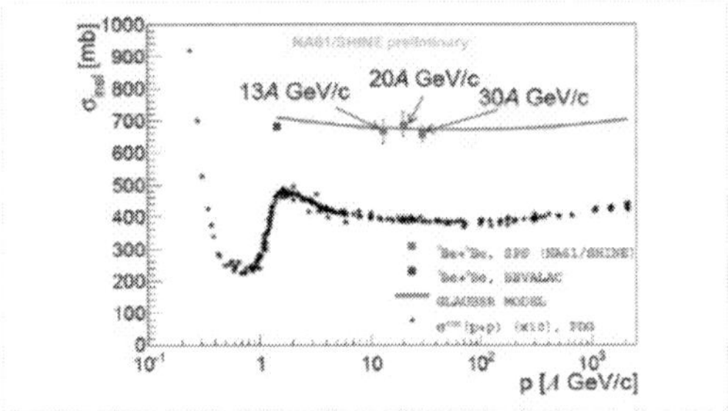

Figure 3. The inelastic cross section for $^{7}Be + {}^{9}Be$ collisions. It is well reproduced by the Glauber model[8]

Pion spectra for $^{7}Be + {}^{9}Be$ collisions

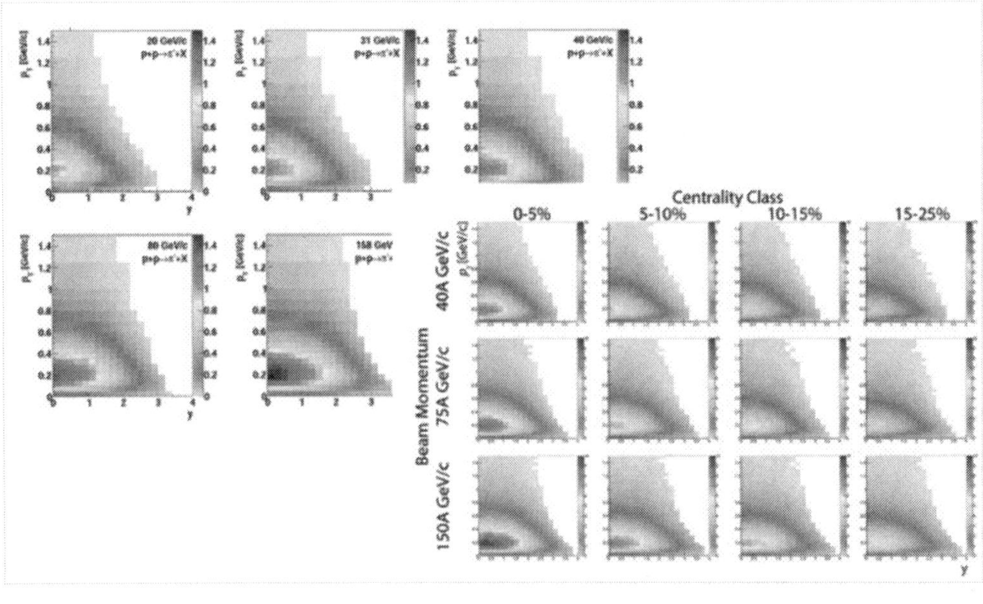

Figure 4. Precise measurements of π^{-} in $p + p$ and $^{7}Be + {}^{9}Be$ collisions

Journal of Physics: Conference Series **623** (2015) 012027 doi:10.1088/1742-6596/623/1/012027

Pion rapidity distribution for $^7Be + {}^9Be$ collisions

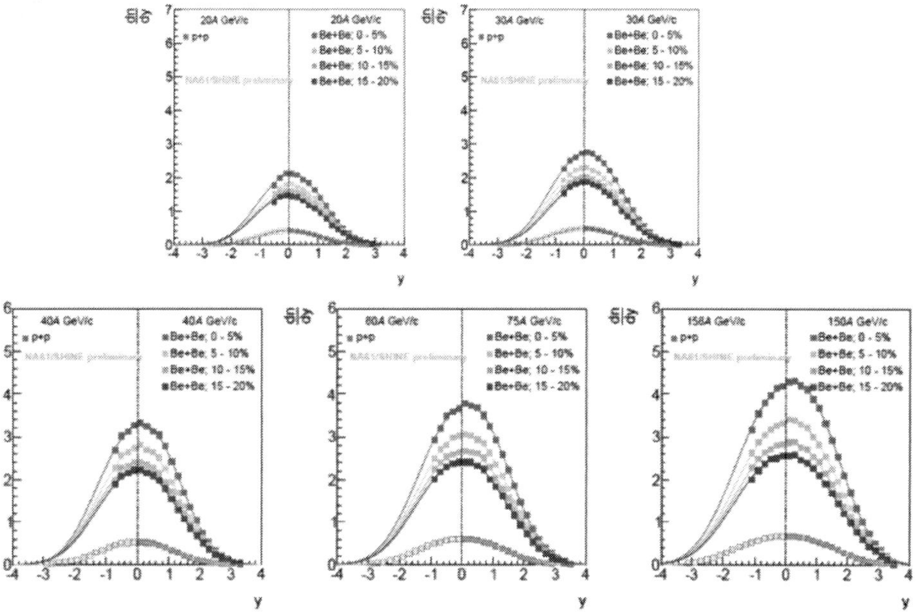

Figure 5. Pion rapidity distribution of π^- in $p + p$ and $^7Be + {}^9Be$ collisions

Inverse slope parameter T

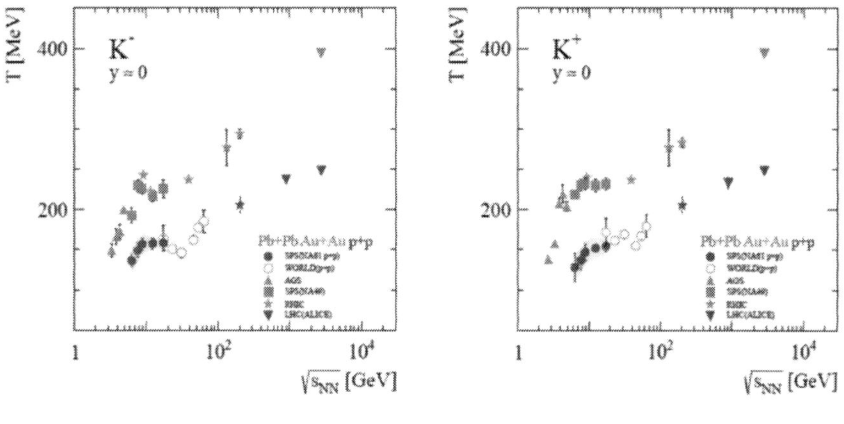

Figure 6. Inverse slope parameter T exhibits rapid changes in the SPS energy range - also in $p + p$ collision

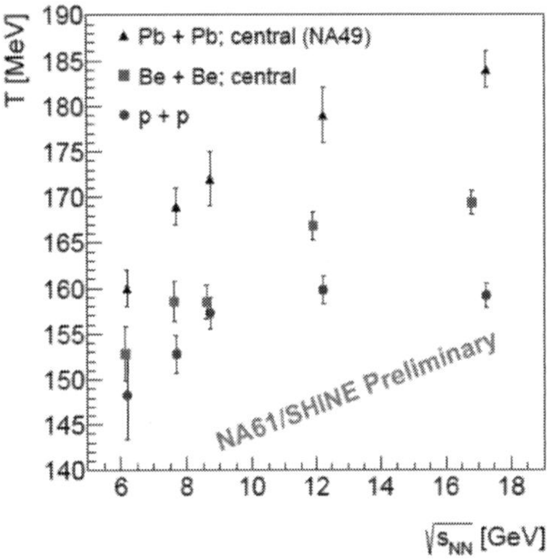

Figure 7. Inverse slope parameter T in $^7Be + ^9Be$ larger than in $p + p$ collision indicates possibility of the transverse collective flow

Transverse momentum fluctuations

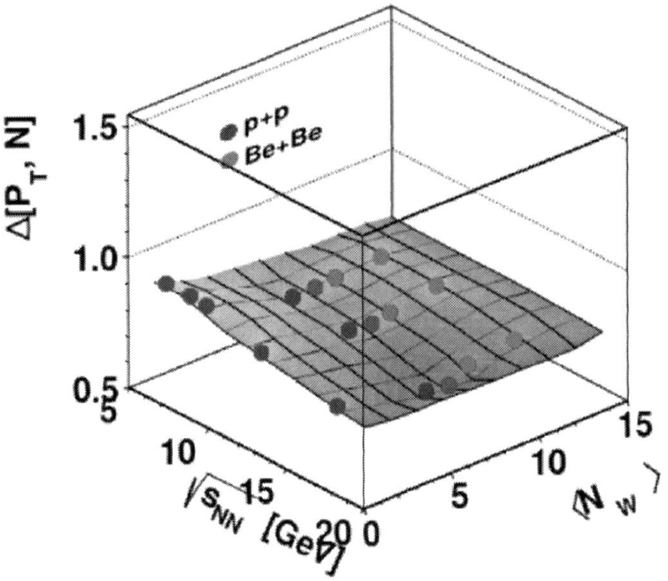

Figure 8. No critical fluctuations neither in $^7Be + ^9Be$ nor in $p + p$ collision

Acknowledgments
This work was supported in part by the Polish National Science Center (NCN) under contract No. DEC-2012/04/M/ST2/00816

[1] N. Antoniou et al. (NA49-future Collaboration), CERN-SPSC- 2006-034, 2006.
[2] N. Abgrall et al. (NA61/SHINE Collaboration), CERN-SPSC- 2008-018, 2008.
[3] N. Abgrall *et al.* [NA61/SHINE Collaboration], Phys. Rev. C **84**, 034604 (2011) [arXiv:1102.0983 [hep-ex]].
[4] J. Abraham et al. (Pierre Auger Collaboration), Nucl. Instrum. Methods, Phys. Res., Sect. A **523**, 50 (2004).
[5] T. Antoni et al. (KASCADE Collaboration), Nucl. Instrum. Methods, Phys. Res., Sect. A **513**, 490 (2003)
[6] C. Alt et al. [NA49 Coll.], Phys. Rev. C **77**, 024903 (2008)
[7] M. Gazdzicki et al. [NA49 Coll.], J. Phys. G **30**, S701 (2004).
[8] W. Broniowski, M. Rybczynski and P. Bozek, Comput. Phys. Commun. **180**, 69 (2009) [arXiv:0710.5731 [nucl-th]]

CERN: A European laboratory for a global project

Rüdiger Voss
CERN, Geneva, Switzerland

Abstract. In the most important shift of paradigm of its membership rules in 60 years, CERN in 2010 introduced a policy of "Geographical Enlargement" which for the first time opened the door for membership of non-European States in the Organization. This short article reviews briefly the history of CERN's membership rules, discusses the rationale behind the new policy, its relationship with the emerging global roadmap of particle physics, and gives a short overview of the status of the enlargement process.

In 2014, CERN celebrates it 60th anniversary. On July 1, 2014, the cycle of events commemorating the birthday of this unique scientific, political and human endeavour was opened in Paris, commemorating the anniversary of the CERN Convention, which was signed at the UNESCO Headquarters by representatives of the twelve founding Members in 1953. These twelve signatures are worth commemorating indeed. For more than half a century, the Convention has stood the test of time as a masterpiece of simple and minimalistic legal language that focuses wisely on the essential cornerstones of CERNs institutional basis and governance. At the same time, it provides for the leeway that is necessary to adapt the Organization to a changing political environment, and to new scientific and technological challenges. The Convention is a testimony to the wisdom and foresight of CERN's founding fathers, on par with their vision of rebuilding peace in Europe by establishing a unique focal point that would foster scientific collaboration on an unprecedented scale, between nations that that had fought war against each other only a few years earlier. On the basis of this Convention, CERN has served as a model for other, successful European science organizations, and most recently for the SESAME synchrotron light source in the Middle East.

Some of the most intriguing aspects of the CERN Convention are in the provisions for Membership in the Organization. Whereas it stipulates that "the Organization shall provide for collaboration among European States in nuclear research of a pure scientific and fundamental character ...", nowhere does it state explicitly that Membership in CERN is restricted to European States. This ambiguity is by no means fortuitous. It reflects the fact that already in the early 1950s, a possible enlargement of Membership beyond Europe was a hotly debated issue on which the provisional Council could not reach agreement. It agreed, however, on a carefully crafted compromise that left a door open to shaping the membership policy of CERN at a later stage, and to adapt it to an evolving scientific and political landscape. Indeed CERN continued to grow and counted 20 European Member States at the end of the 20th century, thanks largely due to the accession of countries from Central and Eastern Europe following the fall of the Berlin Wall.

The Council debated a widening of Membership beyond Europe on several occasions in the past, but repeatedly confirmed a restrictive interpretation of the Convention, whereby Membership remained reserved for European countries. Only in 2010, it approved the most

Figure 1. The 21 Member States of CERN after the accession of Israel in 2014. Romania and Serbia are expected to become Member States in 2015 and 2017, respectively. Turkey joined as an Associate Member State in 2015.

radical shift of paradigm of CERN's membership policy to date, embedded in a policy of "Geographical Enlargement" and opening Membership to non-European States irrespective of their geographical location [1]. At the same time, the Council introduced the new instrument of Associate Membership in order to facilitate the accession of new Members, including emerging countries outside Europe, which may not command sufficient resources to sustain full Membership in the foreseeable future. Under this new policy, Israel joined CERN as 21st, and first non-European Member State, in 2014 (Fig. 1). In parallel, a new policy of "Scientific Enlargement" allows, for the first time, the Laboratory to participate in *global* accelerator projects outside Europe.

CERN's new Membership rules are not a political end in itself but follow a twofold science policy rationale: they underpin the unprecedented migration of the global particle physics community to Europe, reflecting the scientific attractiveness and success of the LHC; and they serve to prepare CERN for its long-term future. They mirror the globalization of particle physics, which in turn has become a prominent paradigm for the globalization of science at large.

For more than fifty years, CERN has supported free movement of scientists across countries,

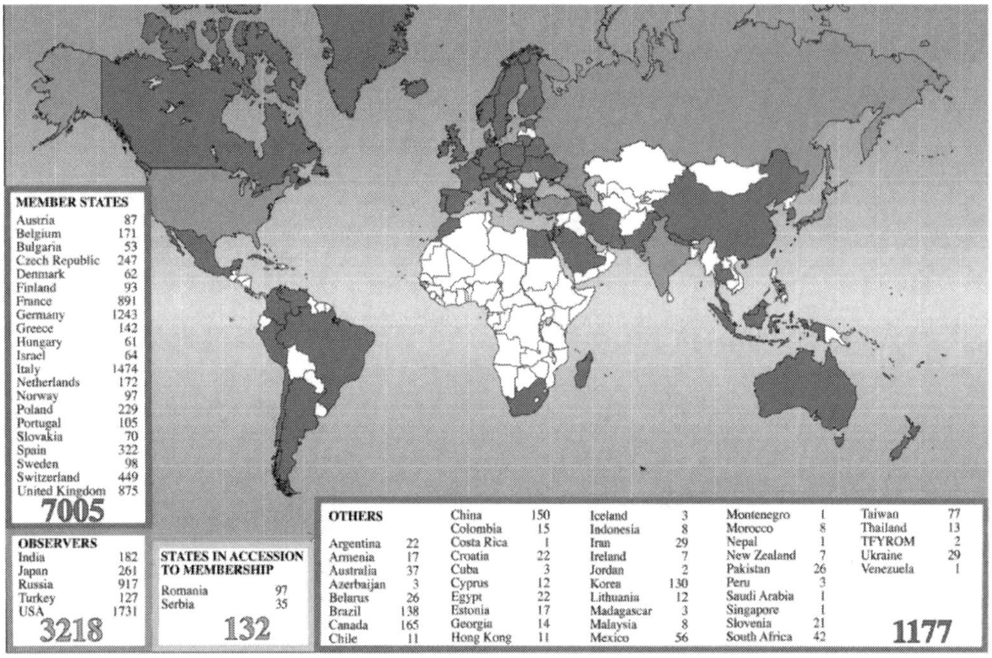

Figure 2. Countries of origin of CERN's visiting scientists (Users) by location of their home institute (January 2015)

regions and continents with a policy of open doors, providing free access to its scientific facilities without discrimination between Member States and non-Member States. This policy had been formalized in 1980 in ICFA guidelines for free, mutual and *balanced* access of physicists from different regions to regional laboratories in other parts of the world [2]. It was shaped at a time when Europe, North America and Japan owned all major facilities for accelerator-based particle physics, when the global HEP community was strongly dominated by scientists from the same three regions, and when exchange between the same regions was healthy and balanced indeed.

Since 1990, however, the landscape has changed dramatically: starting with the demise of the SSC, the shut-down of major accelerators around the globe – or their conversion into applied science facilities – has severely broken the symmetry of scientific exchange. Since 2004, the community of CERN's visiting scientists ("Users") has grown from just above 6000 to more than 11000 physicists and engineers from all continents. Today, these scientists come to CERN from about 650 different institutes in more than 70 countries (Fig. 2) and represent around 100 different nationalities. This growth has been driven by non-Member States (NMS) more than by the Member States: scientists from NMS institutes now account for almost 40% of CERN's Users, and their participation is expected to increase further. Whereas the numbers are dominated by North America, in recent years the most important growth rates are observed for Asia and Latin America, where new players emerge on the field of international science (Fig. 3). These data convincingly illustrate particle physics' strong tradition of defying political, geographical and cultural boundaries.

From a European perspective, the unprecedented non-Member State participation in the LHC has brought about substantial scientific, technical and political benefits. It has contributed to

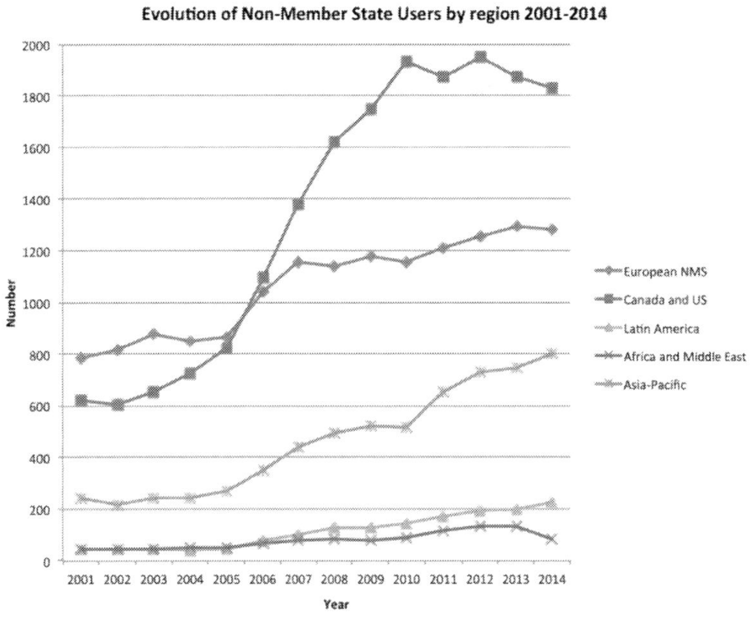

Figure 3. Evolution of the number of CERN Users from Non-Member States (NMS) grouped by region. The apparent decrease for Africa and Middle East reflects the fact that Users from Israel are not included any longer for 2014, when Israel became a Member State.

establishing CERN as the world's leading center at the high energy frontier in the awareness of governments and funding agencies, and of the general public. The LHC has convincingly demonstrated the potential of global partnership in basic science, and is widely perceived as a paradigm of successful world-wide co-operation on megascience projects. To take this co-operation to the next higher level, and to fully exploit its potential to the benefit of all stakeholders, CERN's enlargement policy offers a new institutional framework for a sustained global partnership in particle physics research.

More important, the Geographical Enlargement policy is a first step in preparing CERN's membership and governance for the post-LHC future. Whereas the LHC experiments today are truly global operations, the LHC machine was built as a predominantly European project, with a contribution from outside Europe that was technically and politically significant but represented only about 10% of the construction cost, and was mostly provided in-kind. This model cannot be expected to work for a large, next-generation facility in Europe. With the CLIC and FCC studies, CERN is exploring two different, challenging avenues to prepare its future, and the future of our field, after the LHC. No cost estimate exists yet for either option, but it seems inconceivable that any of them can be approved and build within the same membership, governance and funding structures that worked more than twenty years ago – successfully, but under great labour pains – for the LHC.

In one of the most significant developments in recent years, the three most important *regional* particle physics roadmaps – *i.e.* the proposal of the Japanese community to host the ILC; the 2013 Update of the European Strategy for Particle Physics [3]; and the US P5 report of 2014 [4] – have been developed in a complementary approach and with a strong focus on inter-regional collaboration, such that together they form an important milestone on the way to a *global* roadmap. CERN's enlargement policy is designed to fit seamlessly into this emerging global strategy for accelerator-based particle physics. It opens Membership in the Organization to all States irrespective of their geographical location, and introduces the new scheme of Associate Membership with reduced obligations and benefits. The important instrument of International Co-operation Agreements is maintained for partners who are not willing or able to shoulder the obligations of CERN membership in the near future; almost 50 such Agreements are in place today and provide the formal framework for most of the co-operation with non-European countries.

In a reciprocal move, for the first time CERN is formally allowed to participate in *global* HEP projects outside Europe; this policy, which was confirmed by the 2013 European Strategy Update, will initially enable CERN to serve as a gateway for European participation in the LBNF facility at Fermilab, and later in the ILC.

With ten applications for Membership or Associate Membership received in recent years from countries of different size, and from in- and outside Europe (Brazil, Croatia, Cyprus, Israel, Pakistan, Russia, Serbia, Slovenia, Turkey, Ukraine), the enlargement process has made a promising start; some of the accession procedures have been completed (Israel, Turkey) or are expected to conclude in the near future. Other countries which are perceived as natural candidates, while they acknowledge the promise and potential of a continued scientific and technological partnership, have so far remained absent, or are hesitant on political or financial grounds. More work, stamina, and patience will be needed to enlarge the Membership of CERN to a size that is commensurate with its future ambitions in quantity and quality. Not all States that are obvious candidates for a closer scientific and technical partnership may share today the values of a governance that is excellence-driven and consensus-oriented, and which has prevailed most of the time in CERN's 60-years history. In the long term, broadening the institutional base without sacrificing the traditional values of European co-operation that have been a key ingredient in CERN's past successes is likely to emerge as the true challenge of the enlargement process.

The partnership between CERN, its Member States and non-Member States in building the LHC and harvesting its scientific potential has developed into a solid backbone of successful co-operation of unprecedented, global dimensions. CERN wants this partnership to continue, to grow and to flourish on the basis of an enhanced institutional participation of both European and non-European countries. The policy of Geographical Enhancement is well matched to the unfolding, global roadmap of particle physics.

1. References

[1] CERN Council document CERN/2918/Rev., available at http://cds.cern.ch/record/1289091?ln=en

[2] ICFA Guidelines for the Interregional Utilization of Major Regional Experimental Facilities for High-Energy Particle Physics Research, http://www.fnal.gov/directorate/icfa/icfa_guidelines.html

[3] CERN-Council-S/106, available at http://cern.ch/council/en/EuropeanStrategy/esc-e-106.pdf

[4] Particle Physics Project Prioritization Panel (P5): Building for Discovery: Strategic Plan for U.S. Particle Physics in the Global Context, http://science.energy.gov/ /media/hep/hepap/pdf/May%202014/FINAL_P5_Report_053014.pdf

Dark matter searches in CMS

Nadir Daci, on behalf of the CMS Collaboration
IIHE - VUB, Pleinlaan 2, 1050 Brussels, Belgium
E-mail: nadir.daci@cern.ch

Abstract. The CMS collaboration searched for Dark Matter (DM) particles produced in pairs in proton-proton (pp) collisions performed by the LHC at a centre-of-mass energy of 8 TeV, corresponding to an integrated luminosity of 19.7 /fb. The following signatures were investigated: a single boosted object and a significant transverse momentum imbalance; associated production of DM with a pair of top quarks; decay of a Higgs boson (H) into a pair of DM particles. Exclusion limits were set on the cross sections of DM production and interaction cross sections, as a function of the DM particle mass.

1. Introduction

The Compact Muon Solenoid (CMS) experiment searches for Dark Matter (DM) under the form of Weakly Interacting Massive Particles (WIMPs). This class of DM candidates is constrained by the DM thermal relic abundance: they should have weak-scale mass and interaction cross section with baryonic matter. Thus, they do not interact within the CMS detector and remain undetected. As the LHC accelerates proton beams longitudinally, their transverse momentum (projection in the plane transverse to the beam axis, $\vec{p}_{\mathrm{T}}$) is negligible; thus, the vectorial sum of all final state particles $\vec{p}_{\mathrm{T}}$ cancels out. Therefore, if a WIMP pair is produced and recoils against detectable particles, it causes a transverse momentum imbalance. This signature is used to select WIMP-like events in the recorded data from LHC collisions performed at a centre-of-mass energy of 8 TeV, corresponding to an integrated luminosity of 19.7 fb^{-1}. The monophoton [3] (monojet [4], monolepton [5]) signature is the production of DM with a photon (gluon, leptonically decaying W boson) radiated in the initial state. The monotop [6] signature is the production of a new mediator, connected to a dark sector, in association with a t quark through a Flavour-Changing Neutral Current (FCNC) mechanism. The production of DM in association with a $t\bar{t}$ pair decaying in a dileptonic [7] or a semileptonic [8] final state, is also explored. The "Higgs portal" to DM considers the possible decay of a H boson into DM [9].

2. Particle detection and reconstruction in CMS.

A full description of the CMS detector can be found at [10]. It comprises a tracker, electromagnetic (ECAL) and hadronic (HCAL) calorimeters, and a muon detector. The CMS particle-flow (PF) algorithm reconstructs each individual particle with an optimized combination of information from all CMS sub-detectors. Energy deposits in each calorimeter are clustered ; hits in the tracker (resp. muon detector) are fitted to reconstruct tracks from charged particles (resp. muons). These elements are then matched using geometrical criteria and χ^2 fits. The algorithm distinguishes charged hadrons from electrons, and neutral hadrons from photons by comparing the energy (calorimeters) and the momentum (tracker) of the PF candidates. The analyses reported here require particles originating from the primary pp interaction, defined as the vertex with the largest sum of p_T^2 of all the associated tracks. Distance criteria are applied to select only events with a primary vertex close enough to the centre of the detector. Hadronic jets are clustered from PF candidates with the anti-k_t algorithm (size parameter $R = 0.5$). Jet energy is corrected, to remove the contribution from additional pp interactions, and the effect of the non-linearity and non-uniformity of the detector. A b jet is identified with a $\approx$70% by a b-tagger which exploits track impact parameters, reconstructed decay vertices, clustered leptons.

3. Theoretical models

The monojet, monophoton, monolepton and di-top analyses use Effective Field Theories (EFT): these approximate the DM production mechanism by contact interaction, which can be spin-independent (scalar/vector; *SI*) or spin-dependent (pseudo-scalar/axial-vector; *SD*). For a given operator, only two parameters are scanned: an energy scale $\Lambda = M/\sqrt{g_\chi g_q}$ (M: mediator mass; $g_{\chi,q}$: couplings to DM, quarks); the DM mass M_χ. Thus, the interpretation of the experimental data (referred to as "data" hereafter) stays generic and simple to translate in terms of results on the DM-nuclei scattering cross section [1]. However, EFT interpretations display several limitations [2]. They assume that the interaction between standard model (SM) particles and DM is well represented by a single operator. The EFT is perturbative if $\sqrt{g_\chi g_q} < 4\pi$, which can be unverified in part of the probed phase space. The EFT is realistic only if the mass of the mediator is larger than the centre-of-mass energy of the parton-parton interactions (multi-TeV range). Simplified models involving more search parameters, such as the mass M and width Γ of the mediator, can help exploring more realistic scenarios. The monojet and monophoton analyses include a simplified model interpretation, using an s-channel mediator with vector interactions. The monotop analysis use a specific FCNC model where the new mediator appears explicitely in the interaction lagrangian. The Higgs portal analyses use the framework of searches for the SM H boson, assuming a mass of $m_H = 125$ GeV and explore its possible invisible decays. Background and signal contributions are predicted using Monte Carlo simulations. These are reweighted using scale factors measured in data, to correct for differences in selection efficiencies. Some background estimations use control samples, *i.e.* events from data in a control region, where some signal requirements are replaced by background requirements.

4. Searches based on mono-object final states

4.1. Monojet analysis

The monojet analysis requires a jet with $p_T > 110$ GeV, and $E_T^{\rm miss} > 250$ GeV. As there might be two ISR, the presence of a second softer jet with $p_T > 30$ GeV is allowed; it is required to be close to the first jet in azimuthal angle ($\Delta\phi < 2.5$) to reject QCD dijet processes. The jet energy fractions carried by charged or neutral hadrons, or photons, are used to suppress instrumental and beam-related backgrounds. Events containing a third jet with $p_T > 30$ GeV are vetoed, thus rejecting QCD and $t\bar{t}$ events. Events containing prompt isolated leptons are removed as well, to reject (di-)boson (V, VV) and single top (t) processes.

The dominant backgrounds, estimated using control samples, are: $Z(\nu\nu)$; $W(l\nu)$, where the lepton is out of the detector acceptance, is not reconstructed, or fails the lepton veto definition. The residual contributions from QCD, $Z(ll)$, VV, t and $t\bar{t}$ (4% of the total background) are estimated from simulations. The $Z(\nu\nu)$ background is estimated using a $Z(\mu\mu)$ control sample, where the signal selection (without muon veto) is applied; the muons are removed in the calculation of $\vec{p}_T^{\rm miss}$. The estimation is corrected for: the contamination from other processes (simulated); the selection acceptance; the different branching ratios (BR) for ν and μ. The $W(\mu\nu)$ estimation is similar to the $Z(\nu\nu)$ case, correcting for the probability of loosing the muon from the W decay (due to detector acceptance and reconstruction efficiency). The $W(l\nu)$ ($l = e, \tau$) contributions are obtained using the $W(\mu\nu)$ estimation, correcting for the ratio of

$W(l\nu)$ events to $W(\mu\nu)$ events passing the signal selection (without lepton vetoes), and for the different lepton loss probability. The QCD background is simulated, applying a scale factor computed in a QCD control region (signal selection, no jet veto, $\Delta\phi > 0.3$).

A single-bin event counting is performed for 7 different $E_{\rm T}^{\rm miss}$ thresholds; the cut inducing the best sensitivity to non-SM processes leading to a monojet signature is found to be $E_{\rm T}^{\rm miss} > 500$ GeV. The main uncertainties in this region are: the statistical error from the control samples (11% in $Z(\mu\mu)$, 5.5% in $W(\mu\nu)$); the acceptance and efficiency measurements (6.4% in $W(\mu\nu)$); the renormalization and factorization scales (15%) and the ISR modelling (5%) in the signal prediction. The observation is consistent with the background-only hypothesis.

4.2. Monophoton analysis

The monophoton analysis requires a photon with $p_T > 145$ GeV and $E_{\rm T}^{\rm miss} > 140$ GeV. The photon purity is maximized using requirements on the photon electromagnetic shower shape and the hadronic activity around the photon. The photon must fall in the central part of the ECAL ($|\eta| < 1.44$); the energy in the HCAL channel behind the photon must be below 5% of the photon energy; the photon isolation is ensured by requiring a low energy sum of surrounding particles. The compatibility of the photon energy deposit in the ECAL with hits in the innermost layers of the tracker is used to reject electrons faking photons.

The photon and the $\vec{p}_{\rm T}^{\rm miss}$ are required to be back-to-back ($\Delta\phi(\vec{p}_{\rm T}^{\rm miss}, \gamma) > 2$), to suppress γ+jet events (a mismeasured jet can lead to significant $E_{\rm T}^{\rm miss}$ in the event). Additionnally, a χ^2 function is used to evaluate the quality of the $\vec{p}_{\rm T}^{\rm miss}$ reconstruction and reject events with no genuine $\vec{p}_{\rm T}^{\rm miss}$: this suppresses 80% (35%) of γ+jet (QCD) events, with a 99.5% signal efficiency. Events containing prompt isolated leptons are vetoed to reduce the $W(l\nu) + \gamma$ contribution.

The main backgrounds are $Z(\nu\nu) + \gamma$ and $W(l\nu) + \gamma$, where the lepton is lost, as in the monojet analysis. These contributions, as well as the minor backgrounds γ+jet, $W(\mu\nu)$, $\gamma\gamma$ and $Z(ll) + \gamma$, are predicted from simulations, with a data-driven cross-check for $Z(\nu\nu) + \gamma$. Non-collision backgrounds are estimated in data using the electromagnetic shower shapes and the ECAL energy deposition timing.

The $W(e\nu)$ estimation uses a control sample, where the events pass a modified signal selection, reverting the veto on the compatibility between the ECAL energy deposit and tracker hits. The efficiency of this matching is estimated in data, and it is used to extrapolate the $W(e\nu)$ contribution from this control region to the full signal region.

The QCD background, where a jet fakes a photon, is estimated in the data, using events passing the signal selection with reverted photon isolation cuts. This estimation is then multiplied by the ratio of QCD events with a jet passing the photon selection, to QCD events with a jet failing the photon isolation cuts.

The major uncertainties affect the $Z(\nu\nu) + \gamma$ (12%) and $W(l\nu) + \gamma$ (20%) and are mainly due to Parton Distribution Functions (PDF) (which are correlated accross these two channels). Uncertainties from the QCD control sample and scale factor affect the QCD estimation (31%).

4.3. Monolepton analysis

The monolepton analysis searches for processes where two quark interact: one radiates a W boson which decays leptonically, and the other one radiates a pair of DM particles. This analysis is sensitive to possible interferences between two diagrams with same initial and final states, in which the DM particles couple either to up quarks or to down quarks. Three scenarios are considered here : destructive or constructive interference, and coupling to only one type of quark. Here, the master variable is the transverse mass m_T (lepton-$\vec{p}_{\rm T}^{\rm miss}$ invariant mass).

The monolepton analysis requires a prompt isolated lepton, with p_T above 100 GeV (electron) or 45 GeV (muon). The relative uncertainty on the muon transverse momentum is required to be below 30%, to suppress largely mismeasured muons. Events with two same-flavour leptons,

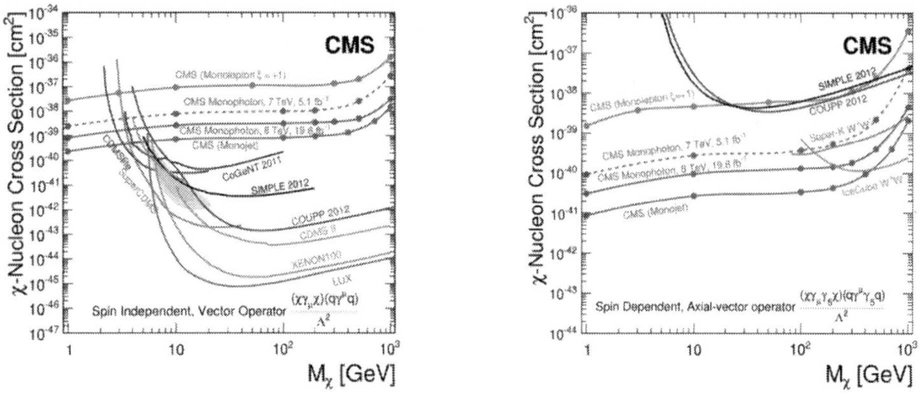

Figure 1. 90% CL upper limits on the DM-nucleon cross section from the monojet, monophoton, and monolepton (destructive interferences) analyses, as a function of the DM mass, for vector (left) and axial-vector (right) interactions.

where the second lepton has p_T above 35 GeV (electron) or 25 GeV (muon), are vetoed, in order to remove dilepton processes (Drell-Yan $Z/\gamma*$). The lepton and the $\vec{p}_{\mathrm{T}}^{\mathrm{miss}}$ are required to be back-to-back ($\Delta\phi(l, \vec{p}_{\mathrm{T}}^{\mathrm{miss}}) > 2.5$) and balanced ($0.4 < p_T/E_{\mathrm{T}}^{\mathrm{miss}} < 1.5$), which discriminates signal and background events (e.g. QCD multijets where the $\vec{p}_{\mathrm{T}}^{\mathrm{miss}}$ is due to mismeasured jets).

The major background in this analysis is the $W(l\nu)$ process, with minor contributions from t, $t\bar{t}$, $Z/\gamma * (ll)$, γ+jets, and VV; all these processes are estimated from simulations. In order to reach an accurate background estimate at very high m_T values, the prediction is fitted using an empirical function: $f(m_T) = \exp\left(a + bm_T + cm_T^2\right) M_T^d$. The $W(l\nu)$ simulation is corrected using NLO k-factors binned in m_T. The QCD background, where a jet fakes an electron, is estimated from data, using events passing the signal selection with reverted electron isolation cuts. This estimation is then extrapolated to the full signal region using the fraction of QCD events with an electron passing the isolation cuts. This ratio is evaluated in a control sample, where events verify: $1.5 < p_T(e)/E_{\mathrm{T}}^{\mathrm{miss}} < 10$. Such a sample is dominated by QCD processes. The ratio is computed in several $p_T(e)$ and $\eta(e)$ bins.

In the e channel, the major uncertainty arise from the PDF, with a subdominant component due to the W k-factors. The total uncertainty in this channel is below 5% at $m_T = 300$ GeV, and reaches 30% at $m_T = 3$ TeV. In the μ channel, the total uncertainty is largely dominated by the μ p_T scale: it ranges from 5% at $m_T = 300$ GeV, to 85% at $m_T = 3$ TeV.

4.4. Interpretation of the results

In the monojet, monophoton and monolepton analyses, the observation is found to be consistent with the background-only hypothesis, and exclusion limits are set. In the monojet and monophoton analyses, single-bin event counting experiments are performed to set limits, whereas a multi-bin counting experiment is performed in the monolepton analysis. These limits on the DM production cross section are then translated in limits on the DM - nucleon interaction cross section, as a function of the DM particle mass (as described in [1]), and displayed on figure 1.

In the monojet and monophoton analysis, a first simplified model interpretation is performed, using an s-channel vector mediator. Figure 2 shows the lower limit on the interaction scale $\Lambda = M/\sqrt{g_\chi g_q}$ as a function of the mediator mass M. Six scenarios are considered: two DM masses (50, 500 GeV) and three mediator widths ($M/8\pi$, $M/10$, $M/3$).

This limit displays a resonant behaviour around the TeV scale, where the mediator can be produced on-shell. At higher mediator masses, the result is consistent with the EFT limit. At intermediate mass, the on-shell production increases the cross section, hence a tighter exclusion with respect to the EFT. At lower mass, the observed limit is below the EFT limit, and the EFT approach becomes too optimistic.

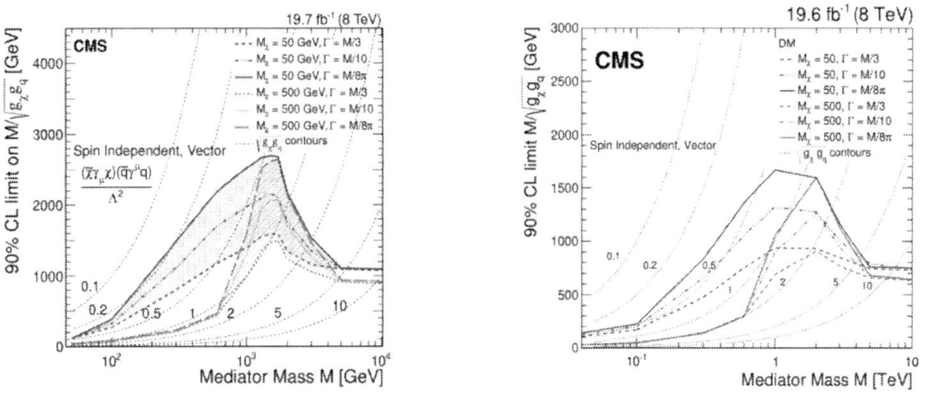

Figure 2. Observed limits on the interaction scale Λ as a function of the mediator mass M, for two DM masses and three mediator widths. The dashed lines show contours of constant coupling $g_\chi g_q$. Left: monojet, right: monophoton.

5. Searches involving top quarks

5.1. *Top pair analyses*

These analyses search for a DM pair produced in association with a $t\bar{t}$ pair, using an EFT with the following interaction lagrangian: $L = \frac{m_q}{\Lambda} q\bar{q}\chi\bar{\chi}$. Given this m_q dependence, the ditop final state enhances the analysis sensitivity. Two channels are defined: dileptonic and semileptonic, whether both or only one W boson(s) from the $t(bW)$ process decays leptonically.

The dileptonic analysis selects two isolated leptons with: $p_T > 20$ GeV; invariant mass $m_{ll} > 20$ GeV and outside of a Z mass window ($m_Z \pm 15$ GeV) in same-flavour dilepton events, thus reducing the Z background. The scalar sum of the two lepton p_T must be above 120 GeV, and their opening azimuthal angle $\Delta\phi$ must be below 2 radians, hence reducing $t\bar{t}$ and other backgrounds. The events must also contain at least two jets with $p_T > 30$ GeV; the scalar jet p_T sum must be below 400 GeV, thus reducing the $t\bar{t}$ background. Finally, as it is a WIMP search, the event must contain significant $E_{\mathrm{T}}^{\mathrm{miss}}$: the cut applied here is 320 GeV. The irreducible processes ($t\bar{t}, t, Z/\gamma^*, W$), which comprise 99% of the total background, are estimated from simulations, requiring leptons to originate from a W/Z boson decay. The residual backgrounds consist in events containing one or two fake leptons (mostly jets reconstructed as leptons); they are estimated using a control sample. The largest source of uncertainty on the total background is the jet energy measurement (scale: 15%, resolution: 5.4%); another important uncertainty arises from the t-quark p_T reweighting applied in $t\bar{t}$ simulated events (11%).

The semileptonic channel selects events with a single isolated lepton with $p_T > 30$ GeV; at least 3 jets with $p_T > 30$ GeV, including at least 1 jet identified as a b quark jet. The $\vec{p}_{\mathrm{T}}^{\mathrm{miss}}$ and the pair of leading jets are requested to have an azimuthal separation $\Delta\phi > 1.2$ rad (reduces $t\bar{t}$ background); the events must verify: $E_{\mathrm{T}}^{\mathrm{miss}} > 320$ GeV (WIMP-like events), $m_T > 160$ GeV (reduces W background). An additional variable m_{T2}^W, characterizing the W decay kinematics, is used to reduce the dileptonic $t\bar{t}$ contribution at large m_T. All backgrounds $(t, Z, VV, W, t\bar{t})$ are estimated from simulations. The dominant processes $(t\bar{t}, W)$ are weighted using scale factors estimated in control samples. The major uncertainties on the total background arises from the W scale factors (13%) and the estimated contamination from sub-dominant backgrounds in the control samples (10%).

In both channels, a single bin counting experiment is performed in order to extract 95% CL lower limits on the interaction scale (denoted here M^*), displayed on figure 3.

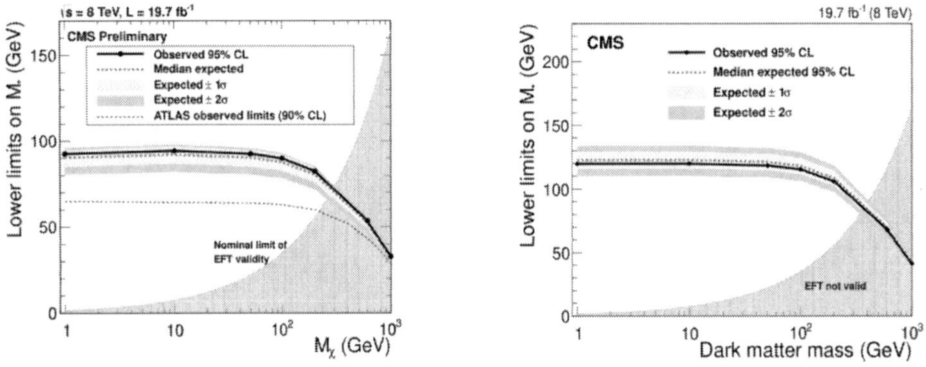

Figure 3. Lower limits on the interaction scale: ditop (left: dileptonic; right: semileptonic).

5.2. Monotop analysis

The monotop analysis searches for a new mediator connected to a dark sector, produced in a FCNC process leading to a t quark ($t \to bW(q\bar{q})$) and significant E_T^{miss}. The event selection requires 3 jets with $p_T > 60, 60, 40$ GeV and a total invariant mass below 250 GeV; $E_T^{\text{miss}} > 350$ GeV. Events containing a fourth jet with $p_T > 35$ GeV, and/or a prompt isolated lepton ($p_T > 20(\mu), 10(e)$ GeV), are vetoed. Two event categories are defined: background-dominated 0 b jet; signal-dominated 1 b jet.

The dominant backgrounds $Z(\nu\nu)$ and $W(l\nu)$ are estimated using control samples, with strategies similar to the monojet analysis. The $Z(\nu\nu)$ evaluation uses $Z(\mu\mu)$+3Jets events, and the $W(l\nu)$ estimation uses $W(l\nu)$+3Jets events. Both control samples are corrected for contaminations from other backgrounds and μ acceptance/efficiency. The $Z(\mu\mu)$ estimation is also corrected for ν/μ BR, while the $W(l\nu)$ is multiplied by the probabilities of: the lepton to remain undetected, a light quark jet to pass b-tagging. All other backgrounds $(t\bar{t}, t, VV)$, excepting QCD, are estimated from simulations.

The numbers of signal, QCD and other background events are fitted to the number of observed events, simultaneously in each category. Signal and QCD events are weighted by the probabilities to tag 0 or 1 b jet (depending on the category) in such events; these probabilities are measured in simulations. This fit allows to estimate the number of QCD and signal events. The sources of uncertainty are: statistical uncertainties in control samples and their contaminations, jet energy, b-tagging performance, PDF, choices of renormalization/factorization scales in simulations. Figure 4 shows the upper limit on the production cross section of a vector mediator, as a function of its mass.

6. The Higgs portal to Dark Matter

These analyses search for invisible decays of a H boson, considering three production modes: Vector Boson Fusion (VBF) and Higgstrahlung: $Z(bb, ll)H(\text{inv})$.

The VBF analysis requires: $E_{\mathrm{T}}^{\mathrm{miss}} > 130$ GeV; two leading jets ($p_T > 50$ GeV) well separated in η ($\eta_1\eta_2 < 0$, $\Delta\eta > 4$) and ϕ ($\Delta\phi_{JJ} < 1$, reduces QCD); vetoes the presence of a central jet ($\eta_1 < \eta_J < \eta_2$, reduces QCD), or a prompt isolated lepton (reduces V). The major backgrounds $Z(\nu\nu)$ and $W(l\nu)$ are estimated in control samples ($Z_{\mathrm{CR}}^{\mu\mu}$, $W_{\mathrm{CR}}^{l\nu}$), subtracting contaminations from other backgrounds (simulated) and correcting for BR (Z), and differences in selection efficiencies between the signal and the control region (simulated). The QCD background is estimated by combining three QCD-enriched control samples, where the $E_{\mathrm{T}}^{\mathrm{miss}}$ cut, the CJV requirement, or both, are reverted. The contamination from electroweak processes (simulated) is subtracted. The major uncertainties affecting the total background are: statistical uncertainties in control samples and simulations (11% each). The jet and $E_{\mathrm{T}}^{\mathrm{miss}}$ energy measurement uncertainty induce a 7% (13%) error on the background (signal) estimations.

The $Z(bb)H$ analysis defines three $E_{\mathrm{T}}^{\mathrm{miss}}$ regions ([100,130], [130, 170], > 170) and exploits the boosted H topology by requiring two b-tagged (reduce V, VV) jets ($p_T > 60, 30$ GeV) with total p_T above 130 GeV (120 GeV in the low-$E_{\mathrm{T}}^{\mathrm{miss}}$ region), and $\Delta\phi(Z, H) > 2$. This jet pair is required to be consistent with a Z boson decay ($M_{JJ} < 250$ GeV); in the low-$E_{\mathrm{T}}^{\mathrm{miss}}$ region, events with more than 1 additional jet are rejected. A lepton veto ($p_T > 15$ GeV) reduces the $WZ, t\bar{t}$ backgrounds. Additional requirements are applied to reject "fake $E_{\mathrm{T}}^{\mathrm{miss}}$" from QCD processes (mismeasured jets): $\Delta\phi(E_{\mathrm{T}}^{\mathrm{miss}}, J) > 0.7, \Delta\phi(E_{\mathrm{T}}^{\mathrm{miss}}, E_{\mathrm{T}}^{\mathrm{miss}\pm}) < 0.5$, where J designates the jet closest to the $\vec{p}_{\mathrm{T}}^{\mathrm{miss}}$, and $E_{\mathrm{T}}^{\mathrm{miss}\pm}$ the $E_{\mathrm{T}}^{\mathrm{miss}}$ computed only from charged particles. The main analysis variable is a Boosted Decision Tree (BDT) trained on simulated signal and background processes. It exploits the event topology and results from the b-tagging algorithm. Control samples are used to validate the BDT input distributions, and to derive scale factors. The major uncertainty affecting the background processes are the b-tagging performance (7 %) and the scale factors (8%); the main signal uncertainty arises from the PDF and the factorization/renormalization scales chosen for the simulations (7%).

The $Z(ll)H$ analysis selects same-flavour opposite-charge lepton pairs, with a total p_T above 20 GeV and invariant mass within a Z window ($m_Z \pm 15$ GeV). Events containing additional e and μ ($p_T > 10$ GeV) are vetoed in order to reduce the WZ background. Two event categories (0-jet, 1-jet) are defined based on the number of jets with $p_T > 30$ GeV, to exploit their very different signal-to-background ratios. Higher multiplicity events are rejected in order to reduce the $Z(ll)$+jets background (where the $E_{\mathrm{T}}^{\mathrm{miss}}$ arises mostly from mismeasured jets). A b-quark veto is applied, based on the presence of a soft μ ($p_T > 7$ GeV) or a b-tagged jet ($p_T > 20$ GeV), to reduce the t background. The $E_{\mathrm{T}}^{\mathrm{miss}}$ is required to be above 120 GeV. The $\vec{p}_{\mathrm{T}}^{\mathrm{miss}}$ and the dilepton systems must be back-to-back ($\Delta\phi > 2.7$) and well balanced ($|E_{\mathrm{T}}^{\mathrm{miss}} - p_T^{ll}|/p_T^{ll} < 0.25$), thus reducing the $Z(ll)$+jets and t contributions. The major backgrounds are diboson processes (ZZ, WZ), estimated in simulations. The $Z(ll)$+jets is estimated using a γ+jets control sample, subtracting contaminations from other backgrounds (simulated) and correcting for different momentum distributions for Z and γ. The other background ($t, WW, W+$jets$, Z(\tau\tau)$) estimations use a [$e^{\pm}\mu^{\mp}$] control sample passing the full signal selection, excepting the same-flavour requirement. Contaminations from other processes (siulated) are subtracted, and the estimation is then extrapolated to the same-flavour region, using yield ratios $N_{ll}/N_{e\mu}$ estimated in sidebands of the Z peak. The main uncertainties are related to the factorization/normalization scales and the PDF used in the simulations (5.0-7.0%), as well as the $Z(ll)$ normalization (4.8%).

The three channels lead to results compatible with the SM prediction. The results are used to compute a 95% CL upper limit on the product of the H production cross section and its invisible branching fraction, as a function of the H mass. After combining the three channels, the upper limits of the invisible branching fraction of the 125 GeV H boson are established; 95% CL: 0.58 (observed) and 0.44 (expected); 90% CL: 0.51 (obs.) and 0.38 (exp.). These limits are in turn translated to limits on the DM-nucleon elastic scattering cross section, for scalar/vector/Majorana fermion DM (for $m_\chi < m_H/2$), as displayed on figure 5.

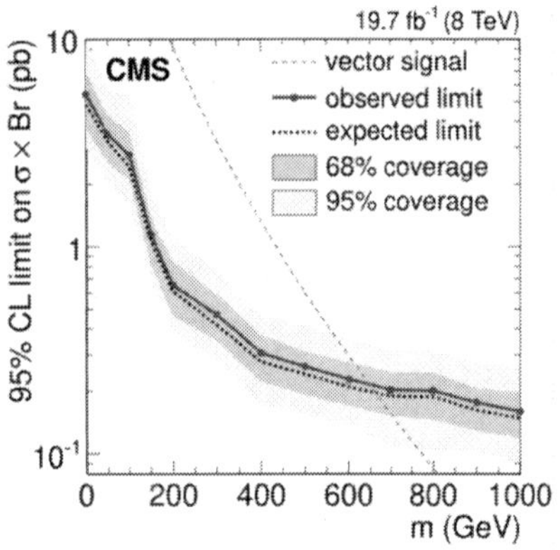

Figure 4. Monotop: upper limit on the new mediator production cross section.

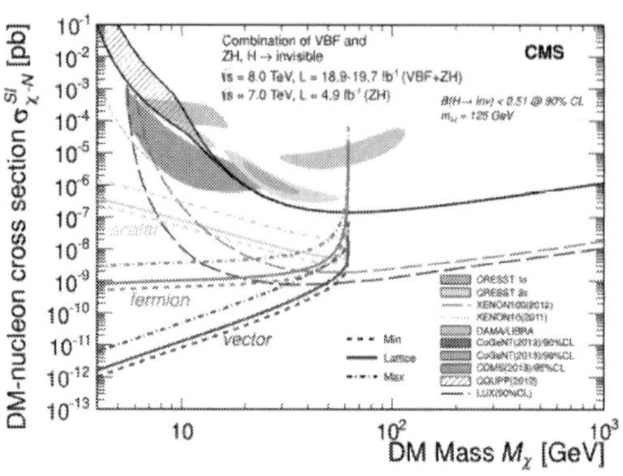

Figure 5. H_{125} upper limits on the DM-nucleon interaction cross section versus DM mass.

7. Conclusion

The CMS DM searches cover a broad panel of final states and scenarios, leading to stringent constraints on the existence of WIMPs. The mono-object analyses provide 90% CL upper limits on the DM-nucleon interaction cross section (90% CL) from 10^{-36} to 10^{-41} cm^2. The monojet and monophoton analyses perform a simplified model cross-check, establishing how realistic the EFT approaches are in this context. The monolepton analysis allows to test flavour-dependent couplings, while the ditop and monotop explore top-DM couplings. These searches provide the only limits at low masses (m_χ <O(1) GeV for SI interactions, m_χ <O(10) GeV for SD interactions), and the most stringent constraints for SD interactions at m_χ <O(100) GeV). The Higgs portal analyses set limits on the invisible decays of a H_{125} boson, then translated in limits on the DM-nucleon SI (vector) interaction cross section. These are even tighter than the limits from other analyses, reaching 10^{-45}cm^2 for Majorana fermion DM with $m_\chi < 20$ GeV.

References

[1] Y. Bai et al. The Tevatron at the Frontier of Dark Matter Direct Detection. *JHEP* 1012:048,2010
[2] O. Buchmueller et al. Beyond Effective Field Theory for Dark Matter Searches at the LHC. *JHEP* 01(2014)025
[3] The CMS Collaboration. Search for new phenomena in monophoton final states in proton-proton collisions at sqrt(s) = 8 TeV. *Submitted to Phys. Lett.* B (arXiv:1410.8812)
[4] The CMS Collaboration. Search for dark matter, extra dimensions, and unparticles in monojet events in proton-proton collisions at sqrt(s) = 8 TeV. *Submitted to Eur. Phys. J.* C (arXiv:1408.3583)
[5] The CMS Collaboration. Search for dark matter in the mono-lepton channel with pp collision events at center-of-mass energy of 8 TeV. (CMS-PAS-EXO-13-004)
[6] The CMS Collaboration. Search for monotop signatures in proton-proton collisions at sqrt(s) = 8 TeV. *Phys. Rev. Lett.* **114**, 101801 (2015)
[7] The CMS Collaboration. Search for the Production of Dark Matter in Association with Top Quark Pairs in the Di-lepton Final State in pp collisions at sqrt(s) = 8 TeV. (CMS-PAS-B2G-13-004)
[8] The CMS Collaboration. Search for Dark Matter production in association with top quark pairs in the semileptonic channel. (CMS-B2G-14-004)
[9] The CMS Collaboration. Search for invisible decays of Higgs bosons in the vector boson fusion and associated ZH production modes. *Eur. Phys. J.* C **74** (2014) 2980
[10] The CMS Collaboration. The CMS experiment at the CERN LHC. *JINST* **3** (2008) S08004

Institute of Physics
Dirac House, Temple Back
Bristol BS1 6BE UK

CURRAN ASSOCIATES INC
proceedings.com